Engineering Materials

This series provides topical information on innovative, structural and functional materials and composites with applications in optical, electrical, mechanical, civil, aeronautical, medical, bio- and nano-engineering. The individual volumes are complete, comprehensive monographs covering the structure, properties, manufacturing process and applications of these materials. This multidisciplinary series is devoted to professionals, students and all those interested in the latest developments in the Materials Science field, that look for a carefully selected collection of high quality review articles on their respective field of expertise.

Indexed at Compendex (2021) and Scopus (2022)

Akash Katoch · Puneet Kaur · Rahul
Editors

2D Materials

Sensing Applications

Springer

Editors
Akash Katoch
Centre for Nanoscience
and Nanotechnology
Panjab University
Chandigarh, India

Puneet Kaur
Department of Physics
National Cheng Kung University
Tainan, Taiwan

Rahul
Department of Physics
National Cheng Kung University
Tainan, Taiwan

ISSN 1612-1317 ISSN 1868-1212 (electronic)
Engineering Materials
ISBN 978-981-97-6257-6 ISBN 978-981-97-6258-3 (eBook)
https://doi.org/10.1007/978-981-97-6258-3

This Springer imprint is published by the registered company Springer Nature Singapore Pte Ltd.
The registered company address is: 152 Beach Road, #21-01/04 Gateway East, Singapore 189721, Singapore

Preface

The emergence of atomically thin 2D materials having a myriad property stands out as a promising platform for highly selective, versatile, and sensitive sensors in diverse dimensions. The atomically thin nature of these materials offers an exceptionally high surface-to-volume ratio, a significant feature that leads to the amplification of their interactions with the surrounding environment to enhance their sensing abilities. Additionally, the possibility of band gap tunability offered by these 2D nanomaterials including semiconductor transition metal chalcogenides, graphene, reduced graphene oxides, transition metal nitrides, carbon nitrides, black phosphorus, boron nitride, etc. opens avenues to tailor sensor's performance to achieve high precision and accuracy.

This book provides a comprehensive overview of the utilization of 2D materials for various sensing applications. It highlights the present state of the art of 2D materials employed as gas sensors, electrochemical sensors, colorimetric sensors for detecting water contaminants, biosensors, biomarkers, textile sensors for human health monitoring, and photodetectors and their mechanisms. Furthermore, it delves into the various top-down and bottom-up synthesis strategies that can be explored to achieve low-cost scalable 2D materials for affluent 'Lab-to-Fab' transition. The layer-by-layer structure of 2D materials and their morphology at the nanoscale have shown the ability to detect gas molecules at lower temperatures or even at room temperature making these materials highly favorable for gas sensing, colorimetric sensing of pollutants. The high electron mobility offered by these materials invokes their role in light sensing and photodetection. A step ahead, the involvement of 2D materials in modern medical technologies by detecting the disease/health condition at an early stage by analyzing exhaled breath, identifying the biomarkers, etc. discusses the engineering aspect of these materials for the development of the wearable biosensors, textile sensor is addressed. Moreover, this book discusses the latest advancements in the designing and construction of electrochemical sensors based on food quality monitoring and biomolecular detection, sensing of pesticides, and heavy metal ion detection at ppb.

"2D Materials-Sensing Applications" provides a comprehensive overview of the synthesis, characterization, and applications of 2D nanomaterials in various sensing technologies. We believe that the broad spectrum of the sensor applications covered in this book will provide invaluable insight to the researchers and scientific community to further advance their understanding of 2D nanomaterials and their practical implications. We extend our sincere appreciation to all the contributors, including scientists, research scholars, and academicians for their invaluable contributions, as well as to the reviewers for their critical evaluation of the manuscripts. A special acknowledgment to Praveen Kumar Boominathan at Springer for their unwavering cooperation and guidance in facilitating the publication of this book.

Chandigarh, India — Akash Katoch
Tainan, Taiwan — Puneet Kaur
Tainan, Taiwan — Rahul

Contents

Scalable and Cost-Effective Synthesis of 2D Materials 1
Krishna K. Yadav, Sunaina, Supriya Rana, and Sujit Kumar Guchhait

2D-MoS_2 and WS_2-Based Chemical Gas Sensor 21
Sajjad Hussain, Rohit, Sandeep Kaur, Akash Katoch, and Deepika Jamwal

Surface Engineered 2D TMD Materials for Advanced Wearable Biosensors 45
Shalu Rani, Sanjay Kumar, Ruchi Singh, and Pawan Kumar

Recent Developments in Various 2D Material-Based Gas Sensors: Diagnostic Perspective of Human Exhalation 73
Levna Chacko and Zdeněk Sofer

Electrochemical Sensors Based on 2D Materials (2DMs) and Their Heterostructures 97
Ranjeet Kaur, Randeep Kaur, Bhupender Singh, Jaiveer Singh, Gurpreet Singh, Shweta Rana, Puneet Kaur, and Deepender Kaushik

2D Materials for Gas Sensing Application 129
Nainsi Sharma, Raeesah Islam, Harish Mudila, and Anil Kumar

Organic Linkers and Their Modified Nanoparticles for Colorimetric Monitoring Devices for Inorganic Water Contaminants 143
Nancy Sidana

2D Material-Based Textile Sensors for Human Health Monitoring Applications 165
Pradeep Kumar, Monika Gupta, Varun Jeoti, and Goran M. Stojanović

2D Materials—Applications in Photo Sensors 187
Juri Kalita, Linkon Bharali, and Siddhartha Sankar Dhar

2D Materials' Sensing Mechanisms 217
Yadagiri L N K Mallela

Editors and Contributors

About the Editors

Akash Katoch is currently working as a UGC Assistant Professor at the Centre for Nanoscience and Nanotechnology (CNSNT), Panjab University, India. He received his M.Tech. in Materials Science and Engineering from Thapar University, India (2009) and a Ph.D. in Materials Science and Engineering from Inha University, Republic of Korea (2014). Later he worked as a postdoctoral researcher at Inha University (2014–2015) and as an Assistant Professor at Seokeyong University, South Korea (2015), and DST INSPIRE faculty at IIT Roorkee (2015–2018). He is a renowned researcher and established academician. He has won many accolades and fellowships nationally and internationally. His current research focus includes chemical gas sensors and heavy metal ion detection. He has published 70 papers, nine book chapters and six patents internationally.

Puneet Kaur received her Ph.D. from Guru Nanak Dev University, Amritsar, following which she has worked as a project associate at CSIR-CSIO, Chandigarh. Currently she is working as a post-doctoral fellow in the Department of Physics, at National Cheng Kung University, Taiwan (R.O.C.). Her research interests include the development of efficient and color-tunable phosphors, ion implantation/irradiation, energy storage and conversion, photo electrocatalysis, VOC degradation, and strongly correlated complex oxide for various device applications. She has published more than 20 research papers in peer-reviewed international journals, and three book chapters. Dr. Kaur also serves as a guest editor and referee of renowned national and international journals. Dr. Kaur has been the recipient of several research grants from India and abroad including the prestigious INSPIRE scholarship and fellowship award by the Department of Science and Technology, Government of India.

Rahul received his doctoral degree from the Centre of Nanoscience and Nanotechnology, Panjab University, Chandigarh, India, and is currently working at National Cheng Kung University, Taiwan as a post-doctoral research scientist funded by NSTC, Taiwan. He has research specialization in the fields of energy storage devices, band alignment studies, water splitting applications, nanomaterials, and thin film synthesis using physical and chemical vapor deposition. He has published more than 19 research papers, and two book chapters in peer-reviewed journals. His research interest primarily lies in interdisciplinary Materials Science. He has served as a reviewer for various journals of National and International repute and has been awarded several recognitions. He has received the prestigious INSPIRE fellowship funded by the Department of Science and Technology, New Delhi, India, and TEEP@AsiaPlus funded by MOST, Taiwan.

Contributors

Linkon Bharali Department of Chemistry, National Institute of Technology, Cachar, Assam, India

Levna Chacko Department of Inorganic Chemistry, University of Chemistry and Technology Prague, Prague 6, Czech Republic

Siddhartha Sankar Dhar Department of Chemistry, National Institute of Technology, Cachar, Assam, India

Sujit Kumar Guchhait Electrochemical Process Division, Sardar Swaran Singh National Institute of Bio Energy, Kapurthala, Punjab, India

Monika Gupta Department of Electrical and Electronic Engineering, Universiti Teknologi PETRONAS, Perak, Malaysia

Sajjad Hussain Smart Nanomaterials and Sensor Laboratory, Centre for Nanoscience and Nanotechnology, Panjab University, Chandigarh, India

Raeesah Islam Department of Chemistry, Lovely Professional University, Phagwara, Punjab, India

Deepika Jamwal Department of Chemistry, University Institute of Sciences, Chandigarh University, Gharuan, Punjab, India

Varun Jeoti Faculty of Technical Sciences, University of Novi Sad, Novi Sad, Serbia

Juri Kalita Department of Chemistry, National Institute of Technology, Cachar, Assam, India

Akash Katoch Smart Nanomaterials and Sensor Laboratory, Centre for Nanoscience and Nanotechnology, Panjab University, Chandigarh, India

Puneet Kaur Department of Physics, National Cheng Kung University, Tainan, Taiwan ROC

Randeep Kaur Department of Chemistry, Panjab University, Chandigarh, India

Ranjeet Kaur University Centre for Research & Development (UCRD), Department of Chemistry, Chandigarh University, Gharuan, Punjab, India

Sandeep Kaur Smart Nanomaterials and Sensor Laboratory, Centre for Nanoscience and Nanotechnology, Panjab University, Chandigarh, India

Deepender Kaushik College of Pharmacy, University of Nebraska Medical Center, Omaha, NE, USA

Anil Kumar Department of Chemistry, Lovely Professional University, Phagwara, Punjab, India

Pawan Kumar Centre for Applied Research in Electronics (CARE), Indian Institute of Technology, Delhi, New Delhi, India

Pradeep Kumar Faculty of Technical Sciences, University of Novi Sad, Novi Sad, Serbia

Sanjay Kumar Department of Electronics Engineering, Indian Institute of Technology (ISM) Dhanbad, Dhanbad, Jharkhand, India

Yadagiri L N K Mallela The University of Texas at Austin, Austin, TX, USA

Harish Mudila Department of Chemistry, Lovely Professional University, Phagwara, Punjab, India

Shweta Rana Department of Chemistry, Panjab University, Chandigarh, India

Supriya Rana Institute of Nano Science and Technology, Mohali, India

Shalu Rani Department of Electronics Engineering, Indian Institute of Technology (ISM) Dhanbad, Dhanbad, Jharkhand, India

Rohit Smart Nanomaterials and Sensor Laboratory, Centre for Nanoscience and Nanotechnology, Panjab University, Chandigarh, India

Nainsi Sharma Department of Chemistry, Lovely Professional University, Phagwara, Punjab, India

Nancy Sidana Department of Applied Sciences, Chandigarh Engineering College-CGC, Landran, Mohali, Punjab, India

Bhupender Singh Department of Chemistry, Panjab University, Chandigarh, India

Gurpreet Singh University College of Dhilwan, Barnala, Punjab, India

Jaiveer Singh Department of Chemistry, Panjab University, Chandigarh, India

Ruchi Singh Department of Electronics and Communication Engineering, Jaypee Institute of Information Technology, Noida, India

Zdeněk Sofer Department of Inorganic Chemistry, University of Chemistry and Technology Prague, Prague 6, Czech Republic

Goran M. Stojanović Faculty of Technical Sciences, University of Novi Sad, Novi Sad, Serbia

Sunaina Department of Chemistry, University of Calgary, Calgary, Canada

Krishna K. Yadav Department of Chemical Engineering, Sami Shamoon College of Engineering, Beer-Sheva, Israel

Scalable and Cost-Effective Synthesis of 2D Materials

Krishna K. Yadav, Sunaina, Supriya Rana, and Sujit Kumar Guchhait

Abstract Nowadays, low-dimensional materials, especially two-dimensional (2D) materials show their importance in a wide range of applications in almost all sectors of science and technology such as energy, electronics, optics, and quantum information. However, their properties are sometimes compromised due to their availability, cost, and high-quality large-scale 2D materials. Earlier, apart from graphene, the literature was limited to the scalable method of production for other 2D materials, which are new in the family; however, shown their potential in various important applications. The scalable and economical synthesis should be at the forefront of a 'Lab-to-Fab' transition. In the past few years, a significant advancement has been carried out. Several researchers attempted the synthesis of high-quality large-scale 2D materials. The present book chapter deals with the synthesis of low-cost 2D materials of different types and their scalability. Along with the synthesis a few real-time example was also included for the synthesis of 2D material and their application.

K. K. Yadav (✉)
Department of Chemical Engineering, Sami Shamoon College of Engineering, 8410802 Beer-Sheva, Israel
e-mail: Krishnay447@gmail.com

Sunaina
Department of Chemistry, University of Calgary, Calgary, Canada

S. Rana
Institute of Nano Science and Technology, Knowledge City, Sector 81, Mohali, India

S. K. Guchhait
Electrochemical Process Division, Sardar Swaran Singh National Institute of Bio Energy, Kapurthala, Punjab 144603, India

A. Katoch et al. (eds.), *2D Materials*, Engineering Materials,
https://doi.org/10.1007/978-981-97-6258-3_1

1 Introduction

The development in science and technology has resulted in remarkable scientific discoveries and inventions, especially in the realm of nano- and atomic-level materials [1–10]. The development of innovative materials is key to improving current applications and boosting elemental output in various fields, including medicine, materials, electronics, energy, and catalysis [11–18]. Nanomaterials are classified based on their dimensions, which range from zero-dimensional (0D) to three-dimensional (3D) materials including quantum dots, nanoribbons, nanotubes, nanowires, nanoballs, and nanocones [19–21]. The size of a material plays a crucial role in determining its properties and behaviour, and when size is reduced to the nanoscale level, materials can exhibit improved electrical, chemical, mechanical, and optical properties [21–27]. For instance, 2D materials have a higher strength and a greater surface area-to-volume ratio than 3D materials, which can enhance the rate of reactions [20, 28]. Out of a large number of elements presented in the periodic table, carbon plays a significant role as a pioneering material in the field of nanotechnology [22, 29–32]. The various bonding patterns in carbon compounds result in a large number of allotropes that can form various structure compounds such as 3D graphite, 2D graphene, 1D nanotubes, and 0D fullerenes [31, 33]. Each of these structures has its own set of physicochemical features and the properties of these carbon-based materials are affected by their shape, dimensionality, and chemical bonds [34–36]. Out of various forms of carbon, the discovery of graphene sparked an unprecedented interest in 2D materials [23, 36]. Graphene exhibits sheet-like atomic structures among the other studied low-dimensional materials [24]. The arrangements of carbon in graphene tune their physical and chemical properties, and they possess enormous potential in various fields of science and technology such as optoelectronics, biosensing, photonics, and energy conversion and storage [22, 37, 38]. However, despite the successful synthesis of wafer-scale graphene single crystals, their semimetallic nature and large dark current render them unsuitable for electronic switches [39]. Apart from this, traditionally used materials often have low mobility, high driving voltage, complex technological processes, and large leakage current whereas 2D materials have strong covalent bonds in the plane, which provides high mechanical robustness, resulting in high fracture strains, making them ideal for flexible electronic products [40]. Additionally, 2D materials exhibit remarkable electrical properties, as well as transparency and flexibility, which make them suitable for use in flexible electronic products that are subject to mechanical strain and cyclic stress [41–44]. Graphene is well used 2D material however due to its semimetallic nature scientists are now shifting their focus towards 2D materials beyond graphene, such as TMDs, h-BN, graphitic carbon nitride, h-MO, TMOs, metallic glass, 2D metals, and MXenes and can be found in Fig. 1 [45–50].

It is interesting to note that graphene is a well-explored 2D material that can be scaled economically based on top-down or bottom-up approaches [52]. In the author's opinion, the term scalability can vary based on the application where materials would

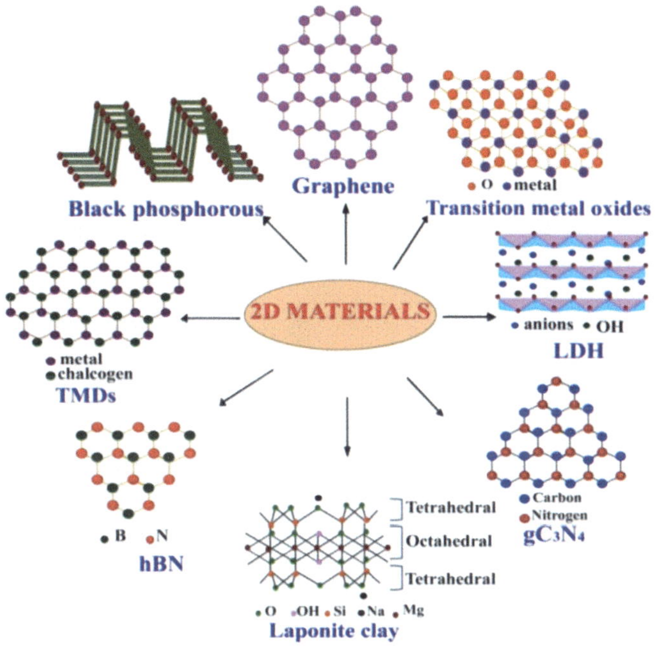

Fig. 1 Different types of 2D material structures. Reprinted with permission Ashtami et al. [51]. *Copyright* 2023 Elsevier

be used. If 2D materials are used in semiconductor industries, then printing or transferring the materials over a millimetre to centimetres to inch size wafers can be regarded as large scale whereas in the case of energy storage materials which can be produced in kilograms will be considered as scalable. Mainly two types of synthesis strategies have been used for 2D materials synthesis; (1) top-down, (2) bottom-up [52]. The primary top-down techniques involve mostly exfoliation processes, which can present challenges in generating high-quality 2D materials when compared to other methods. In these processes basically, 3D building blocks will be broken into 2D by pealing or off lamination. Whereas in the case of a bottom-up process, small units will be arranged to form 2D materials such as Chemical Vapor Deposition (CVD). It can be noted that the bottom-up approach is widely employed for 2D materials synthesis on wafer [53]. In the scientific community generally, liquid exfoliation and wet chemical synthesis are named as scalable synthesis processes whereas chemical vapour deposition (CVD), pulsed laser deposition (PLD), atomic layer deposition (ALD), and molecular beam epitaxy (MBE) large-scale thin-film scalable techniques for 2D materials [53]. CVD is regarded as the most practical method for industrial production due to its ability to generate uniform and crystalline 2D films while meeting high throughput, cost-effectiveness, and scalability requirements. A commonly used technique for the 2D materials synthesis has been shown in Fig. 2. Nowadays, biowaste-derived large-scale synthesis of graphene using

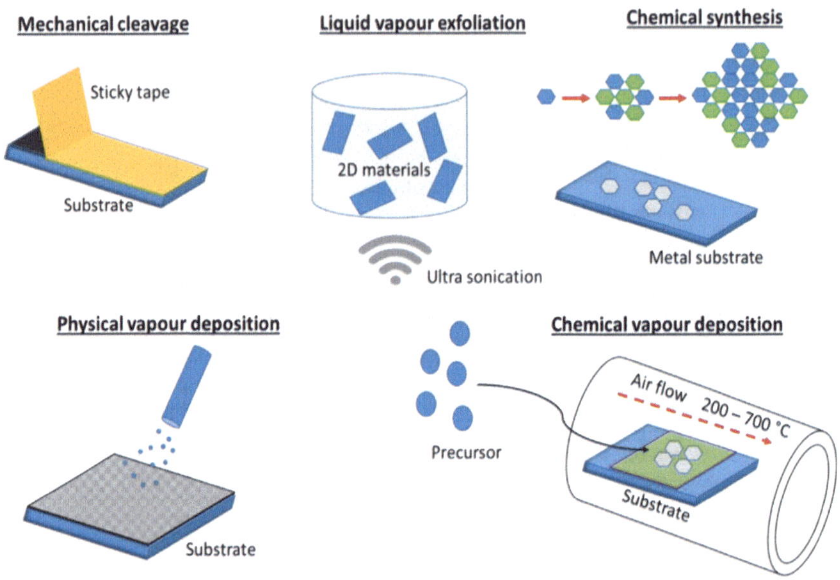

Fig. 2 Various synthesis routes of 2D materials [1]

a combination of wet chemical and exfoliation methods is also very popular for 2D materials synthesis on a large scale; however, the properties of synthesized materials can vary based on the initial biowaste.

2 Synthesis of 2D Materials

2.1 Top-Down Approach

Various top-down methods can be found in the literature such as mechanical exfoliation, electrochemical exfoliation, and hydrothermal exfoliation. The most commonly used method is mechanical cleavage also known as the scotch tape method, where bulk sheets are delaminated and transferred over large-scale wafers; however, scalability is not effective and lack of control over the number of layers. Apart from that few other approaches such as the electrochemical approach play a very significant role in the large-scale production of 2D materials.

2.1.1 Mechanical Exfoliation

The process of tape-assisted graphene exfoliation can separate nanosheets from bulk crystals using mechanical force. However, this process is time-consuming and not scalable due to the layer-by-layer approach of tape peeling, therefore for scale-up of the production, shear exfoliation and ball milling methods have developed, which can effectively produce large quantities of exfoliated materials in a single run. Materials have been placed into liquid for shear exfoliation at high speed. The efficiency of exfoliation is influenced by the choice of solvent, with the best results obtained when the surface tension of the solvent matches that of the 2D products to minimize restacking of nanosheets.

Paton et al. in 2014 showed large-scale shear exfoliated graphene and achieved a 1.44 g/h production rate, which is higher than the sonication method [54]. Exfoliation of 2D materials other than graphene has been also demonstrated by various researchers [55, 56]. In one case, WS_2 nanosheets were exfoliated with a 0.95 g/h production rate by Biccai et al. whereas the exfoliation yield was just 1%, requiring additional post-treatments to increase efficiency [57]. Apart from the scotch tape method, Ball milling methods have been shown their importance in increasing the yield of exfoliated 2D materials [58, 59]. In a typical ball milling process, a cylindrical shell is filled with continuously rotating ceramic balls, subjecting layered materials to both horizontal and vertical forces that cause interlayer sliding and intralaminar breakdown. This method has better efficiency and control over the morphology of the exfoliated products compared to shear exfoliated products. Overall the efficiency of shear exfoliation is low whereas the ball milling process is a time-consuming method [57].

2.1.2 Sonication

Sonication is a common liquid-phase exfoliation (LPE) method that involves the use of ultrasonic energy to exfoliate layered materials. Recent literature suggests that this is only suitable for graphene due to several drawbacks such as low product yield, extended processing time, poor uniformity, and limited parameters for process control [60]. The sonication method can also cause breakage of readily exfoliated nanosheets, leading to wide lateral size distribution and uncontrollable morphology, while most bulk materials remain unaffected [61]. Often, liquid cascade centrifugation requires getting relatively uniform 2D materials through sonication, which sometimes acts as a limiting factor for scalability [62]. It has very little control over exfoliation and can be influenced by the operator's skill, instrument and position of the vessels [63]. To overcome the above issue, Coleman et al. have prepared a comprehensive recommendation and shown a way for large-scale 2D materials exfoliation [64]. In this method, they demonstrated the exfoliation of bulk transition metal dichalcogenides (TMDs) and boron nitride crystals in a common solvent for the production of mono- and few-layer nanosheets.

2.1.3 Laser-Assisted Exfoliation

Laser energy can be utilized for exfoliation by irradiating a material with high-density photons, resulting in a dramatic increase in temperature at the material/laser interface. The detachment of surface species occurs due to high-pressure cavities, vapourization, recondensation, shockwaves and plasma formation on the material's surface and also executive for exfoliating layered materials like MoS_2 and $MoTe_2$ [65–67]. For successful exfoliation using lasers, a moderate range of energy should be provided for the promotion of detachment of few-layer flakes, not individual atoms [67]. The important fact is that the laser will finish the exfoliation in a part of seconds and no other method can beat this in terms of time. Qian et al. first demonstrated the use of laser-assisted exfoliation to produce few-layer graphene, utilizing a photothermal mechanism and a protective Ar gas environment to achieve precise control over photon energy and limit side reactions [68]. Later, they extended this method to a liquid phase, achieving higher efficiency and ease of nanosheet suspension formation. Other researchers have also utilized laser irradiation to exfoliate TMDs, with water molecules playing a key role in promoting the peeling process. Solvent choice has been shown to have a significant impact on the resulting layer number, with low-boiling-point solvents such as hexane producing the thinnest products. The efficiency and quality of laser-assisted exfoliation are determined not only by the solvent used but also by the laser parameters, such as wavelength and energy. This method offers several advantages over others, including quick exfoliation, the ability to use various starting materials and contamination-free products. However, the limited number of materials exfoliated by laser due to the challenge of delivering the right amount of energy without causing structural damage as well as some time change in phase that also occurs during exfoliation [69]. Thus, this approach requires specific thermal stability and careful selection of laser energy, wavelength and liquid environment when applied to other layered materials. A schematic representation of the experimental setup for the production of $MoTe_2$ has been shown in **Fig. 3**.

2.1.4 Electrochemical Exfoliation

Compared to conventional methods, electrochemical exfoliation is a user-friendly process that can be performed under ambient conditions, eliminating the need for multiple processing steps. This method is also environmentally friendly as it allows for the recycling of electrolytes, thereby reducing chemical waste [70]. The exfoliation reaction is typically completed within a few minutes to hours, resulting in the production of solution-processable 2D materials at the gram-scale level in the laboratory. By engineering the electrochemical cells, it may be possible to increase the throughput to kilogram-scale levels [2, 71]. Furthermore, the broad range of electrochemical reactions (e.g., oxidation or reduction) provides ample opportunity to customize the thickness, crystal structure and chemical properties of the exfoliated materials to meet specific requirements based on the application. Despite being a relatively new method, electrochemical exfoliation has garnered considerable interest

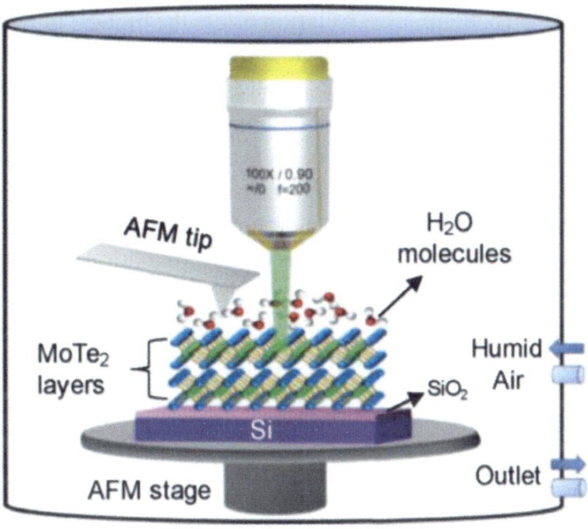

Fig. 3 Schematic illustration of the experimental setup used for thinning and in situ imaging of $MoTe_2$ [65]

from both academia and industry. Recently, it has been that electrochemical exfoliation will be an alternative powerful exfoliation technique for graphene and other 2D semiconductors [72–74]. Electrochemical exfoliation has a strategy for scalability and offers a good balance between cost, quality, and designed functionality [53]. Yang et al. exfoliated different 2D materials such as graphene, graphene oxide, phosphorene and transition metal dichalcogenides (TMDs) and found electrolyte has a crucial role in the large-scale production of 2D materials [75, 76]. A typical electrochemical cell is shown in Fig. 4. The electrochemical needs the following configuration.

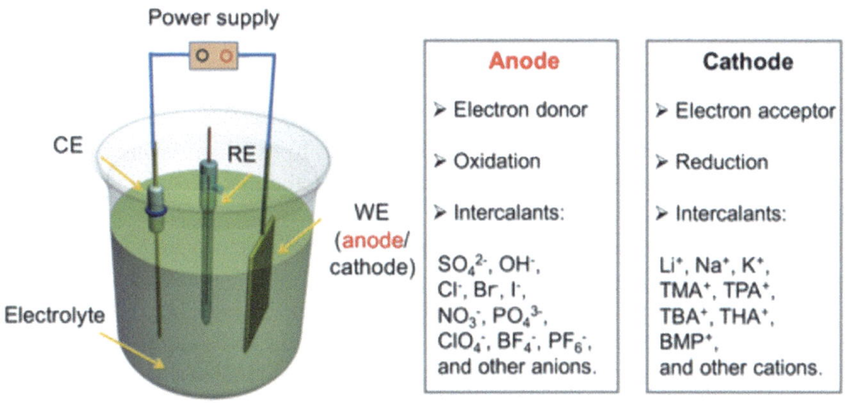

Fig. 4 A typical 3-electrode electrochemical cell for exfoliation of various working electrode materials. WE refer to the working electrode, RE refers to the reference electrode, and CE refers to the counter electrode [75]

Electrochemical exfoliation offers the highest uniformity due to its unique configuration, which applies a voltage bias directly to the starting material, promoting uniform ion intercalation due to the homogeneous potential of its surface. This method allows well-defined control over quality by tuning potential and electrolyte whereas choice of electrolyte is crucial in the exfoliation process. The electrolyte should possess excellent conductivity and have solvated ions that match with interlayer spacing of the electrode materials, which will facilitate intercalation. In addition, gaseous species evolve during oxidation or reduction making the exfoliated sheets damage free and similar surface tension of electrolytes prevents the restacking of exfoliated flakes. It allows to removal of solvents and electrolytes easily, especially in the fabrication of electronic devices and at the same time water or organic solvents have limitations of inevitable oxidation or insufficient intercalation [77]. The materials which will be exfoliated also play a crucial role. Layered materials that possess strong in-plane bonding and weak interlayer connection are generally suitable for the exfoliation process. Apart from that the essential criteria, the starting materials should be conducting or semiconducting [78].

2.2 *Bottom-Up Approach*

The bottom-up methods offer excellent compatibility with a broad spectrum of applications, taking advantage of their diverse principles. These methods encompass various techniques such as magnetron sputtering, chemical vapour deposition (CVD) and wet chemical methods [79, 80]. The bottom-up approaches offer versatility and flexibility, making them well-suited for a wide range of applications. The bottom-up approach is mostly utilized when 2D materials are needed on a wafer scale, especially for semiconductor industries [81–83].

2.2.1 Hydrothermal Exfoliation

Researchers have been looking for alternative ways to exfoliate layered materials due to the limitations of mechanical methods. Hydrothermal exfoliation is one of the methods that has been proposed, and it involves treating a suspension of the layered material in water with chemical and thermal disturbance where intercalation or chemical reaction strategies have been used [84]. In the intercalation approach, intercalants diffused into two layers of bulk materials under hydrothermal conditions causing slight lattice expansion and then treated for the exfoliation process. For MXenes formation from the MAX phase, chemical reaction or chemical etching is another hydrothermal approach where an acid or salt mixture is used for MAX etching. Other 2D materials were also exfoliated using a similar method [85–89]. In hydrothermal exfoliation generally, Li^+, Na^+ and K^+ are used for intercalation [90]. Metal sulphide can be exfoliated very fast with high yield in the presence of the combination of Li^+ ions and isopropyl alcohol as a solvent during hydrothermal.

Chemical reaction can give ~100% yield for the MXene nanosheets exfoliation from MAX [53].

2.2.2 Chemical Vapour Deposition (CVD)

Chemical vapour deposition (CVD) is a versatile and well-explored technique for nanostructure fabrication. It has been a prominent technique in the microelectronics industry for several decades and continues to be widely used today to overcome challenges associated with modern technologies [91]. In this process, simple materials were used to prepare nanostructures where a thin layer will be deposited over the substrate from gaseous reactants, which undergoes a chemical reaction upon contact with the heated substrate within a reaction chamber [92]. Consequently, a thin film forms on the surface of the substrate, which can then be selectively etched and utilized for various applications. For proper implementation of CVD, requires several operating requirements to be met such as gas-phase reactants, inert or carrier gas-containing reaction chamber, gas discharge technique, management of pressure inside the reaction chamber, environmental compliance and so on [93]. Meeting these requirements enables the efficient and controlled deposition of thin films using CVD, facilitating its widespread adoption in various industries and research fields.

The CVD process begins with the delivery of reactant gases into the reactor and these gases undergo homogeneous reactions to create intermediates and by-products or can be directly diffused over the substrates [94]. As the growing species evaporate, they are transported to the substrate with the assistance of a carrier gas. Upon reaching the substrate, the species assemble and undergo growth through heterogeneous reactions at the gas–solid interface. This continuous thin-film production process involves coalescence, growth, and nucleation, resulting in the formation of the desired thin film. Alongside the thin-film production, there may also be the generation of reaction by-products.

Nand et al. utilized the CVD method for synthesizing the large-scale MoS_2 fabrication over the silicon substrate [95]. They explored the monolayer synthesis of high-quality MoS_2 in an economical way. Large area scale ($\approx 5 \times 1$ mm^2) monolayer of MoS_2 on SiO_2/Si substrate were obtained via chemical vapour deposition (CVD). The CVD process has some unique advantages over other methods such as finest control, large-scale fabrication, good crystal quality, high purity, low cost and few substrate flaws [96, 97]. However, it has some disadvantages such as it is not cost-effective and can release toxic gases during the process. The typical synthesis of 2D WSe_2 layers using the CVD route has been illustrated in Fig. 5.

2.2.3 Molecular Beam Epitaxy (MBE)

Molecular Beam Epitaxy (MBE), is commonly used to grow 2D films at wafer scale. In the process, a substrate is placed inside an ultra-high vacuum (UHV) chamber

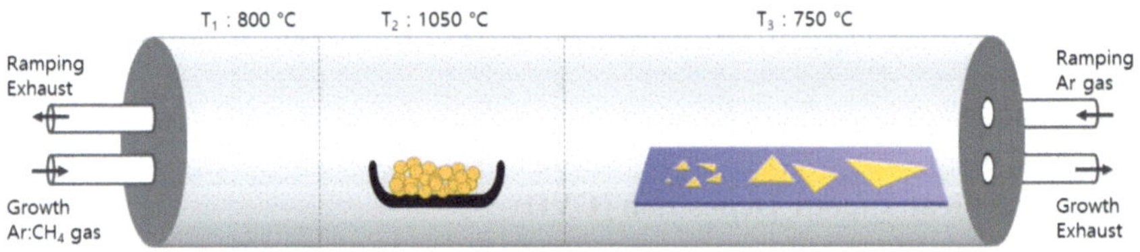

Fig. 5 Schematic diagram of the CVD process for the growth of WSe_2 [98]

on a stage held at the desired growth temperature. The UHV chamber is a specialized enclosure designed to achieve and maintain extremely low-pressure levels by removing all possible contaminants and gases which leads to the growth of high-quality 2D films [99, 100]. By analyzing the diffraction patterns, researchers can obtain real-time information about the structure and quality of the growing film. This capability allows for precise control and optimization of the growth parameters. By carefully controlling the evaporation of these precursors, researchers can achieve precise layer-by-layer growth, leading to highly controlled film thickness. Apart from this, it has the disadvantage of a very slow deposition rate (typically less than 5 Å min^{-1}) and costly technique as compared to CVD. Briefly the UHV environment, in situ monitoring, high-purity precursors and slow deposition rate, make it a promising technique for achieving large-scale growth of high-quality 2D films [99, 100]. For instance, a 2-inch SiO_2/Si wafer was utilized for the growth via the MBE method and 100% monolayer film coverage of the 2H phase of $MoTe_2$ has been successfully developed by the Yao group [101]. Apart from that Loh et al. achieved a significant breakthrough in the growth of highly oriented monolayer MoS_2 films on 2-inch h-BN/sapphire substrates using Molecular Beam Epitaxy (MBE) [102]. They were able to achieve a remarkable 96.3% zero-aligned domain orientation of the films. This achievement was made possible by employing a low Mo precursor flux during the growth process [102].

3 Evaluation and Criteria of Scalability in Exfoliation Method

2D materials can be synthesized via various methods as discussed previously in the section; however, in most cases, the trade-off between quality and quantity occurs. To evaluate the scalability of exfoliation methods, several factors including production rate, exfoliation yield, quality uniformity, reproducibility, cost and tunability should be taken into consideration. In this section exfoliation-based has been discussed for the scalability. Nowadays due to reducing the harsh environmental impact, we have to keep in mind that methods should not have any negative impact on the environment. These factors play a crucial role in determining the feasibility and scalability of a particular 2D material production process and these parameters can be kept in mind during scalable production.

3.1 Production Rate

It is interesting to note that production rates are defined as the main factor when evaluating and comparing different exfoliation methods for 2D materials in recent reports. However, the production rate may have influenced by process time, amount of starting materials, and exfoliation yield. The production rate (mostly measured in grams per hour) is non-linear with time, i.e., longer treatment times may lead to decreased production rates due to loss of materials and in most of the exfoliation process significant exfoliation often occurs early in the process. The production rate can be increased by increasing the starting materials even if the quality of those materials is not consistent. The higher end product results in a higher production rate, assuming that other parameters remain constant. Therefore, for scalable 2D materials synthesis relying solely on production rate may not accurately reflect the efficiency or quality of an exfoliation method [1].

3.2 Exfoliation Yield

Exfoliation yield is another crucial factor even sometimes regarded as the most important factor during scalability. Only when the exfoliation yield is very high then only one can say the methods are scalable. Microwave-assisted exfoliation has been found to have surprisingly high yields, with graphite exfoliated at a yield of 93% within 30 min. Electrochemical exfoliation typically delivers yields ranging from 30 to 80%, but the treatment duration can vary widely. Mechanical exfoliation can achieve yields ranging from less than 1% to 95%, but it is highly dependent on the setup and engineering parameters used. Post-treatments such as product selection, separation and purification are required to obtain 2D materials with desired size and thickness.

3.3 Reproducibility

The reproducibility with the consistent manner of 2D materials is another important criterion for scaling up the synthesis otherwise the performance of the end devices can be compromised. Standard operating procedures (SOPs) can help eliminate uncontrollable factors and define well-established parameters to ensure reproducibility. However, some methods, such as sonication, can be challenging to standardize due to the complexity of factors affecting sonication energy, making it highly dependent on operator skills and experience. In contrast, mechanical and electrochemical exfoliation methods are more reproducible as their parameters are well-defined and less influenced by operator experience. Laser and microwave-assisted exfoliation methods are manageable when worked for specific layered materials with fixed laser power and microwave energy whereas less controllable with various layered materials. Nonetheless, these methods can provide high yields and speed up the exfoliation resulting in a higher production rate.

3.4 Cost

When considering the commercialization of a laboratory-scale method for producing 2D materials, cost is a crucial factor to consider. However, the cost of raw materials is often overlooked in research, as many researchers use high-quality crystals of layered materials, such as Highly Ordered Pyrolytic Graphite (HOPG), for exfoliation. Using powders instead of bulk crystals as the starting material for exfoliation methods is much more cost-effective and important beyond graphite, as large crystals of these materials are extremely expensive. The production cost of exfoliation methods also includes chemicals and post-treatment. Therefore, low-cost aqueous exfoliation techniques should be preferred over organic solvents as well as ionic liquids. However, it is interesting to note that ionic liquid will have a high exfoliation yield as compared to aqueous exfoliation with lower thickness. Therefore, laser-assisted and mechanical exfoliation are one of the prime techniques for exfoliation, however; exfoliated nanosheets should be filtered from their bulk form which will again require the solvents. In this regard, electrochemical exfoliation has the advantage of automatically separating products whereas hydrothermal exfoliation requires excessive chemicals and results in higher cost. Overall, choosing the most cost-effective and environmentally friendly method for producing 2D materials at a commercial scale is essential.

3.5 Tunability

For industrial-scale manufacturing of 2D materials, the adjustability of production methods is essential. Electrochemical exfoliation stands out ahead of other methods due to its versatility in potential sequences and electrolyte selection. Hydrothermal and microwave-assisted exfoliation introduce diverse intercalants, yielding 2D materials with varying properties. Laser-assisted exfoliation offers customizable parameters like energy, wavelength, frequency and solvent, impacting material characteristics. Mechanical exfoliation through ball milling achieves diverse layer counts and lateral sizes.

3.6 Quality Uniformity

When it comes to scaling up the production of chemicals, maintaining product uniformity is a significant challenge. This is because reactions that perform well on a small scale may not be as successful on an industrial scale. One major problem with hydrothermal exfoliation is the difficulty in achieving a uniform spreading of thermal energy and ingredients. Due to uneven energy distribution different exfoliated thicknesses of 2D materials can be obtained and complex post-treatments are necessary to achieve products with similar lateral sizes and layer numbers. In addition, hydrothermal exfoliation requires more chemicals than other methods, which makes purification challenging. This is particularly problematic for electronic applications where ion intercalant residues can contaminate 2D materials. Apart from that mechanical force, laser and microwave exfoliation also faced energy distribution issues; however, fewer chemicals are required for post-treatment than hydrothermal methods, which is an advantage. Direct electrochemical exfoliation offers the highest uniformity thanks to its unique configuration. This method applies a voltage bias directly to the starting material, promoting uniform ion intercalation due to the homogeneous potential of its surface. The doped/functionalized/defective products obtained after the electrochemical exfoliation are not suitable for electronics; however, can fit energy-related applications, where structurally modified 2D materials are popular choices. Until now, most exfoliation procedures for 2D materials have taken place in liquid environments, commonly known as liquid-phase exfoliation (LPE).

4 Prospects and Conclusions

The prospects for 2D materials synthesis are promising and continue to evolve as researchers explore advanced manufacturing processes, expand beyond naturally occurring materials such as graphene, and investigate the synthesis of specific materials like TMDs, BN and bio-based graphene. These efforts contribute to a deeper understanding of the synthesis techniques and open up exciting possibilities for the applications of 2D materials in diverse fields.

The current chapter offers a comprehensive analysis of various synthesis and exfoliation methods for 2D materials. These methods include hydrothermal, mechanical, electrochemical, CVD and laser- and microwave-assisted exfoliation. The scalability, production rate, exfoliation yield, quality uniformity, reproducibility, cost and tunability of each method are also assessed in this chapter. The chapter thoroughly discussed the general principles, representative examples, progress, advantages and disadvantages of each method. It is worth noting that scaling up any exfoliation method depends on the specific application scenario and finding a balance between quality and quantity. Electrochemical and mechanical exfoliation techniques offer the best scalability; however, laser- and microwave-assisted methods have higher production rates and efficiency for various applications. These methods show promise but require further development and refinement. The high costs of raw materials, manpower, and post-treatments, along with the limited lab-to-fab transition caused by small batch production, pose obstacles. Researchers are encouraged to consider the possibility of scaling up their inventions when developing new methods or adopting new processes.

References

1. Shanmugam, V., Mensah, R.A., Babu, K., Gawusu, S., Chanda, A., Tu, Y., Neisiany, R.E., Försth, M., Sas, G., Das, O.: A review of the synthesis, properties, and applications of 2D materials. Part. Part. Syst. Charact. **39**, (2022)
2. Choi, S.H., Yun, S.J., Won, Y.S., Oh, C.S., Kim, S.M., Kim, K.K., Lee, Y.H.: Large-scale synthesis of graphene and other 2D materials towards industrialization. Nat. Commun. **13**, (2022)
3. Devi, S., Sunaina, Wadhwa, R., Yadav, K.K., Jha, M.: Understanding the origin of ethanol oxidation from ultrafine nickel manganese oxide nanosheets derived from spent alkaline batteries. J. Clean Prod. **376**, 134147 (2022)
4. Castro Neto, A.H., Guinea, F., Peres, N.M.R., Novoselov, K.S., Geim, A.K.: The electronic properties of graphene. Rev. Mod. Phys. **81**, 109–162 (2009)
5. Yadav, K.K., Sunaina, Sreekanth, M., Ghosh, S., Ganguli, A.K., Jha, M.: Excellent field emission from ultrafine vertically aligned nanorods of NdB6 on silicon substrate. Appl. Surf. Sci. **526**, 146652 (2020)
6. Shi, B., Qi, P., Jiang, M., Dai, Y., Lin, F., Zhang, H., Fang, Z.: Exotic physical properties of 2D materials modulated by moiré superlattices. Mater. Adv. **2**, 5542–59 (2021)
7. Kan, X., Fan, C., Wu, C., Sun, C., Li, Z., Zhao, Y.: Graphdiyne-supported atomic catalysts: synthesis and applications. Chempluschem **85**, 2570–2579 (2020)
8. Ingason, A.S., Mockute, A., Dahlqvist, M., Magnus, F., Olafsson, S., Arnalds, U.B., Alling, B., Abrikosov, I.A., Hjörvarsson, B., Persson, P.O.Å., Rosen, J.: Magnetic self-organized atomic laminate from first principles and thin film synthesis. Phys. Rev. Lett. **110**, 195502 (2013)
9. Rohilla, D., Kaur, N., Shanavas, A., Chaudhary, S.: Microwave mediated synthesis of dopamine functionalized copper sulphide nanoparticles: an effective catalyst for visible light driven degradation of methlyene blue dye. Chemosphere **277**, 130202 (2021)
10. Kaur, G., Kaur, V., Kaur, N., Kaur, C., Sood, K., Shanavas, A., Sen, T.: Design of Silica@Au hybrid nanostars for enhanced SERS and photothermal effect. ChemPhysChem (2023)
11. Rana, S., Yadav, K.K., Sunaina, Mehta, S.K., Jha, M.: Structurally engineered anisotropic cobalt-based nanostructures for efficient chlorine and oxygen evolution. Adv. Mater. Interfaces **9**, 2200740 (2022)
12. Ankush, Yadav, K.K., Guchhait, S.K., Rana, S., Jha, M.: Excellent hydrogen generation from ultrathin nanosheets of cobalt cyclotetraphosphate. Mater. Sci. Eng. B Solid State Mater. Adv. Technol. **265**, 114983 (2021)
13. Yadav, K.K., Kumar. G., Rana, S., Sunaina, Ghosh, S., Jha, M.: Understanding the role of sheet thickness on field emission from engineered hexagonal tin disulphide. Appl. Surf. Sci. **606**, 154816 (2022)
14. Rana, S., Yadav, K.K., Guchhait, S.K., Nishanthi, S.T., Mehta, S.K., Jha, M.: Insights of enhanced oxygen evolution reaction of nanostructured cobalt ferrite surface. J. Mater. Sci. **56**, 8383–8395 (2021)
15. Kaur, N., Puppala, E.R., Srivastava, R., Naidu, V.G.M., Shanavas, A.: Gold-capsuled polymeric nanomedicine for synergistic breast cancer photo-chemotherapy. J. Nanopart. Res. **24**, 1–14 (2022)

16. Guchhait, S.K., Sutradhar, D., Nandi, R., Sarma, A.K.: Biomass derived metal free hierarchical porous activated carbon for efficient oxygen evolution reaction (2023). https://doi.org/10.1080/15567036.2023.2209540455957-69
17. Kumar, S., Battula, V.R., Sharma, N., Samanta, S., Rawat, B., Kailasam, K.: A metal-free heptazine–porphyrin based porous polymeric network as an artificial leaf for carbon-free solar fuels. J. Mater. Chem. A Mater. **10**, 14568–75 (2022)
18. Battula, V.R., Rawat, B., Kailasam, K.: Ultrahigh photocatalytic oxygen reduction to hydrogen peroxide by a true oxygen-bridged heptazine polymer. ACS. Appl. Polym. Mater. (2022)
19. Wadhwa, R., Arora, A., Yadav, K.K.: Fiber materials: design, fabrication and applications. In: Aslam, J., Verma, C.: In: Fiber materials: design, fabrication and applications ed. Walter de Gruyter GmbH & Co KG, p. 191 (2023)
20. Wadhwa, R., Arora, A., Rana, S., Yadav, K.K.: Overview of advanced fiber materials. In: Aslam, J., Verma, C. (ed.), Fiber materials: design, fabrication and applications. Walter de Gruyter GmbH & Co KG, p. 1 (2023)
21. Yadav, K.K., Arora, A., Jangra, S., Jha, M.: Chemically modified carbon nanotubes in 3D and 4D printing. Chemically Modified Carbon Nanotubes for Commercial Applications 419–39 (2023)
22. Yadav, K.K., Khan, N., Jha, M.: Application of surface modified carbon nanotubes in energy. ACS Symp. Ser. **1425**, 101–119 (2022)
23. Yadav, K.K., Singh, H., Rana, S., Sunaina, Sammi, H., Nishanthi, S.T., Wadhwa, R., Khan, N., Jha, M.: Utilization of waste coir fibre architecture to synthesize porous graphene oxide and their derivatives: An efficient energy storage material. J. Clean Prod. 124240 (2020)
24. Yadav, K.K., Wadhwa, R., Khan, N., Jha, M.: Efficient metal-free supercapacitor based on graphene oxide derived from waste rice. Curr. Res. Green Sustain. Chem. **4**, 100075 (2021)
25. Khan, N., Yadav, K.K., Wadhwa, R., Ankush, Jha, M.: One-pot green method for fabrication of MoS2 decorated graphite rod and their application in H2 evolution. Mater. Lett. **313**, 131784 (2022)
26. Mahajan, A., Khan, N., Yadav, K.K., Jha, M., Ghosh, S.: Efficient field emission from ultrafine nanostructured lanthanum sulfide synthesized by chemical route. Appl. Surf. Sci. **623**, 156996 (2023)
27. Bagheri, M., Komsa, H.-P., Bagheri, M., Komsa, H.-P.: Screening 0D materials for 2D nanoelectronics applications. Adv. Electron Mater. **9**, 2200393 (2023)
28. An, Y., Tian, Y., Wei, C., Tao, Y., Xi, B., Xiong, S., Feng, J., Qian, Y.: Dealloying: An effective method for scalable fabrication of 0D, 1D, 2D, 3D materials and its application in energy storage. Nano Today **37**, 101094 (2021)
29. Fischer, R.A., Cottrell, E., Hauri, E., Lee, K.K.M., Le Voyer, M.: The carbon content of Earth and its core. Proc. Natl. Acad. Sci. USA **117**, 8743–8749 (2020)
30. Noon, M.L., Goldstein, A., Ledezma, J.C., Roehrdanz, P.R., Cook-Patton, S.C., Spawn-Lee, S.A., Wright, T.M., Gonzalez-Roglich, M., Hole, D.G., Rockström, J., Turner, W.R.: Mapping the irrecoverable carbon in Earth's ecosystems. Nature Sustain. **5**, 37–46 (2021)
31. Karthik, P.S., Himaja, A.L., Singh, S.P.: Carbon-allotropes: synthesis methods, applications and future perspectives. Carbon letters **15**, 219–37 (2014)
32. Ku, S.H., Lee, M., Park, C.B.: Carbon-based nanomaterials for tissue engineering. Adv. Healthc. Mater. **2**, 244–260 (2013)
33. Pradhan, B., Das, S., Li, J., Chowdhury, F., Cherusseri, J., Pandey, D., Dev, D., Krishnaprasad, A., Barrios, E., Towers, A., Gesquiere, A., Tetard, L., Roy, T., Thomas, J.: Ultrasensitive and ultrathin phototransistors and photonic synapses using perovskite quantum dots grown from graphene lattice. Sci. Adv. **6**, 5225–5237 (2020)
34. Yu, W., Gong, K., Li, Y., Ding, B., Li, L., Xu, Y., Wang, R., Li, L., Zhang, G., Lin, S.: Flexible 2D materials beyond graphene: synthesis, properties, and applications. Small **18** (2022)
35. Slepička, P., Hubáček, T., Kolská, Z., Trostová, S., Kasálková, N.S., Bačáková, L., Švorčík, V., Slepička, P., Hubáček, T., Kolská, Z., Trostová, S., Kasálková, N.S., Bačáková, L., Švorčík, V.: The properties and application of carbon nanostructures. Polym. Sci. (2013)

36. Yang, G., Li, L., Lee, W.B., Ng, M.C.: Structure of graphene and its disorders: a review. Sci. Technol. Adv. Mater. **19**, 613–648 (2018)
37. Chandra, Y., Mukhopadhyay, T., Adhikari, S., Figiel, L.: Size-dependent dynamic characteristics of graphene based multi-layer nano hetero-structures. Nanotechnology **31**, 145705 (2020)
38. Pawlyta, M., Smykała, S., Liszka, B., Blacha-Grzechnik, A., Krzywiecki, M., Jurkiewicz, K., Jakóbik-Kolon, A.: Influence of carbon support structure on cathode catalysts durability. Appl. Surf. Sci. **611**, 155637 (2023)
39. Ogawa, S., Shimatani, M., Fukushima, S., Okuda, S., Kanai, Y., Ono, T., Matsumoto, K.: Broadband photoresponse of graphene photodetector from visible to long-wavelength infrared wavelengths. Opt. Eng. **58**, 057106 (2019)
40. Hassan, A., Azam, M., Ahn, Y.H., Zubair, M., Cao, Y., Khan, A.A.: Low Dark Current and Performance Enhanced Perovskite Photodetector by Graphene Oxide as an Interfacial Layer. Nanomaterials **12**, 190 (2022)
41. Kou, J., Nguyen, E.P., Merkoçi, A., Guo, Z.: 2-Dimensional materials-based electrical/optical platforms for smart on-off diagnostics applications. 2d Mater. **7**, 032001 (2020)
42. Su, C.Y., Lu, A.Y., Xu, Y., Chen, F.R., Khlobystov, A.N., Li, L.J.: High-quality thin graphene films from fast electrochemical exfoliation. ACS Nano **5**, 2332–2339 (2011)
43. Liu, F., Wang, C., Sui, X., Riaz, M.A., Xu, M., Wei, L., Chen, Y.: Synthesis of graphene materials by electrochemical exfoliation: recent progress and future potential. Carbon Energy **1**, 173–199 (2019)
44. Lowe, S.E., Shi, G., Zhang, Y., Qin, J., Jiang, L., Jiang, S., Al-Mamun, M., Liu, P., Zhong, Y.L., Zhao, H.: The role of electrolyte acid concentration in the electrochemical exfoliation of graphite: mechanism and synthesis of electrochemical graphene oxide. Nano Mater. Sci. **1**, 215–223 (2019)
45. Alam, S., Asaduzzaman Chowdhury, M., Shahid, A., Alam, R., Rahim, A.: Synthesis of emerging two-dimensional (2D) materials—Advances, challenges and prospects. FlatChem **30** (2021)
46. Bhimanapati, G.R., Glavin, N.R., Robinson, J.A.: 2D boron nitride: synthesis and applications. Semicond Semimetals **95**, 101–47 (2016)
47. Gogotsi, Y., Anasori, B.: The rise of MXenes. ACS Nano **13**, 8491–8494 (2019)
48. Wadhwa, R., Yadav, K.K., Ankush, Rana, S., Sunaina, Jha, M.: Cobalt decorated graphitic carbon nitride photoanode for electrochemical ethanol oxidation: a sustainable way towards clean energy. Int. J. Hydrogen Energy (2023)
49. Ou, C.H., Lin, Y.C., Keikoin, Y., Ono, T., Esashi, M., Tsai, Y.C.: Two-dimensional MEMS Fe-based metallic glass micromirror driven by an electromagnetic actuator. Jpn. J. Appl. Phys. **58**, SDDL01 (2019)
50. Manzeli, S., Ovchinnikov, D., Pasquier, D., Yazyev, O.V., Kis, A.: 2D transition metal dichalcogenides. Nat. Rev. Mater. **2**, 8 **2**, 1–15 (2017)
51. Jayakumar, A., Surendranath, A., PV, M., 2D materials for next generation healthcare applications. Int. J. Pharm. **551**, 309–21 (2018)
52. Zavabeti, A., Jannat, A., Zhong, L., Haidry, A.A., Yao, Z., Ou, J.Z.: Two-dimensional materials in large-areas: synthesis, properties and applications. Nanomicro. Lett. **12** (2020)
53. Zheng, W., Lee, L.Y.S.: Beyond sonication: advanced exfoliation methods for scalable production of 2D materials. Matter **5**, 515–45 (2022)
54. Paton, K.R., Varrla, E., Backes, C., Smith, R.J., Khan, U., O'Neill, A., Boland, C., Lotya, M., Istrate, O.M., King, P., Higgins, T., Barwich, S., May, P., Puczkarski, P., Ahmed, I., Moebius, M., Pettersson, H., Long, E., Coelho, J., O'Brien, S.E., McGuire, E.K., Sanchez, B.M., Duesberg, G.S., McEvoy, N., Pennycook, T.J., Downing, C., Crossley, A., Nicolosi, V., Coleman, J.N.: Scalable production of large quantities of defect-free few-layer graphene by shear exfoliation in liquids. Nat. Mater. **13**, 624–630 (2014)
55. Pan, S., Valencia-Acuna, P., Kong, W., Liu, J., Ge, X., Xie, W., Zhao, H.: Efficient interlayer electron transfer in a MoTe2/WS2/MoS2 trilayer heterostructure. Appl. Phys. Lett. **118** (2021)

56. Lin, H., Wang, J., Luo, Q., Peng, H., Luo, C., Qi, R., Huang, R., Travas-Sejdic, J., Duan, C.G.: Rapid and highly efficient chemical exfoliation of layered MoS2 and WS2. J. Alloys Compd. **699**, 222–9 (2017)
57. Biccai, S., Barwich, S., Boland, D., Harvey, A., Hanlon, D., McEvoy, N., Coleman, J.N.: Exfoliation of 2D materials by high shear mixing. 2d Mater **6**, 015008 (2018)
58. Chu, T., Liu, D., Tian, Y., Li, Y., Liu, W., Li, G., Song, Z., Jian, Z., Cai, X.: Cationic hexagonal boron nitride, graphene, and MoS2Nanosheets heteroassembled with their anionic counterparts for photocatalysis and sodium-ion battery applications. ACS Appl. Nano Mater. **3**, 5327–34 (2020)
59. Huang, J., Songfeng, E., Li, J., Jia, F., Ma, Q., Hua, L., Lu, Z.: Ball-milling exfoliation of hexagonal boron nitride in viscous hydroxyethyl cellulose for producing nanosheet films as thermal interface materials. ACS Appl. Nano Mater. **4**, 13167–13175 (2021)
60. Alzakia, F.I., Tan, S.C.: Liquid-exfoliated 2D materials for optoelectronic applications. Adv. Sci. **8** (2021)
61. Telkhozhayeva, M., Teblum, E., Konar, R., Girshevitz, O., Perelshtein, I., Aviv, H., Tischler, Y.R., Nessim, G.D.: Higher Ultrasonic frequency liquid phase exfoliation leads to larger and monolayer to few-layer flakes of 2D layered materials Langmuir **37**, 4504–14 (2021)
62. Backes, C., Hanlon, D., Szydlowska, B.M., Harvey, A., Smith, R.J., Higgins, T.M., Coleman, J.N.: Preparation of liquid-exfoliated transition metal dichalcogenide nanosheets with controlled size and thickness: a state of the art protocol. J. Vis. Exp. **2016**, 54806 (2016)
63. Wang, J., Li, G., Li, L., Wang, J., Li, G., Li, L.: Synthesis strategies about 2D materials. In: *Two-dimensional Materials - Synthesis, Characterization and Potential Applications* (IntechOpen) (2016)
64. Coleman, J.N., Lotya, M., O'Neill, A., Bergin, S.D., King, P.J., Khan, U., Young, K., Gaucher, A., De, S., Smith, R.J., Shvets, I.V., Arora, S.K., Stanton, G., Kim, H.Y., Lee, K., Kim, G.T., Duesberg, G.S., Hallam, T., Boland, J.J., Wang, J.J., Donegan, J.F., Grunlan, J.C., Moriarty, G., Shmeliov, A., Nicholls, R.J., Perkins, J.M., Grieveson, E.M., Theuwissen, K., McComb, D.W., Nellist, P.D., Nicolosi, V.: Two-dimensional nanosheets produced by liquid exfoliation of layered materials. Science **1979**(331), 568–571 (2011)
65. Karthik Nagareddy, V., Octon, T.J., Townsend, N.J., Russo, S., Craciun, M.F., David Wright, C., Nagareddy, V.K., Octon, T.J., Townsend, N.J., Russo, S., Craciun, M.F., Wright, C.D.: Humidity-controlled ultralow power layer-by-layer thinning, nanopatterning and bandgap engineering of MoTe2. Adv. Funct. Mater. **28** 1804434 (2018)
66. Castellanos-Gomez, A., Barkelid, M., Goossens, A.M., Calado, V.E., Van Der Zant, H.S.J., Steele, G.A.: Laser-thinning of MoS 2: On demand generation of a single-layer semiconductor. Nano Lett. **12**, 3187–92 (2012)
67. Han, G.H., Chae, S.J., Kim, E.S., Güneş, F., Lee, I.H., Lee, S.W., Lee, S.Y., Lim, S.C., Jeong, H.K., Jeong, M.S., Lee, Y.H.: Laser thinning for monolayer graphene formation: Heat sink and interference effect. ACS Nano **5**, 263–268 (2011)
68. Qian, M., Zhou, Y.S., Gao, Y., Feng, T., Sun, Z., Jiang, L., Lu, Y.F.: Production of few-layer graphene through liquid-phase pulsed laser exfoliation of highly ordered pyrolytic graphite. Appl. Surf. Sci. **258**, 9092–9095 (2012)
69. Le, T.H., Oh, Y., Kim, H., Yoon, H.: Exfoliation of 2D materials for energy and environmental applications. Chem—A Eur. J. **26**, 6360–401 (2020)
70. Xia, Z.Y., Pezzini, S., Treossi, E., Giambastiani, G., Corticelli, F., Morandi, V., Zanelli, A., Bellani, V., Palermo, V.: The exfoliation of graphene in liquids by electrochemical, chemical, and sonication-assisted techniques: a nanoscale study. Adv. Funct. Mater. **23**, 4684–4693 (2013)
71. Yang, Y., Hou, H., Zou, G., Shi, W., Shuai, H., Li, J., Ji, X.: Electrochemical exfoliation of graphene-like two-dimensional nanomaterials. Nanoscale **11**, 16–33 (2018)
72. Hu, Z., Liu, Q., Chou, S.-L., Dou, S.-X., Hu, Z., Liu, Q., Chou, S.-L., Dou, S.-X.: Advances and challenges in metal sulfides/selenides for next-generation rechargeable sodium-ion batteries. Adv. Mater. **29**, 1700606 (2017)

73. Fang, Y., Li, X., Li, J., Yao, C., Hoh, H.Y., Hai, X., Lu, J., Su, C.: Janus electrochemical exfoliation of two-dimensional materials. J. Mater. Chem. A Mater. **7**, 25691–25711 (2019)
74. Hashemi, P., Sabaghi, D., Yang, S., Nia, A.S., Feng, X.: Electrochemical exfoliation to produce high-quality black phosphorus. J. Vis. Exp. **2022** (2022)
75. Yang, S., Zhang, P., Nia, A.S., Feng, X.: Emerging 2D materials produced via electrochemistry. Adv. Mater. **32** (2020)
76. Zhang, P., Yang, S., Pineda-Gómez, R., Ibarlucea, B., Ma, J., Lohe, M.R., Akbar, T.F., Baraban, L., Cuniberti, G., Feng, X.: Electrochemically exfoliated high-quality 2H-MoS2 for multiflake thin film flexible biosensors. Small **15** 1901265 (2019)
77. Kim, J., Kwon, S., Cho, D.H., Kang, B., Kwon, H., Kim, Y., Park, S.O., Jung, G.Y., Shin, E., Kim, W.G., Lee, H., Ryu, G.H., Choi, M., Kim, T.H., Oh, J., Park, S., Kwak, S.K., Yoon, S.W., Byun, D., Lee, Z., Lee, C.: Direct exfoliation and dispersion of two-dimensional materials in pure water via temperature control. Nature Commun. **6**:1 **6** 1–9 (2015)
78. Abdelkader, A.M., Cooper, A.J., Dryfe, R.A.W., Kinloch, I.A.: How to get between the sheets: a review of recent works on the electrochemical exfoliation of graphene materials from bulk graphite. Nanoscale **7**, 6944–6956 (2015)
79. Baig, N., Kammakakam, I., Falath, W., Kammakakam, I.: Nanomaterials: a review of synthesis methods, properties, recent progress, and challenges. Mater Adv **2**, 1821–71 ((2021))
80. Abid, N., Khan, A.M., Shujait, S., Chaudhary, K., Ikram, M., Imran, M., Haider, J., Khan, M., Khan, Q., Maqbool, M.: Synthesis of nanomaterials using various top-down and bottom-up approaches, influencing factors, advantages, and disadvantages: a review. Adv. Colloid Interface Sci **300** 102597 (2022)
81. Geim, A.K., Novoselov, K.S. (2007) The rise of graphene. Nat. Mater. 2007 6:3 **6** 183–91
82. Sun, Z., Martinez, A., Wang, F.: Optical modulators with 2D layered materials. Nat. Photon. 10:4 **10** 227–38 (2016)
83. Moss, D.J.: Fabrication methods for integrating 2D materials
84. Wang, N., Yang, G., Wang, H., Yan, C., Sun, R., Wong, C.P.: A universal method for large-yield and high-concentration exfoliation of two-dimensional hexagonal boron nitride nanosheets. Mater. Today **27**, 33–42 (2019)
85. Shuck, C.E., Sarycheva, A., Anayee, M., Levitt, A., Zhu, Y., Uzun, S., Balitskiy, V., Zahorodna, V., Gogotsi, O., Gogotsi, Y.: Scalable synthesis of Ti3C2Tx MXene. Adv. Eng. Mater. **22** 1901241 (2020)
86. Zhou, J., Palisaitis, J., Halim, J., Dahlqvist, M., Tao, Q., Persson, I., Hultman, L., Persson, P.O.Å., Rosen, J.: Boridene: Two-dimensional Mo4/3B2-x with ordered metal vacancies obtained by chemical exfoliation. Science **373**, 801–5 (2021)
87. Naguib, M., Mashtalir, O., Carle, J., Presser, V., Lu, J., Hultman, L., Gogotsi, Y., Barsoum, M.W.: Two-dimensional transition metal carbides. ACS Nano **6**, 1322–1331 (2012)
88. Ding, Y., Chen, Y.P., Zhang, X., Chen, L., Dong, Z., Jiang, H.L., Xu, H., Zhou, H.C.: Controlled intercalation and chemical exfoliation of layered metal-organic frameworks using a chemically labile intercalating agent. J. Am. Chem. Soc. **139**, 9136–9139 (2017)
89. Ren, L., Qi, X., Liu, Y., Hao, G., Huang, Z., Zou, X., Yang, L., Li, J., Zhong, J.: Large-scale production of ultrathin topological insulator bismuth telluride nanosheets by a hydrothermal intercalation and exfoliation route. J. Mater. Chem. **22**, 4921–4926 (2012)
90. Cheng, Z.L., Kong, Y.C., Liu, Z.: Li+/Na+ Co-assisted hydrothermal exfoliation for graphite into few-layer graphene nanosheets and their excellent friction-reducing performance. ACS Sustain. Chem. Eng. **7**, 19770–8 (2019)
91. Sayago, I., Hontañón, E., Aleixandre, M.: Preparation of tin oxide nanostructures by chemical vapor deposition. Tin Oxide Mater. 247–80 (2020)
92. Bhaviripudi, S., Mile, E., Steiner, S.A., Zare, A.T., Dresselhaus, M.S., Belcher, A.M., Kong, J.: CVD synthesis of single-walled carbon nanotubes from gold nanoparticle catalysts. J. Am. Chem. Soc. **129**, 1516–1517 (2007)
93. Sun, L., Yuan, G., Gao, L., Yang, J., Chhowalla, M., Gharahcheshmeh, M.H., Gleason, K.K., Choi, Y.S., Hong, B.H., Liu, Z.: Chemical vapour deposition. Nat. Rev. Methods Primers 1:1 **1** 1–20 (2021)

94. Choy, K.L.: Chemical vapour deposition of coatings. Prog. Mater. Sci. **48**, 57–170 (2003)
95. Kumar, N., Tomar, R., Wadehra, N., Devi, M.M., Prakash, B., Chakraverty, S.: Growth of highly crystalline and large scale monolayer MoS2 by CVD: the role of substrate position. Crystal Res. Technol. **53**, 1800002 (2018)
96. Adachi, M., Tsukui, S., Okuyama, K.: Nanoparticle synthesis by ionizing source gas in chemical vapor deposition. Jpn. J. Appl. Phys., 2 Lett. **42**, L77 (2003)
97. Pottathara, Y.B., Grohens, Y., Kokol, V., Kalarikkal, N., Thomas, S.: Synthesis and processing of emerging two-dimensional nanomaterials. Nanomater. Synthes. Des. Fabric. Appl. 1–25
98. Jang, H.S., Lim, J.Y., Kang, S.G., Hyun, S.H., Sandhu, S., Son, S.K., Lee, J.H., Whang, D.: Methane-mediated vapor transport growth of monolayer WSe2 crystals. Nanomaterials 9, 1642 (2019)
99. Xu, X., Guo, T., Kim, H., Hota, M.K., Alsaadi, R.S., Lanza, M., Zhang, X., Alshareef, H.N.: Growth of 2D Materials at the Wafer Scale. Adv. Mater. **34** (2022)
100. Walsh, L.A., Hinkle, C.L.: van der Waals epitaxy: 2D materials and topological insulators. Appl. Mater. Today **9**, 504–15 (2017)
101. He, Q., Li, P., Wu, Z., Yuan, B., Luo, Z., Yang, W., Liu, J., Cao, G., Zhang, W., Shen, Y., Zhang, P., Liu, S., Shao, G., Yao, Z., He, Q., Li, P., Wu, Z., Luo, Z., Yang, W., Liu, J., Cao, G., Shen, Y., Zhang, P., Shao, G., Yao, Z., Yuan, B., Zhang, W., Liu, S.: Molecular beam epitaxy scalable growth of wafer-scale continuous semiconducting monolayer MoTe2 on inert amorphous dielectrics. Adv. Mater. **31**, 1901578 (2019)
102. Fu, D., Zhao, X., Zhang, Y.Y., Li, L., Xu, H., Jang, A.R., Yoon, S.I., Song, P., Poh, S.M., Ren, T., Ding, Z., Fu, W., Shin, T.J., Shin, H.S., Pantelides, S.T., Zhou, W., Loh, K.P.: Molecular beam epitaxy of highly crystalline monolayer molybdenum disulfide on hexagonal boron nitride. J. Am. Chem. Soc. **139**, 9392–9400 (2017)

2D-MoS_2 and WS_2-Based Chemical Gas Sensor

Sajjad Hussain, Rohit, Sandeep Kaur, Akash Katoch, and Deepika Jamwal

Abstract Two-dimensional (2D) nanomaterials have been explored in gas sensing applications because of their excellent physical and chemical properties. In recent years, 2D-transition metal dichalcogenide (TMDs) and their hybrids have shown a propitious potential in the gas sensing application due to their excellent properties like high surface area, abundant adsorption site, and tunable electrical properties. The 2D morphology and layer-by-layer formation of TMDs at the nanoscale have shown the ability to detect gas molecules at lower temperatures or even at room temperature. This chapter will briefly discuss the gas sensing performance of 2D-MoS2 and WS2 nanomaterial and their hybrids towards different gas molecules in the parts-per-million range. We also discussed thoroughly the different strategies to enhance the gas sensing performance of 2D-TMDs. The decoration of zero-dimensional (0D) nanomaterial such as metal nanoparticles (Au, Ag, Pt, Pd), metal oxide nanoparticles (SnO_2, ZnO), and other metal chalcogenide quantum dots (Pbs). Heterostructure formation with different nanomaterials including metal oxide, metal dichalcogenide, Mxenes, and a carbon-based material such as reduced graphene oxide enhances the gas sensing performance due to the synergistic effect. Also, the impact of light illumination on the gas-sensing performance was discussed.

Keywords Gas sensor · 2D nanomaterial · Transition metal dichalcogenide · MoS_2 · WS_2

S. Hussain · Rohit · S. Kaur · A. Katoch (✉)
Smart Nanomaterials and Sensor Laboratory, Centre for Nanoscience and Nanotechnology, Panjab University, Chandigarh 160014, India
e-mail: akatoch@pu.ac.in

D. Jamwal (✉)
Department of Chemistry, University Institute of Sciences, Chandigarh University, Gharuan, Punjab 140413, India
e-mail: jamwaldeepika@gmail.com

A. Katoch et al. (eds.), *2D Materials*, Engineering Materials,
https://doi.org/10.1007/978-981-97-6258-3_2

1 Introduction

Atmospheric pollution, a reservoir of all toxic, harmful, and flammable gases, becomes a serious issue nowadays. Due to rapid industrialization and development in transportation, the use of other pollution-emitting technologies has become a serious threat to humans as well as to the ecology [1]. In 2019, the WHO declared air pollution a "new tobacco" because it causes severe effects on the human respiratory system and other organs [2]. To minimize the damage caused by these toxic gases, monitoring or detection of these harmful and toxic gases is necessary specifically within the limit of standard norms mentioned by different authorities. Therefore, real-time detection, cost-effective, reliable, low power consumption, and higher selectivity are required. For this, the semiconductor gas sensor is widely investigated in monitoring the gases having health hazards, as these materials fulfill the requirements of gas sensors [3, 4].

Different kinds of materials are explored in semiconductor-based gas sensors, such as transition metal oxide (zinc oxide (ZnO), tin oxide (SnO_2), titanium oxide (TiO_2), and tungsten oxide (WO_3)) and transition metal dichalcogenides (TMDs) (WS_2, MoS_2, WSe_2, $MoSe_2$, etc.,) including other materials of carbon-based (carbon nanotubes (CNTs), graphene, reduced graphene oxide(rGO)) and polymers (polypyrrole (PPy), polyaniline(PANI)) [5–7]. All these materials have their advantages and disadvantages in sensing applications. Metal oxides are the most widely explored and commercialized gas sensors. However, the sensor usually faces some concerns of high operating temperature and detection of multiple gases at the same time, which limits their application as real-time detecting gas sensors.

TMDs can detect gas analytes at low temperatures even at room temperature (RT), which reduces power consumption and cost, making them suitable materials for the fabrication of real-time detecting gas sensors [8, 9]. TMDs are layered inorganic semiconductors with the chemical composition of MX_2, M represents transition metals and X represents chalcogen (S, Se, and Te) exist in the form of X-M-X. In a single layer of TMDs, the transition metal (M) lies in between consecutive chalcogen (X) layers and is arranged periodically in a hexagonal structure. The interlayer M-X bond is covalent whereas the sandwiched layers are coupled with van der Waals force. The M-M bond in TMDs varies between 3.15 Å and 4.03 Å, depending on the size of the M and X, respectively. The transition metal M in MX_2 is either trigonal prismatic or octahedral coordinated with chalcogen atoms depending on the coordination between them. TMDs exhibit different types of polymorphs and polytypes. The common polymorphs that exhibit in TMDs are 1 T, 2 H, and 3R, where T, H, and R represent trigonal, hexagonal, and rhombohedral, and 1, 2, and 3 represent the number of the layers in the unit cell respectively as shown in Fig. 1 [10]. Depending on the stacking sequence, the 2 H polymorph has 3 types of polytypes: AbACaCBcB [11–13].

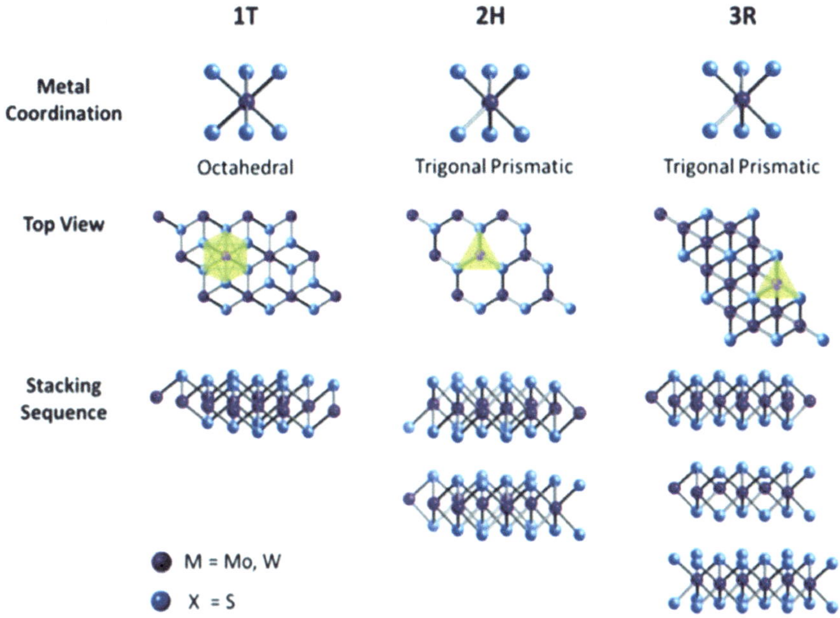

Fig. 1 Metal coordination and stacking sequence of 2D-TMDs structure in a unit cell. Reproduced with permission from Ref. [10]. Copyright 2017, Royal Society of Chemistry

The atomic thin-layered 2D-TMDs have excellent electronic properties along with a high surface-to-volume ratio, having more active sites, layer-dependent properties and the ability to detect gas analyte at low temperatures usually at RT offers opportunities for real-time.

monitoring. The modification in the properties of 2D-TMDs by exposing edge sites, doping or functionalizing with metal or metal oxide nanoparticles and hybrid formation with metal oxide, graphene, and its derivatives and with other 2D-TMDs enhance the sensor performance [14, 15]. The gas detection mechanism generally depends on two important factors, i.e., recognition and transduction. In 2D-TMDs, the sensor response is the estimation of charge transfer between the surface adsorbed gas molecules and the 2D-TMD surface of the sensing material. When the analyte gas molecules are exposed to sensing material (2D-TMDs), the number of gas molecules available on the surface of 2D-TMDs and charge carriers are exchanged between them. The total resistance modulation depends on the types of gas used i.e., oxidizing or reducing gas. The transfer of charge changes the conductivity (resistivity) of the sensing material (2D-TMDs). The sensor recovered to its baseline by the desorption of gas molecules which happens on stoppage of the gas supply. The defects that arise in 2D-TMDs during the synthesis process increase the active sites for gas adsorption. The adsorption of gas molecules on such defective sites or edge sites has higher adsorption energy (chemisorbed) on the surface, which as a result leads to more charge modulation and hence enhances the sensitivity on the sensor [16–18].

In this chapter, we discuss the sensors based on selective transition metal dichalcogenides. The discussion is mainly focused on the performance of molybdenum disulfide (MoS_2) and tungsten disulfide (WS_2) and their composite gas sensors.

2 Molybdenum Disulfide (MoS_2) Based Gas Sensor

MoS_2 is widely explored transition metal dichalcogenide in different applications from energy storage devices, catalytic reduction, photovoltaic, photodetector, and sensors due to their excellent stability for electron transport applications [19]. In a single-layer MoS_2, a molybdenum atom (M) is sandwiched between two sulfur (S) atoms with a covalent in-plane bonding between them. The band gap of the MoS_2 increased from 1.29 eV in bulk to 1.9 eV in monolayer MoS_2. MoS_2 was the first among the 2D-TMDs family widely explored in gas sensing applications.

2.1 Pristine MoS_2-Based Gas Sensor

Huu et al. [20] synthesized a few layered MoS_2 nanosheets via a probe sonication technique using ethanol and water in the volume ratio of 2:1 as solvent. The drop-casting method was used for sensor preparation using the exfoliated nanosheet on the Si substrate with a Pt electrode. To enhance the adhesion of the MoS_2 nanosheet on the Pt electrode, the sensor was vacuum heated at 300 °C. The sensing performance was measured at different temperatures from RT to 150 °C and the sensing response decreased with increasing the operating temperature. This type of trend is accredited to the surface defect reduction and more removal of gaseous species at high temperatures. The sensor was found highly selective toward NO_2 at RT and showed a good response and sluggish recovery time of 54 s and 1250 s, respectively.

The post-synthesis annealing of nanomaterial in air enhances the overall performance of the devices. To study the effect of annealing, Donarelle et al. [21] annealed the MoS_2 nanoflakes after exfoliation at different temperatures and studied the gas detection ability toward NO_2 gas. The post-annealing treatment of MoS_2 in the presence of air amends the electrical properties of the exfoliated nanosheet from p-type at 150 °C to n-type at 250 °C. The variation from p-type to n-type is attributed to the creation of more S vacancy as well as thermal partial oxidation of exfoliated MoS_2 at 250 °C. The concentration of sulfur vacancy increased from $2.3 \pm 0.5\%$ at 150 °C to $5.0 \pm 0.5\%$ at 250 °C and also the ratio of MoO_3/MoS_2 increased from $0.09 \pm 0.05\%$ at 20 °C to $0.97 \pm 0.05\%$ at 250 °C. Both p and n-type MoS_2 showed good sensing performance, the former showed better response and recovery kinetics and the latter showed sensible response with a good detection limit of up to 20 ppb at 200 °C. Since MoS_2 is widely investigated among the TMDs in the gas sensing application, their gas sensing performance at RT can be enhanced by vacancy, functionalization of their surface with metal or metal oxide and also by nanocomposite formation. Light illumination such as ultraviolet light (UV) [22], visible light [23], and near-infrared light (NIR) is also an effective strategy to enhance the gas sensing performance by creating more charge carriers that act as active sites for gas molecule adsorption [24]. Xia et al. [25] fabricated a gas sensor to study the impact of S vacancy, heterostructure, and light illumination on gas sensing performance and studied their gas sensing performance toward NO_2 gas. The presence of sulfur vacancy in MoS_2 (SV-MoS_2) improves the gas sensing performance, response is enhanced by almost 6.5 times compared to defect-free MoS_2 at RT. In the presence of NIR light, the response of SV-MoS_2 increased to 45% for 200 ppb at RT. The improved gas response is ascribed to (i) the formation of sulfur vacancy which acts as an additional active site for gas adsorption and reduces the adsorption energy for gas molecules. (ii) The creation of a more photogenerated charge carrier under NIR illumination. The major setback of both defect-free MoS_2 and SV-MoS_2 is that the reversibility of the sensor is very poor which leads to poor stability and repeatability. To overcome the poor recoverability of SV-MoS_2, a heterostructure is created by functionalizing the surface of SV-MoS_2 with ZnO quantum dots (ZnO QD). The SV-MoS_2/ZnO nanocomposite exhibited a gas sensing response of 226% for 200 ppb of NO_2 at RT and also reduced the response and recovery time to 75 s and 111 s from 201 and 247 s in the case of SV-MoS_2. The formation of heterostructure between SV-MoS_2 and ZnO QD created an additional active site for gas adsorption at the interface, resulting in more gas adsorption thus increasing the response. The ZnO QD acts as a passivation layer between gas molecules and SV-MoS_2 which raises the recovery rate by inhibiting the direct adsorption of atmospheric oxygen to the MoS_2 surface.

2.2 Nanoparticle Decorated/Functionalized MoS_2-Based Gas Sensor

Surface decoration/functionalization of TMDs with noble metal nanoparticles (Au, Pt, Ag, Pd) is also an effective strategy to enhance the gas sensing performance like sensitivity, stability, and selectivity due to their catalytic nature [26, 27]. The gas sensing performance is enhanced after functionalization with metal nanoparticles because of the two significant effects; (a) Chemical sensitization: the dissociation of gas molecules by the metal nanoparticle i.e., spillover effect, and (b) electronic sensitization: the formation of the junction at the interface between the metal atom and host material [28]. Theoretical studies also found that the decoration of host sensing material with metal nanoparticles enhances the sensing performance by enhancing the adsorption of target gas species by the catalytic effect [29–31]. Lee et al. [32] fabricated an H_2 gas sensor by decorating Pt nanoparticles (Pt NPs) on MoS_2. The morphological observation suggests that the Pt NPs decorated with the atomic layer deposition (ALD) technique grown well-ordered and densely on the MoS_2 surface with the plasma treatment as shown in Fig. 2a. The plasma treatments induce phase transformation on the top surface of the MoS_2 layer i.e., the formation of MoO_3 which creates a dangling bond, which facilitates the nucleation of Pt NPs. The presence of a sufficient amount of dangling bond in plasma-treated MoS_2 is responsible for the uniform, dense distribution of Pt NPs. The Pt/MoS_2 sensor showed a response of 440 at 250 °C for 1000 ppm of H_2. The sensing performance depends on the ALD cycle, the excessive Pt NPS on the MoS_2 surface lowers the response due to the hindrance in the transportation of dissociated H_2 atoms toward the MoS_2 surface. The morphology of the nanostructure also shows a significant impact on the device's performance. Different types of MoS_2 morphology (such as nanoflakes [33], nanosheets [34], nanoflowers [35] and nanoworms [36]) have been investigated in the gas sensing application. In most of this morphology basal plane which is less catalytically active is exposed to the outer environment. To investigate the gas sensing performance of edge-exposed MoS_2 Jaiswal et al. [37] fabricated a Pd-decorated edge-exposed MoS_2 thin film. The hybrid Pd/MoS_2 sensor displayed a 33.7% sensor response with a response and recovery time of 16 s and 38 s for 500 ppm of H_2 at 30 °C compared to pure MoS_2 which exhibits a response of 1.2%. The enhanced H_2 response of hybrid Pd/MoS_2 is attributed to three factors: (i) catalytic activity of Pd NPs which dissociate H_2 atoms and also increase the surface area. (ii) Edge-oriented MoS_2 which leads to the formation of porous nanostructure leads to increased surface area, and (iii) the formation of Schottky junction. The H_2 gas sensing mechanism of the Pd/MoS_2 gas sensor was explained as the electronic and chemical sensitization between H_2 gas and Pd/MoS_2 sensing material which is demonstrated in Fig. 2b.

The decoration MoS_2 of Zero-dimensional (0D) nanomaterial increases the gas sensing performance. To take advantage of 0D nanomaterial and to overcome the problem of slow recovery time of MoS_2 at RT. Liu et al. [38] used a decorated MoS_2 surface with PbS QD to enhance the sensitivity and selectivity of sensor toward NO_2 gas. The sensor design and fabrication process are illustrated in Fig. 3a and b. The PbS/MoS_2 is highly selective to NO_2 at RT and exhibits a response five times more than the pristine MoS_2 and also fast response/recovery kinetics of 15/16 s respectively as shown in Fig. 3c and d. The author also discussed the effect of humidity on the sensing parameter and observed that the presence of humidity enhances the gas performance. The response/recovery kinetics become faster in the presence of humidity. The enhanced response in the presence of humidity is credited to the increased resistance by the adsorption of dissociated H_2O molecules, i.e., the H^+ and OH^-. The NO_2 molecules adsorbed on the surface by knocking out the weakly physisorbed water molecules, further increasing the resistance, and thus enhancing the response. The PbS QD occupies the high adsorption binding site on the MoS_2 surface and increases the sensitization of MoS_2 with PbS QD by increasing the surface-to-volume ratio. This increment in sensitization of MoS_2 creates more defective sites (dangling bonds, PbS sites, and S defects). The selective NO_2 receptor, favorable charge transport at the interface, good transducer and receptor function as well as utility factors responsible for enhanced performance to NO_2.

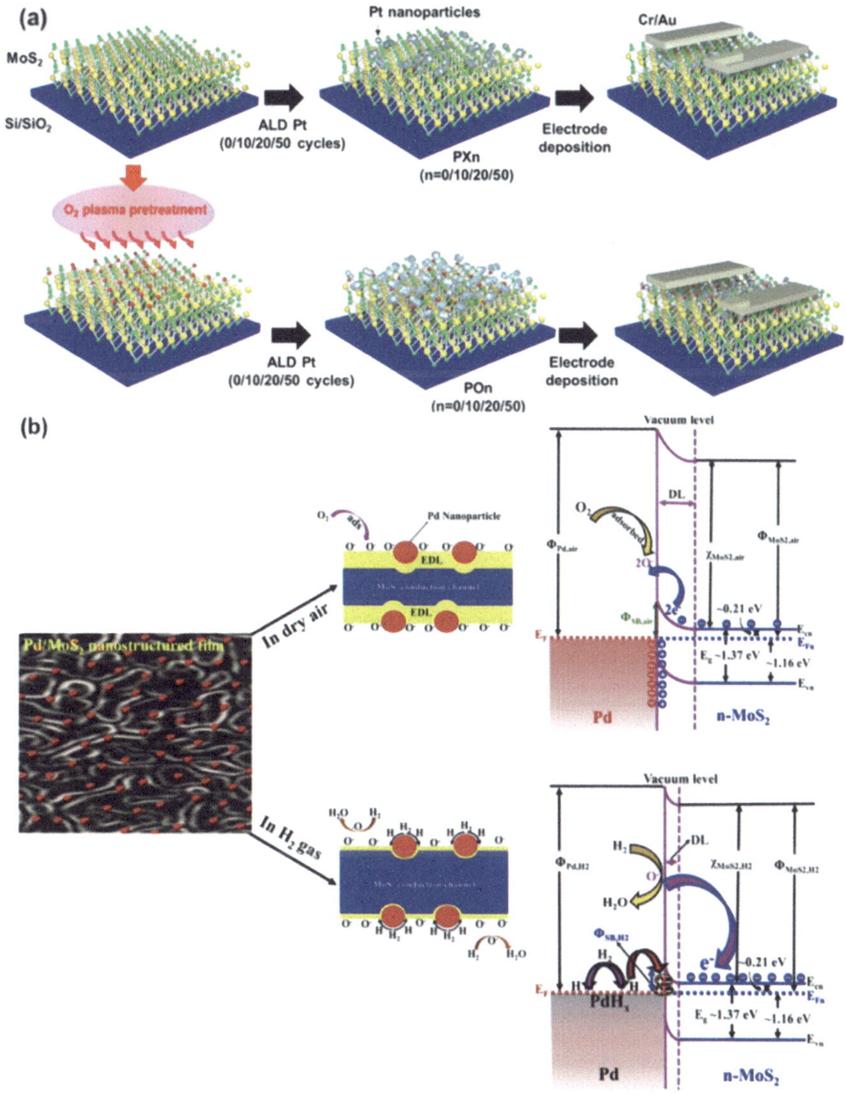

Fig. 2 **a** Schematic of ALD Pt-on-MoS_2 hydrogen sensor fabrication process. Reproduced with permission from Ref. [32]. Copyright 2022, Elsevier. **b** Schematic diagram of H_2 gas sensing mechanism of Pd functionalized vertically aligned edge-oriented MoS_2 sensor. Reproduced with permission from Ref. [37] Copyright 2020, Elsevier

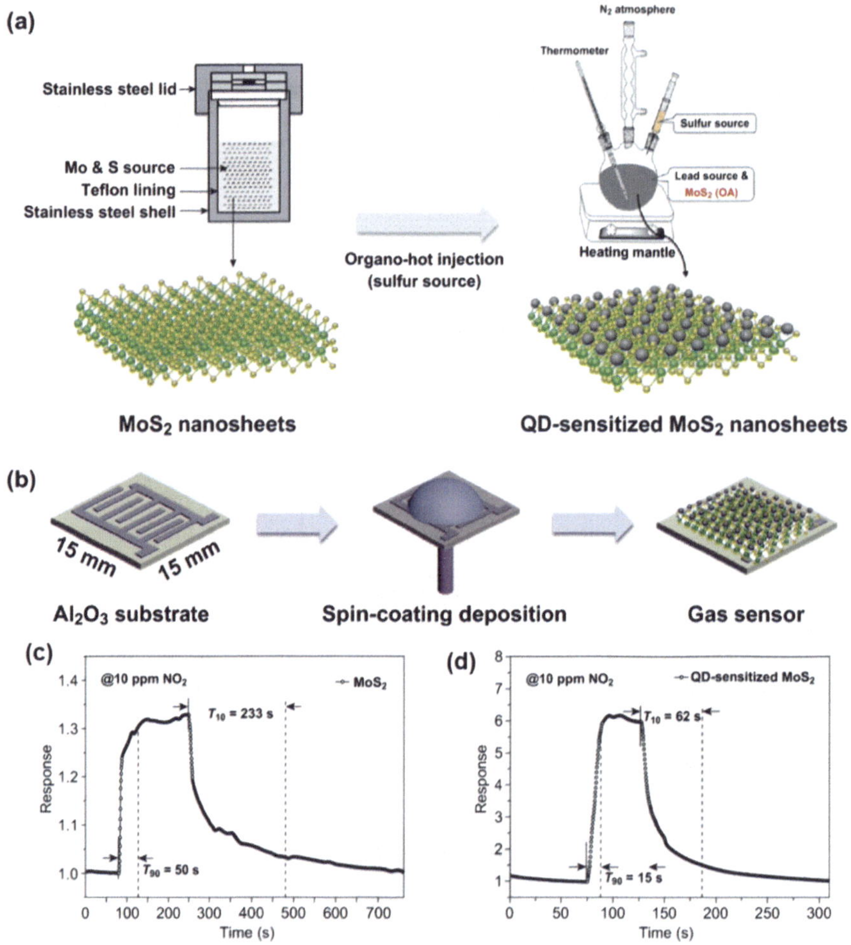

Fig. 3 Schematic illustration of **a** Synthesis of MoS_2 nanosheets and QDs-sensitized MoS_2 nanosheets and **b** Fabrication process of the gas sensor. The transient response curve of **c** MoS_2 nanosheets and **d** QDs sensitized MoS_2 nanosheets gas sensor to 10 ppm of NO_2 at room temperature. Reproduced with permission from Ref. [38] Copyright 2020, Springer

2.3 *MoS_2 Hybrids and Heterostructures*

The stacking and aggregation of TMDs nanosheet reduce the number of active sites resulting in a reduction in sensing response. The decoration of metal nanoparticles is an effective solution to improve the sensor performance however cost offsets their utilization. The integration of TMDs with metal oxide nanoparticles is another reasonable approach to enhance sensor performance such as sensor response, stability, and recoverability. In the hybridization of metal oxide and TMDs, the synergetic effect such as the geometrical aspect, electronic aspect, and chemical aspect comes into play. The hybrid of TMDs with a metal oxide such as WO_3, ZnO, SnO_2, etc. not only enhances the gas adsorption site but also works as a channel for electron transport which as a result enhances the sensitivity, recoverability, and adsorption/desorption time. Singh et al. [39] fabricated a hybrid MoS_2/SnO_2 sensor to detect ammonia (NH_3) at RT. The sensor showed outstanding detection of NH_3 in the concentration of 0.4 to 200 ppm. The performance of the sensor depends on the choice of substrate and its surface roughness. The surface roughness of the substrate was found effective in enhancing the active surface roughness of the material. In this case, the response of the sensor is almost identical to the sensor fabricated on alumina and quartz substrate. However, a significant difference in the response and recovery was observed on different substrates. The theoretical study supports the experimental result of higher sensitivity of the sensor toward NH_3. The density functional theory showed that NH_3 gas is strongly adsorbed on the surface of the MoS_2/SnO_2 hybrid and the absorption energy for NH_3 is −0.4354 eV in comparison to 0.0939 eV, −0.2680 eV, and −0.0789 eV for H_2S, CH_3OH and HCHO, respectively.

In addition to SnO_2, the hybrid nanostructure of MoS_2 was formed with other metal oxides such as ZnO [40], CuO [41], PtO_2 [42], WO_3 [43], and TiO_2 [44] to detect various toxic gas molecules, which enhance the gas sensing response and improve the selectivity [45]. To improve the selectivity of the sensor towards particular analyte gas molecules Singh et al. [46] fabricated a gas sensor of MoS_2/WO_3 nanocomposite by coating the hydrothermally synthesized nanocomposite on a gold electrode. They studied their gas sensing performance towards H_2S and ethanol at different temperatures to enhance selectivity. The operating temperature for the sensor device is significant in determining sensor performance because different decisive factors like reaction barrier energy, adsorption of gas molecules, and the reaction of gas molecules with preadsorbed oxygen species on the sensor surface all depend on the operating temperature. Thus, operating the sensor at the optimal working temperature is necessary to attain maximum sensing performance and enhance sensitivity. The prepared MoS_2/WO_3 is selective toward ethanol at 260 °C and at 320 °C it is selective towards H_2S gas and has shown a response of 17 for 5 ppm of ethanol and 15 for 0.5 ppm of H_2S. The detection of ethanol at lower temperatures is because the LUMO (lowest unoccupied molecular orbit) energy of ethanol (0.125 eV) lies lower than that of H_2S (0.685 eV). The lower LUMO energy enhances the electron-capturing capability and facilitates gas molecule detection at lower temperatures. The gas sensing performance of the MoS_2/WO_3 is due to the synergetic effect due to the formation of heterostructure which increases the surface area and creates more active sites.

Graphene and its derivatives, i.e., GO and RGO are a key class of carbon family and possess excellent conducting and semiconducting properties. The large surface-to-volume ratio, good electrical conductivity, mobility, and a large number of active sites in the form of functional groups, etc. attributed to the good sensing performance of graphene-based nanomaterial. The hybrid of transitional metal dichalcogenide (TMDs) with graphene-based material enhances the sensing performance because of their synergistic effect (Table 1).

Table 1 Summary of the various MoS_2 and their hybrids for gas sensing application

S. no	Material	Synthesis method	Gas	Operating temperature (°C)	Response (concentration)	Response/Recovery time	Limit of detection (LOD)	Reference
1	MoS_2 NS	Exfoliation	NO_2	RT	88% (5 ppm)	60 s/1250 s	27 ppb	[20]
2	SV-MoS_2	Hydrothermal	NO_2	RT	226% (0.2 ppm)	75 s/111 s	0.1 ppb	[25]
3	MoS_2/WO_3	Hydrothermal	H_2S	320	15 (0.5 ppm)	49 s/57 s (50 ppm)	–	[46]
			Ethanol	260	17 (5 ppm)	80–100 s/–		
4	Pd/MoS_2	DC Sputtering	H_2	RT (30)	~33.7% (500 ppm)	~16 s/38 s	–	[37]
5	PbS/MoS_2	Hydrothermal	NO_2	RT (25)	6.15 (10 ppm)	15 s/62 s	94 ppb	[38]
6	MoS_2/SnO_2	Hydrothermal	NH_3	RT (30)	10% (0.4 ppm)	2 s/10 s	–	[39]
7	MoS_2/rGO	Hydrothermal	HCHO	RT	~60% (10 ppm)	17 s/–	–	[23]
8	MoS_2/SnS	Hydrothermal	NO_2	RT	2.8 (250 ppb)	74 s/140 s	–	[47]
9	MoS_2/CuO	Hydrothermal	NO_2	RT	8.98	18 s/53 s	50 ppb	[41]
10	MoS_2/GaN	CVD	NO_2	27	98.3% (50 ppm)	184 s/369 s	–	[48]

A visible-light-illuminated rGO/MoS_2 hybrid sensor was prepared by Wang et al. [23] to demonstrate the formaldehyde detection at RT. The rGO/MoS_2 hybrid shows good sensitivity to formaldehyde (HCHO) and their response increases 8.2 times under visible light illumination in comparison to under dark. The photocatalytic oxidation of HCHO to H_2O and CO_2 on the MoS_2 surface is responsible for increasing the response under visible light illumination. The photocatalytic oxidation of HCHO is triggered by the formation of chemisorbed oxygen species due to the electrons available on the surface. The photocatalytic activity and the band alignment energy at the rGO/MoS_2 interface are responsible for the enhancement of gas sensing performance at RT.

3 Tungsten Disulfide (WS_2) Based Gas Sensor

Tungsten disulfide (WS_2) is the other most widely explored TMDs in gas sensing applications because of its nearly similar semiconducting properties to MoS_2 due to the identical crystal structure. In WS_2, the tungsten atom (W) is sandwiched between two sulfur (S) atoms and is covalently bonded. The band gap of WS_2 varies from 1.2 eV to 2.1 eV with varying the number of layers in WS_2.

3.1 Pristine WS_2 Based Gas Sensor

Similar to MoS_2, WS_2 has also shown good sensing performance to different harmful gases at low temperatures. Li et al. [49] prepared an NH_3 sensor using liquid exfoliated WS_2 flakes and operated at RT. The sensor exhibits a good sensing response to NH_3 at RT and has fast adsorption and desorption kinetics. The superior gas detection ability for NH_3 was likely to be due to the presence of Lewis acid centre $SO_4{}^{2-}$ which acts as an excellent binding site for NH_3 gases. The fast desorption kinetics of WS_2 sensor is because of the weak binding energy of NH_3 of WS_2. In another study, Patrick et al. [50] fabricated a NO_2 gas sensor using a WS_2 nanosheet (NS) through a liquid exfoliation technique. The sensor exhibits a sensitivity of 27% for 10 ppm of NO_2 at RT with rapid response and recovery time. The underlying sensing mechanism for the detection of NO_2 gas molecules on WS_2 is described as follows:

$$WS_2 + NO_2(gas) = WS_2^+ - NO_2^-(ads)$$

$$WS_2^+ - NO_2^-(ads) = WS_2 - NO_2^-(ads) + h_{VB}^+$$

WS_2 sensor showed excellent performance towards different harmful gas analytes at RT. However, their slow recovery at RT or low temperature limits their application. In a layered material, the thickness of the film or the number of layers has a substantial impact on the recovery. Qin et al. [51] observed the recovery rate and direct relationship with layer number. To investigate the relationship, an NH_3 gas sensor was fabricated using WS_2 nanosheets (NSs) with different layer numbers on the alumina substrate by the drop-casting method. The recovery time for NH_3 gas decreased with the decrease in layer number from ~3342 s for bulk (40-layer WS_2) to ~271 s for monolayer WS_2 (1L WS_2) but the response shows the opposite behaviour. NH_3 molecules were adsorbed on the surface of WS_2 NS and donated back the electrons to WS_2. In the case of 1L WS_2 the adsorption energy is low and thus easy to desorb. However, in multilayer or bulk WS_2, NH_3 gas adsorbs on the surface as well as in the interlayer and interacts with both the adjacent layers. Thus, NH_3 molecules diffused into the inner layer have higher adsorption energy so difficult to desorb at RT. The first principal study supports the experimental result showing that NH_3 adsorbed on the surface of 1L WS_2 has a binding energy of −0.196 eV, whereas in the case of bulk, the binding energy was −0.356 eV. Thus, with an increase in the layer number, the ratio of NH_3 intercalated to the NH_3 adsorb on the surface increases thus longer the recovery time.

The gas sensing mechanism of TMDs is acknowledged as similar to the metal oxide semiconductor-based sensors, where the electron movement among the gas molecules/preadsorbed oxygen species on the surface of sensing material takes place, however, it is still in its infancy. To understand the sensing mechanism Liu et al. [52] fabricated a WS_2 sensor and investigated their sensing behaviour to both oxidizing (NO_2) and reducing (H_2S) gas. The sensor shows a good sensing response for both NO_2 and H_2S; however, it is selective toward NO_2 gas. It was reported that the response of NO_2 gas decreases in the presence of oxygen as background gas, the response decreases with increasing concentration of oxygen. For H_2S sensing the presence of oxygen is advantageous, i.e., the response of H_2S increases with increased concertation of oxygen but up to a certain limit after that response starts decreasing. Thus, it may be concluded that for oxidizing gas the presence of preadsorbed oxygen species may interfere with the incoming NO_2 gas, and for reducing gas the presence of oxygen is required to enhance the response.

3.2 *Metal and Metal Oxide Decorated WS_2-Based Gas Sensor*

Similar to MoS_2, the pristine WS_2 sensor shows the same hindrance of poor stability and incomplete recovery at RT. Decoration of metal nanoparticles is an effective strategy to enhance the sensor response and the recovery at RT, as the catalytic nature of noble metal nanoparticles enhances the surface activity, provides more adsorption sites on TMDs surface, and increases the active surface area. Kim et al. [53] fabricated an Au-decorated WS_2 gas sensor on SiO_2 grown on a Si substrate. The decoration of Au nanoparticles (Au NPs) not only increases the sensor detection ability but also improves the selectivity of the gas sensor. The sensor response increases from 1.05 in pristine WS_2 to 1.48 by decorating with Au NPs. The decoration of the WS_2 sensor with Au NPs enhances the selectivity towards CO, which is attributed to the fact that

Au is a good catalyst for CO oxidation. The enhanced response of the fabricated sensor at RT is due to the following reasons: (i) the spillover effect caused by the metal NPs, (ii) the catalytic effect, (iii) the smaller kinetic diameter of CO gas and (iv) the development of the junction at Au and WS_2 interface. It is observed that the response depends on the size and density of NPs. The same group [54] studied the gas sensing performance of Au-SnO_2 co-decorated WS_2 sensor at RT under self-heating mode. In this case, a flexible substrate (polyamide) is employed to make a wearable, flexible, lightweight, and portable gas sensor. The co-decoration of WS_2 by Au and SnO_2 increased the response of the sensor by almost 2.5 times compared to the Au-WS_2 gas sensor, and the maximum response was achieved at 4.7 V and the power consumed by the sensor is 4.24 μW. Such low-power-consuming gas sensors are essential for practical applications. The flexibility of the sensor is demonstrated by bending, tilting, and stretching and it was observed that both bending and tilting have a negligible effect on the sensor response. The response was reduced to 2.9 from 3.6 after binding and tilting 1000 times as shown in Fig. 4a–d. Stretching the sensor up to 1 nm shows a nominal effect on response, and response/recovery time; however, when stretched above 1 nm, the sensor performance degrades significantly as shown in Fig. 4e and f. Humidity is another factor that affects the performance of the sensor. Humidity generates a conducting path and its ionic conducting nature decreases resistance also the adsorption of humidity on the active surface reduces the active sites for gas adsorption, which reduces sensor performance. The favorable sensing response of the Au-SnO_2 co-decorated WS_2 sensor for CO gas is due to the following reasons: (i) The high sensing property of SnO_2. (ii) the spillover effect of Au NPs and also the excellent catalytic effect of Au on CO gas, (iii) the formation of Schottky junction between Au/WS_2 and Au/SnO_2 and the formation heterojunction between SnO_2/WS_2 and Au/SnO_2/WS_2.

Implantation is another widely explored approach in metal oxide sensors and is a promising approach to enhancing gas sensing parameters. In implantation ions doped into the host material induce different kinds of defects such as interstitial defects, substitutional and vacancies that affect the gas sensing performance. The sensing performance in ion-implanted material depends on the doses of implantation. Kim et al. [55] studied the impact of antimony (Sb) implantation at different doses on the gas sensing performance of WS_2 NSs. The implantation with Sb changes the electrical properties from p-type to n-type, which is due to the creation of more Sulfur (S) vacancies. The Sb implanted with a dose of 2×10^{13} ions/cm^2 displayed a reasonable response to CO gas at RT. Annealing is also an important parameter that modulates the sensing parameter. The post-thermal annealing of the Sb-implanted WS_2 NSs at 500 °C further enhanced the CO response at RT, which is either due to electron generation or improved mobility. The optimal working temperature is the temperature where the sensor exhibits the best sensing performance compared to the other temperatures on lower and higher sides. The self-heating technique is a cost-effective technique to achieve the optimal working temperature. In the self-healing technique, heat is generated within the material by applying a voltage, the generated heat increases the working temperature of the sensor. Sb-implanted WS_2 NSs sensor

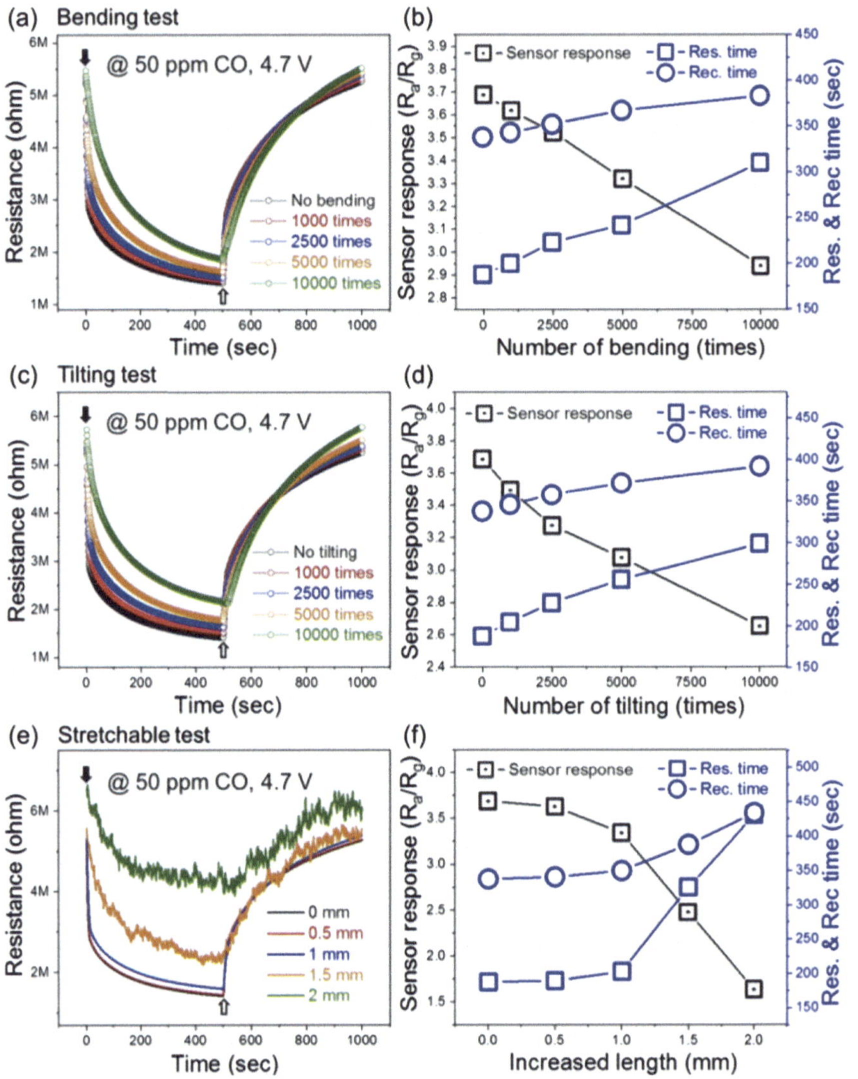

Fig. 4 a and b Dynamic resistance graphs and corresponding sensor response, response time and recovery time by varying the number of bending cycles. **c** and **d** Dynamic resistance plots and CO gas for various tilting cycles Corresponding sensor response, response time and recovery time with varying the number of tilting cycles. **e** and **f** Dynamic resistance plots and corresponding sensor parameters versus the increased length. For Au- SnO_2-co-decorated WS_2 to 50 ppm CO gas at RT (24 °C). Reproduced with permission from Ref. [54] Copyright 2021, Elsevier

achieved maximum response with a voltage supply of 4.8 V. The selectivity of the sensor towards CO is improved after decorating Sb implanted WS_2 by Au NPs.

3.3 WS_2 Hybrids and Heterostructures-Based Gas Sensor

The development of heterostructures between metal chalcogenides is also another methodology to enhance the sensing performance and stability. The heterostructure interface plays a substantial role in modulating the recital of the sensor device. A clear interface at the junction is necessary to attain a reliable sensing performance. However, for the formation of a clear interface at the junction lattice matching between the material should be necessary, which is quite challenging. Jha et al. [56] prepared a simple and cost-effective $WS_2/WO_3.H_2O$ heterostructure and well-matched lattice heterostructure. Heterostructure of $WS_2/WO_3.H_2O$ prepared by partial oxidation of WS_2 NSs prepared using the exfoliation technique. The prepared $WS_2/WO_3.H_2O$ NSs were drop cast on the interdigitated electrode (IDE) to fabricate a gas sensor. The sensor has shown high performance toward ammonia (NH_3) gas at RT. The sensor has shown a response of 11.3% for 50 ppb of NH_3 at RT and exhibits a reasonable response recovery time of 55 s and 23 s for 2 ppm of NH_3 at RT. The high sensitivity and selectivity towards NH_3 gas were attributed to the following points: (i) the presence of more active sites such as WS_2 surface, $WO_3{\cdot}H_2O$ surface, and the interface of $WS_2/WO_3.H_2O$, (ii) the presence of Lewis acid centre (SO_4^{2-}), which is an active surface for Lewis bases molecule NH_3. The decoration or functionalization approach was found effective in enhancing the gas-sensing performance of TMDs gas sensors. The formation of 2D-0D and 2D-1D hybrid heterostructures of TMDs is highly feasible to increase the sensing parameters of TMD sensors. The construction of a hierarchical heterostructure is another approach to enhance the gas sensing performance, where the advantage of geometry is used. Han et al. [58] prepared a quasi 2D-2D WS_2-WO_3 hybrid heterostructure by using a controllable sonochemical oxidation process. Sonochemical oxidation, i.e., oxidation in an aqueous medium is a controlled oxidation process and it leads to the formation of a higher number of uniform heterojunctions and is advantageous than the thermal oxidation process. The oxidized counterparts are well aligned on the exfoliated WS_2 nanoflakes, which construct a hierarchical heterostructure with a large number of heterojunctions. The formation of hierarchical nanostructures utilizes both geometrical and electronic effects together and reduces the probability of stacking WS_2 nanoflakes. The presence of a uniform network at the junction of hierarchical heterostructures facilitates the adsorption process of gas molecules. The presence of both W^{4+} and W^{6+} peaks in XPS confirmed the formation of $WS_2@WO_3$ hybrid heterostructure. The $WS_2@WO_3$ hybrid heterostructure shows excellent sensing characteristics to NO_2 gas RT with a response value of 16.7 for 5 ppm of NO_2, a response time of 34 s, and exhibits excellent stability only degrades the 8% response value after 6 months. The sensor showed an impressive detection limit of 50 ppb. The concentration of H_2O_2 oxidizing agent in an acidic medium has a strong influence over the sensing performance of $WS_2@WO_3$ hybrid heterojunction. With a higher concentration of oxidizing agent excessive WO_3 NSs are formed, which act as a conductive path and reduce the interaction of NO_2 gas molecules with WS_2 NS thus reducing the response value. The formation of a heterostructure modulates the charge carrier, thus generating an energy barrier and providing additional active sites at the junction are considered to enhance the gas sensing performance of $WS_2@WO_3$ hybrid heterostructure to NO_2 gas.

The performance of the gas sensor is related to the morphology of the active material at the nanoscale. For instance, the porous nature of the nanomaterial also affects the gas sensor performance. The porosity increases the surface area and spontaneously enhances the active sites for incoming gas moieties and limits the agglomeration of the active material, all of which contribute to enhancing the sensor performance. Considering the above advantages Qu et al. [58] synthesized a mesoporous WS_2/MoO_3 hybrid heterostructure by a hydrothermal method to overcome the instability and stacking issue of WS_2 nanoflakes. MoO_3 in the hybrids acts as a spacer which helps to limit the stacking of WS_2 nanoflakes. The sensing performance of the hybrid WS_2/MoO_3 heterostructure was investigated with different gases at RT and it shows high sensitivity to NH_3. The hybrid heterostructure exhibits a response of 31.58% for 3 ppm of NH_3 at RT, which is 17.7 and 57.4 times larger than pure MoO_3 and WS_2, respectively. The sensor exhibits excellent recoverability and stability. The enhancement in sensor performance of the WS_2/MoO_3 hybrid heterostructure is attributed to the porous nature and the synergetic effect of WS_2/MoO_3 hybrid heterostructure, where MoO_3 provides a more active site for gaseous species and WS_2 provides a conducting channel.

MXenes are a novel member of 2D material having the general formula $M_{n+1}X_nT_x$ and are widely explored in different applications because of their excellent properties. Currently, MXenes attract significant importance in gas sensing applications and have shown favorable sensing performance because of their good electrical properties, and surface-rich properties. The layered 2D feature of MXenes is beneficial for their hybridization with other 2D layered materials like TMDs and is an effective strategy to enhance the gas sensor performance [59]. Based on this, Quan et al. [60] prepared a flexible paper-based gas sensor by integrating $Ti_3C_2T_x$—MXene electrode (ME) and $Ti_3C_2T_x/WS_2$ gas sensing film. The choice of electrode has a significant impact on the gas sensing performance as the poor electrical conductivity between conventional metal electrodes and semiconducting sensing material becomes a hindrance in sensing performance. The performance of $Ti_3C_2T_x/WS_2$ on ME exhibits superior sensing performance than on the Au interdigitate electrode (AuE). The response of ME + $Ti_3C_2T_x/WS_2$ showed a response 3.2 times higher than AuE + $Ti_3C_2T_x/WS_2$ for 1 ppm of NO_2 at RT. The recovery time of the sensor is also improved on the ME electrode. The enhancement of the sensing performance of the ME + $Ti_3C_2T_x/WS_2$ gas sensor is attributed to the formation of a linear ohmic junction at the interface of ME and $Ti_3C_2T_x/WS_2$. The theoretical limit of detection is also improved in ME + $Ti_3C_2T_x/WS_2$ gas sensor, which is 11 ppb compared to 15 in AuE + $Ti_3C_2T_x/WS_2$ sensor. The ME + $Ti_3C_2T_x/WS_2$ sensor exhibits excellent performance even under mechanical deformation. After bending at 60° and then again flat for 500 times there is no significant degradation in response but its eventual response increases slightly which is due to the defect creation or the formation of micro-cracks during bending that provides additional active sites for gas adsorption.

Recently van der Waals heterostructure has considerable attraction among the various other heterostructures. The layered nature of 2D TMDs was advantageous in the construction of van der Waals junctions either by simply transferring or by growing directly over the other TMDs. The 2D Van der Waals heterojunction-based gas sensor has some significant properties and is tuned precisely to improve the gas sensing properties [61]. The heterostructure formed among 2D-TMDs has shown excellent sensing properties due to the synergetic effect of both properties. Moumen et al. [62] for the first time reported a p-p heterostructure of WS_2-WSe_2 synthesized by liquid exfoliation method after modification with ambient pressure CVD as shown in Fig. 5a. The band diagram of the p-p WS_2-WSe_2 heterojunction is shown in Fig. 5b. The WS_2-WSe_2 heterostructure shows a response of 57% towards 700 ppb of NO_2 at RT as shown in Fig. 5c and exhibits excellent stability for more than 2 months and exhibits high selectivity to NO_2 gas molecules. The enhanced response is due to the contribution of the following factors: (i) Creation of more active sites because of increased specific area. (ii) Creation of more defects in the form of S and Se vacancies and the p–p heterojunction influence to increase the overall properties of the sensor (Table 2).

4 Conclusion

TMDs are proven to be a reliable sensor to detect toxic, harmful, and flammable gases at low temperatures. The unique physical and chemical properties of TMDs such as layer-dependent electronic properties, high adsorption site, and high surface-to-volume ratio are favorable features for gas sensors. The reasonable sensing performance of TMDs at low operating temperatures provides a setup for fabrication of real-time monitoring and detection of gas at an expanse of low power consumption and low cost. The pristine TMDs (MoS_2 and WS_2) sensor showed good sensing performance; however, their poor recoverability at RT restrained their particle application. The recoverability as well as sensing performance in TMDs sensor is enhanced by decoration with metal nanoparticles, metal oxide nanoparticles, and also heterostructure or hybrid with TMDs or other materials like reduced graphene oxide, metal oxide, MXenes. However, the lattice matching and adhesion in heterostructure and adhesion of metal nanoparticles need to be improved to achieve maximum response. It was found that adhesion is enhanced by annealing but it costs the oxidation of TMDs. The sensing performance is also improved by illuminating the sensing material with UV and visible light which creates complexity in the fabrication.

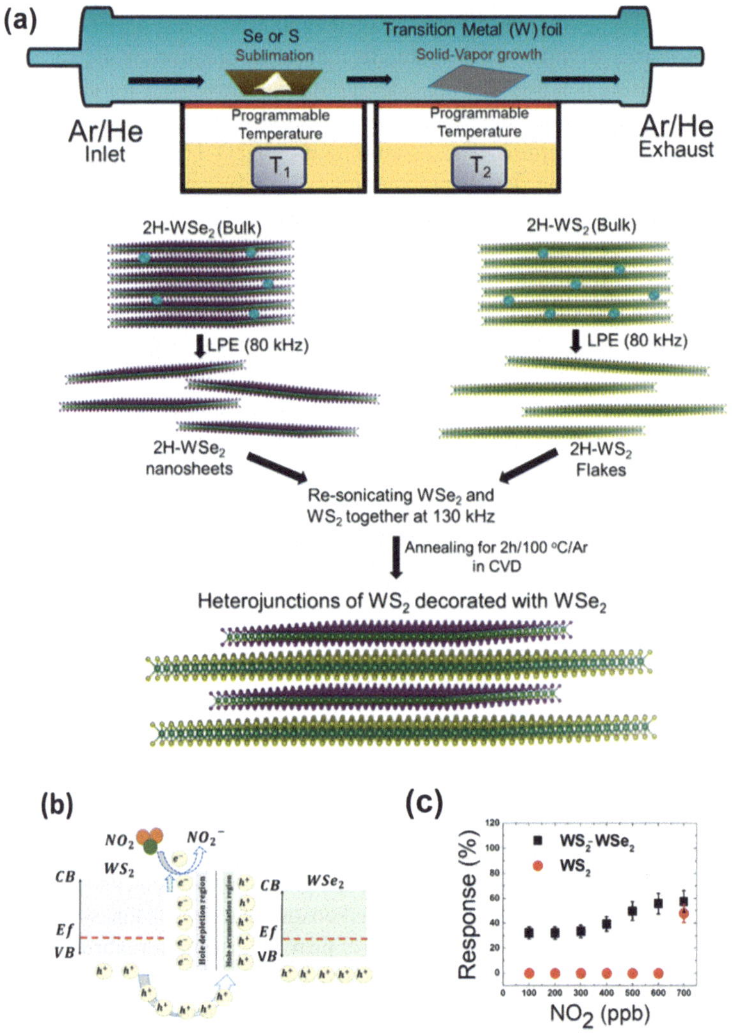

Fig. 5 **a** Schematic of WS2-WSe2 nanocomposite heterojunction synthesis using ambient pressure CVD furnace. **b** Sensing mechanism of p–p WS_2-WSe_2 heterojunction and **c** Response of pristine WS_2 and WS_2-WSe_2 heterojunction sensor toward NO_2 at different concentrations. Reproduced with permission from Ref. [62] copyright 2023, American Chemical Society.

Table 2 Summary of the various WS_2 and their heterostructure for gas sensing application

S.no	Material	Synthesis method	Gas	Operating temperature (°C)	Response (Concentration)	Response/recovery time	LOD	Ref
1	Au-Sb-WS_2	–	CO	20	3.90 (50 ppm)	23 s/410 s	–	[55]
2	$WS_2 \cdot WO_3/H_2O$	Exfoliation	NH_3	RT	254.6% (2 ppm)	55 s/23 s	50 ppb	[56]
3	WS_2-WO_3	Exfoliation	NO_2	25	16.7 (5 ppm)	34 s/–	50 ppb	[57]
4	WS_2/MoO_3	Hydrothermal	NH_3	22 ± 3	31.58% (3 ppm)	57 s/226 s	9 ppb	[58]
5	$Ti_3C_2T_x/WS_2$	CVD & Exfoliation	NO_2	RT	15.2% (1 ppm)	–/60 s	11 ppb	[60]
6	SnO_2/WS_2	Exfoliation	NH_3	25	850% (10 ppm)	–/–	10 ppb	[63]
7	WS_2/SnO_2 UV-light	Exfoliation	NO_2	25	340% (5 ppm)	10 s/9 s	–	[64]
8	rGO/WS_2	Hydrothermal	NH_3	33.5	121% (10 ppm)	~60 s/300 s	–	[65]
9	$T_3C_2T_x/WS_2$ Visible light	Exfoliation	NO_2	RT	~60% (2 ppm)	56 s/53 s	10 ppb	[66]
10	WS_2/PdO WS_2/PtO	CVD	NO_2	RT	26.5 27.5 (800 ppb)	–/–	25 ppb	[67]
11	WS_2/PbS	Exfoliation	NO_2	RT	~12 (10 ppm)	–/–	10 ppb	[68]

LOD: Limit of detection

Conflict of Interest Statement The author declares no competing interests.

References

1. Dai, J., Ogbeide, O., Macadam, N., Sun, Q., Yu, W., Li, Y., Su, B.L., Hasan, T., Huang, X., Huang, W.: Printed gas sensors. Chem Soc Rev **49**, 1756–1789 (2020)
2. Dhall, S., Mehta, B.R., Tyagi, A.K., Sood, K.: A review on environmental gas sensors: materials and technologies. Sensors Int. **2** (2021)
3. Singh, P., Hu, L.L., Zan, H.W., Tseng, T.Y.: Highly sensitive nitric oxide gas sensor based on ZnO-nanorods vertical resistor operated at room temperature. Nanotechnology **30** (2019)
4. Naganaboina, V.R. Anandkumar, M., Deshpande, A.S., Singh, S.G.: Single-phase high-entropy oxide-based chemiresistor: toward selective and sensitive detection of methane gas for real-time applications. Sens. Actuators B Chem. **357** (2022)
5. Liu, X., Zheng, W., Kumar, R., Kumar, M., Zhang, J.: Conducting polymer-based nanostructures for gas sensors. Coord. Chem. Rev. **462** (2022)
6. Joshi, N., Hayasaka, T., Liu, Y., Liu, H., Oliveira, O.N., Lin, L.: A review on chemiresistive room temperature gas sensors based on metal oxide nanostructures, graphene and 2D transition metal dichalcogenides. Microchim. Acta **185** (2018)
7. Liu, L., Wang, Y., Liu, Y., Wang, S., Li, T., Feng, S., Qin, S., Zhang, T.: Heteronanostructural metal oxide-based gas microsensors. Microsyst. Nanoeng. **8** (2022)
8. Kumar, S., Pavelyev, V., Mishra, P., Tripathi, N., Sharma, P., Calle, F.: A review on 2D transition metal di-chalcogenides and metal oxide nanostructures based NO2 gas sensors. Mater. Sci. Semicond. Process **107** (2020)
9. Choi, S.J., Kim, I.D.: Recent developments in 2D nanomaterials for chemiresistive-type gas sensors. Electr. Mater. Lett. **14** 221–60 (2018)
10. Toh, R.J., Sofer, Z., Luxa, J., Sedmidubský, D., Pumera, M.: 3R phase of MoS2 and WS2 outperforms the corresponding 2H phase for hydrogen evolution. Chem. Commun. **53** 3054–7 (2017)
11. Chhowalla, M., Shin, H.S., Eda, G., Li, L.J., Loh, K.P., Zhang, H.: The chemistry of two-dimensional layered transition metal dichalcogenide nanosheets. Nat. Chem. **5**, 263–275 (2013)
12. Wang, Q.H., Kalantar-Zadeh, K., Kis, A., Coleman, J.N., Strano, M.S.: Electronics and optoelectronics of two-dimensional transition metal dichalcogenides. Nat. Nanotechnol. **7**, 699–712 (2012)
13. Manzeli, S., Ovchinnikov, D., Pasquier, D., Yazyev, O.V., Kis, A.: 2D transition metal dichalcogenides. Nat. Rev. Mater. **2** (2017)
14. Hermawan, A., Septiani, N.L.W., Taufik, A., Yuliarto, B., Suyatman, Yin, S.: Advanced strategies to improve performances of molybdenum-based gas sensors Nanomicro Lett. **13** (2021)
15. Luo, P., Zhuge, F., Zhang, Q., Chen, Y., Lv, L., Huang, Y., Li, H., Zhai, T.: Doping engineering and functionalization of two-dimensional metal chalcogenides. Nanoscale Horiz **4**, 26–51 (2019)
16. Zeng, Y., Lin, S., Gu, D., Li, X.: Two-dimensional nanomaterials for gas sensing applications: the role of theoretical calculations. Nanomaterials **8** (2018)
17. Meng, Z., Stolz, R.M., Mendecki, L., Mirica, K.A.: Electrically-transduced chemical sensors based on two-dimensional nanomaterials. Chem. Rev. **119**, 478–598 (2019)
18. Mathew, M., Shinde, P.V., Samal, R., Rout, C.S.: A review on mechanisms and recent developments in p-n heterojunctions of 2D materials for gas sensing applications. J. Mater. Sci. **56** 9575–604 (2021)
19. Nguyen, T.K., Jeong, S., Youn, J.S., Ahn, S., Nam, K.H., Park, C.M., Jeon, K.J.: Insight into mechanism of temperature-dependent limit of NO2 detection using monolayer MoS2. Sens. Actuators B Chem. **329** (2021)

20. Hau, H.H., Duong, T.T.H., Man, N.K., Thi Viet Nga, T., Thi Xuan, C., Thi Thanh Le, D., Van Toan, N., Hung, C.M., Van Duy, N., Van Hieu, N., Hoa, N.D.: Enhanced NO2 gas-sensing performance at room temperature using exfoliated MoS2 nanosheets. Sens. Actuators A Phys. **332** (2021)
21. Donarelli, M., Prezioso, S., Perrozzi, F., Bisti, F., Nardone, M., Giancaterini, L., Cantalini, C., Ottaviano, L.: Response to NO2 and other gases of resistive chemically exfoliated MoS2-based gas sensors. Sens. Actuators B Chem. 602–13 (2015)
22. Kumar, M., Agrawal, A.V., Kumar, R., Venkatesan, S., Zakhidov, A., Yang, G., Bao, J., Kumar, M.: Photoactivated mixed in-plane and edge-enriched p-type MoS2 flake-based NO_2 sensor working at room temperature. ACS Sens. **3** 998–1004 (2018)
23. Wang, J., Deng, H., Li, X., Yang, C., Xia, Y.: Visible-light photocatalysis enhanced room-temperature formaldehyde gas sensing by MoS2/rGO hybrids. Sens. Actuators B Chem. **304** (2020)
24. Suh, J.M., Eom, T.H., Cho, S.H., Kim, T., Jang, H.W.: Light-activated gas sensing: a perspective of integration with micro-leds and plasmonic nanoparticles. Mater. Adv. **2**, 827–844 (2021)
25. Xia, Y., Hu, C., Guo, S., Zhang, L., Wang, M., Peng, J., Xu, L., Wang, J.: Sulfur-vacancy-enriched MoS2 nanosheets based heterostructures for near-infrared optoelectronic NO_2. ACS Appl. Nano Mater. **3** 665–73 (2020)
26. Kim, T., Lee, T.H., Park, S.Y., Eom, T.H., Cho, I., Kim, Y., Kim, C., Lee, S.A., Choi, M.J., Suh, J.M., Hwang, I.S., Lee, D., Park, I., Jang, H.W.: Drastic Gas sensing selectivity in 2-dimensional MoS2 nanoflakes by noble metal decoration. ACS Nano **17** 4404–13 (2023)
27. Wang, Z., Zhu, L., Wang, J., Zhuang, R., Mu, P., Wang, J., Yan, W.: Advances in functional guest materials for resistive gas sensors. RSC Adv **12**, 24614–24632 (2022)
28. Suh, J.M., Shim, Y.S., Kwon, K.C., Jeon, J.M., Lee, T.H., Shokouhimehr, M., Jang, H.W.: Pd- and Au-decorated MoS2 gas sensors for enhanced selectivity. Electr. Mater. Lett. **15** 368–76 (2019)
29. Salih, E., Ayesh, A.I.: First principle study of transition metals codoped MoS_2 as a gas sensor for the detection of NO and NO_2 gases. Phys. E Low Dimens. Syst. Nanostruct. **131** (2021)
30. Xu, S., Zhang, Y., Xu, F., Chen, C., Shen, Z.: Theoretical study of the adsorption behaviors of gas molecules on the Au-functionalized MoS2 nanosheets: a search for highly efficient gas sensors. Comput. Theor. Chem. **1188** (2020)
31. Wang, Z., Zhang, Y., Ren, Y., Wang, M., Zhang, Z., Zhao, W., Yan, J., Zhai, C., Yun, J.: NO gas adsorption properties of MoS2 from monolayer to trilayer: a first-principles study. Mater. Res. Express **8** (2021)
32. Lee, S., Kang, Y., Lee, J., Kim, J., Shin, J.W., Sim, S., Go, D., Jo, E., Kye, S., Kim, J., An, J.: Atomic layer deposited Pt nanoparticles on functionalized MoS2 as highly sensitive H2 sensor. Appl. Surf. Sci. **571** (2022)
33. Gorthala, G., Ghosh, R.: Ultra-fast NO2 detection by MoS2 nanoflakes at room temperature. IEEE Sens. J. **22** 14727–35 (2022)
34. Bharathi, P., Harish, S., Mathankumar, G., Krishna Mohan, M., Archana, J., Kamalakannan, S., Prakash, M., Shimomura, M., Navaneethan, M.: Solution processed edge activated Ni-MoS2 nanosheets for highly sensitive room temperature NO2 gas sensor applications. Appl. Surf. Sci. **600** (2022)
35. Thang, N.T., Hong, L.T., Thoan, N.H., Hung, C.M., Van Duy, N., Van Hieu, N., Hoa, N.D.: Controlled synthesis of ultrathin MoS2 nanoflowers for highly enhanced NO2 sensing at room temperature. RSC Adv. **10** 12759–71 (2020)
36. Tyagi, S., Kumar, A., Kumar, A., Gautam, Y.K., Kumar, V., Kumar, Y., Singh, B.P.: Enhancement in the sensitivity and selectivity of Cu functionalized MoS2 nanoworm thin films for nitrogen dioxide gas sensor. Mater. Res. Bull. **150** (2022)
37. Jaiswal, J., Tiwari, P., Singh, P., Chandra, R.: Fabrication of highly responsive room temperature H2 sensor based on vertically aligned edge-oriented MoS2 nanostructured thin film functionalized by Pd nanoparticles. Sens. Actuators B Chem. **325** (2020)
38. Liu, J., Hu, Z., Zhang, Y., Li, H.Y., Gao, N., Tian, Z., Zhou, L., Zhang, B., Tang, J., Zhang, J., Yi, F., Liu, H.: MoS2 nanosheets sensitized with quantum dots for room-temperature gas sensors. Nanomicro. Lett. **12** (2020)

39. Singh, S., Sattigeri, R.M., Kumar, S., Jha, P.K., Sharma, S.: Superior room-temperature ammonia sensing using a hydrothermally synthesized MoS2/SnO2 composite. ACS Omega **6** 11602–13 (2021)
40. Han, Y., Huang, D., Ma, Y., He, G., Hu, J., Zhang, J., Hu, N., Su, Y., Zhou, Z., Zhang, Y., Yang, Z.: Design of hetero-nanostructures on mos2 nanosheets to boost NO2 roomerature sensing. ACS Appl. Mater. Interfaces **10** 22640–9 (2018)
41. Bai, H., Guo, H., Feng, C., Wang, J., Liu, B., Xie, Z., Guo, F., Chen, D., Zhang, R., Zheng, Y.: Light-activated ultrasensitive NO2 gas sensor based on heterojunctions of CuO nanospheres/MoS2 nanosheets at room temperature. Sens. Actuators B Chem. **368** (2022)
42. Duong, T.T.H., Hau, H.H., Hong, L.T., Vu, L.A., Hung, C.M., Duy, N.V., Hieu, N.V., Hoa, N.D.: PtO2-decorated MoS2 ultrathin nanostructures for enhanced NH3 gas sensing properties. Mater. Sci. Semicond. Process **151** (2022)
43. Singh, S., Deb, J., Sarkar, U., Sharma, S.: MoS2/WO3 nanosheets for detection of ammonia. ACS Appl. Nano Mater. **4** 2594–605 (2021)
44. Zhang, D., Jiang, C., Zhou, X.: Fabrication of Pd-decorated TiO2/MoS2 ternary nanocomposite for enhanced benzene gas sensing performance at room temperature. Talanta **182** 324–32 (2018)
45. Bhati, V.S., Kumar, M., Banerjee, R.: Gas sensing performance of 2D nanomaterials/metal oxide nanocomposites: a review. J. Mater. Chem. C Mater. **9** 8776–808
46. Singh, S., Sharma, S.: Temperature-based selective detection of hydrogen sulfide and ethanol with MoS2/WO3 composite. ACS Omega **7** 6075–85 (2022)
47. Kumar, U., Hsieh, H.W., Liu, Y.C., Deng, Z.Y., Chen, K.L., Huang, W.M., Wu, C.H.: revealing a highly sensitive sub-ppb-level NO2 gas-sensing capability of novel architecture 2D/0D MoS2/SnS heterostructures with DFT interpretation. ACS Appl. Mater. Interfaces **14** 32279–88 (2022)
48. Reddeppa, M., Park, B.G., Murali, G., Choi, S.H., Chinh, N.D., Kim, D., Yang, W., Kim, M.D.: NOx gas sensors based on layer-transferred n-MoS2/p-GaN heterojunction at room temperature: Study of UV light illuminations and humidity. Sens. Actuators B Chem. **308**
49. Li, X., Li, X., Li, Z., Wang, J., Zhang, J.: WS2 nanoflakes based selective ammonia sensors at room temperature. Sens. Actuators B Chem. **240** 273–7 (2017)
50. Simon Patrick, D., Bharathi, P., Krishna Mohan, M., Muthamizchelvan, C., Harish, S., Navaneethan, M.: Liquid phase exfoliated WS2 nanosheet-based gas sensor for room temperature NO2 detection. J. Mater. Sci.: Mater. Electr. **33** 9235–45 (2022)
51. Qin, Z., Zeng, D., Zhang, J., Wu, C., Wen, Y., Shan, B., Xie, C.: Effect of layer number on recovery rate of WS 2 nanosheets for ammonia detection at room temperature. Appl. Surf. Sci. **414** 244–50
52. Liu, D., Tang, Z., Zhang, Z.: Comparative study on NO2 and H2S sensing mechanisms of gas sensors based on WS2 nanosheets. Sens. Actuators B Chem. **303** (2020)
53. Kim, J.H., Mirzaei, A., Kim, H.W., Kim, S.S.: Realization of Au-decorated WS2 nanosheets as low power-consumption and selective gas sensors. Sens, Actuators B Chem. **296** (2019)
54. Kim, J.H., Kim, J.Y., Mirzaei, A., Kim, H.W., Kim, S.S.: Synergistic effects of SnO2 and Au nanoparticles decorated on WS2 nanosheets for flexible, room-temperature CO gas sensing. Sens. Actuators B Chem. **332** (2021)
55. Kim, J.H., Sakaguchi, I., Hishita, S., Ohsawa, T., Suzuki, T.T., Saito, N.: Self-heated CO gas sensor based on Au-decorated Sb-implanted WS2 nanosheets Sens. Actuators B Chem. **382** (2023)
56. Jha, R.K., Nanda, A., Bhat, N.: Ultrasonication assisted fabrication of a tungsten sulfide/tungstite heterostructure for ppb-level ammonia detection at room temperature. RSC Adv **10**, 21993–22001 (2020)
57. Han, Y., Liu, Y., Su, C., Chen, X., Li, B., Jiang, W., Zeng, M., Hu, N., Su, Y., Zhou, Z., Zhu, Z.G., Yang, Z.: Hierarchical WS2-WO3 Nanohybrids with P-N Heterojunctions for NO2Detection. ACS Appl. Nano Mater. **4** 1626–34 (2021)
58. Ou, Y., Niu, W., Zhou, Y., Guo, Y., Gao, C., Wang. Y.: Mesoporous WS2/MoO3 hybrids for high-performance trace ammonia detection. ACS Appl. Mater. Interfaces (2022)
59. He, T., Liu, W., Lv, T., Ma, M., Liu, Z., Vasiliev, A., Li, X.: MXene/SnO2 heterojunction based chemical gas sensors. Sens. Actuators B Chem. **329** (2021)

60. Quan, W., Shi, J., Luo, H., Fan, C., Lv, W., Chen, X., Zeng, M., Yang, J., Hu, N., Su, Y., Wei, H., Yang, Z.: Fully flexible MXene-based gas sensor on paper for highly sensitive room-temperature nitrogen dioxide detection. ACS Sens. **8**, 103–113 (2023)
61. Zheng, W., Liu, X., Xie, J., Lu, G., Zhang, J.: Emerging van der Waals junctions based on TMDs materials for advanced gas sensors. Coord. Chem. Rev. **447** (2021)
62. Moumen, A., Konar, R., Zappa, D., Teblum, E., Nessim, G.D., Comini, E.: Room-temperature NO2 sensing of CVD-modified WS2-WSe2 heterojunctions. ACS Appl. Nano Mater. **6** 7323–9 (2023)
63. Sharma, I., Kumar, K.N., Choi, J.: Highly sensitive chemiresistive detection of NH3 by formation of WS2 nanosheets and SnO2 quantum dot heterostructures. Sens. Actuators B Chem. **375** (2023)
64. Xia, Y., Xu, L., He, S., Zhou, L., Wang, M., Wang, J., Komarneni, S.: UV-activated WS2/SnO2 2D/0D heterostructures for fast and reversible NO2 gas sensing at room temperature. Sens. Actuators B Chem. **364** (2022)
65. Wang, X., Gu, D., Li, X., Lin, S., Zhao, S., Rumyantseva, M.N., Gaskov, A.M.: (2019) Reduced graphene oxide hybridized with WS2 nanoflakes based heterojunctions for selective ammonia sensors at room temperature. Sens. Actuators B Chem. **282** 290–9 (2022)
66. Xia, Y., He, S., Wang, J., Zhou, L., Wang, J., Komarneni, S.: MXene/WS2hybrids for visible-light-activated NO2sensing at room temperature. Chem. Commun. **57** 9136–9 (2021)
67. Alagh, A., Annanouch, F.E., Youssef, K. Al, Bittencourt, C., Güell, F., Martínez-Alanis, P.R., Reguant, M., Llobet, E.: PdO and PtO loaded WS2 boosts NO2 gas sensing characteristics at room temperature. Sens. Actuators B Chem. **364** (2022)
68. Chen, X., Hu, J., Chen, P., Yin, M., Meng, F., Zhang, Y.: UV-light-assisted NO2 gas sensor based on WS2/PbS heterostructures with full recoverability and reliable anti-humidity ability. Sens. Actuators B Chem. **339** (2021)

Surface Engineered 2D TMD Materials for Advanced Wearable Biosensors

Shalu Rani, Sanjay Kumar, Ruchi Singh, and Pawan Kumar

Abstract In the past decade, the 2D materials-based electronic products market is rapidly growing especially in the areas of sensing, energy storage, biomedical, or ultrafast optoelectronics. Out of many 2D materials, transition-metal-dichalcogenides (TMDs) are one of the promising 2D materials which have weak van der-Waals force between each covalently bonded layer (X-M-X) of transition metals (M) and chalcogens (X). These individual-layer of TMDs undergo the transition from in-direct to direct band gap and have remarkable and fascinating electronic and optical properties. Moreover, TMDs have tremendous potential in several applications such as light emission, photodetection, gas sensing, biosensors, bio-imaging, and drug delivery because of their high density of surface sites. Furthermore, TMDs also act as associative biosensing elements for the real-time detection of biomarkers and, hence, exhibit great potential in biosensors and healthcare devices. The surface engineering and modification of TMDs can further enhance their electronic and sensing properties. Therefore, understanding the interfacial properties and atomic structures, and surface engineering of TMDs are thus of significant importance for the advancement of biosensors. The presented book chapter, comprehensively discusses the several deposition techniques for TMDs, their advantages, and disadvantages, the growth quality of deposited thin layers of TMDs, their morphological and structural analysis, several parameters for performance analysis of fabricated 2D-TMDs-based wearable biosensors and the effect of surface modification on device sensitivity, selectivity, and responsivity. Here, the various challenges and their solutions during bandgap tuneability and functionalization with various opportunities for application in advanced wearable biosensors have been discussed.

S. Rani · S. Kumar (✉)
Department of Electronics Engineering, Indian Institute of Technology (ISM) Dhanbad, Dhanbad, Jharkhand, India
e-mail: sanjaykumar@iitism.ac.in

R. Singh
Department of Electronics and Communication Engineering, Jaypee Institute of Information Technology, Noida, India

P. Kumar
Centre for Applied Research in Electronics (CARE), Indian Institute of Technology, Delhi, New Delhi, India

A. Katoch et al. (eds.), *2D Materials*, Engineering Materials,
https://doi.org/10.1007/978-981-97-6258-3_3

Keywords 2D-TMDs · Surface engineering · Functionalization · Wearable biosensors · Selectivity · Responsivity

1 Introduction

Two-dimensional (2D) TMD (transition metal dichalcogenide) materials have recently gained significant attention due to their unique electronic, optical, and mechanical properties, making them promising candidates for advanced wearable biosensors. However, the surface chemistry of these materials can significantly impact their biocompatibility, stability, and sensitivity towards biomolecules. 2D TMDs are a class of materials that belong to the family of layered materials [1–5]. They are composed of individual layers of transition metal atoms sandwiched between layers of chalcogen atoms (typically sulfur, selenium, or tellurium). In their bulk form, TMDs have an alternating stacking of metal and chalcogen atoms, forming a three-dimensional crystal lattice. However, when these materials are thinned down to a single layer or a few layers, they exhibit unique properties and are often referred to as 2D TMDs. The most well-known and extensively studied 2D TMD is molybdenum disulfide (MoS_2), but there are numerous other TMDs with similar layered structures, such as tungsten disulfide (WS_2), molybdenum diselenide ($MoSe_2$), and tungsten diselenide (WSe_2), among others [5–8]. 2D TMDs possess interesting electrical, optical, and mechanical properties due to quantum confinement effects and their atomically thin nature. They have a direct bandgap in the monolayer form, making them attractive for optoelectronic applications. These materials can exhibit strong light-matter interactions, high carrier mobility, and efficient energy transfer, making them promising candidates for various electronic and photonic devices, such as transistors, photodetectors, and solar cells. Furthermore, 2D TMDs can be combined with other materials, such as graphene or other 2D materials, to create heterostructures with tailored properties, opening up possibilities for novel device functionalities and applications. The field of 2D TMDs is still actively evolving, with ongoing research focused on understanding their fundamental properties, developing synthesis methods, exploring device applications, and investigating their potential in areas like energy storage, catalysis, and sensing [8–10].

Surface engineering of 2D TMD materials involves modifying their surface chemistry to achieve desirable properties such as enhanced biocompatibility, improved stability, and higher sensitivity towards biomolecules [12–14]. One of the popular techniques for surface engineering is functionalization with various biomolecules, such as proteins and DNA, to improve their selectivity toward target analytes. Another approach for surface engineering of 2D TMD materials is the deposition of a thin layer of another material, such as gold or graphene, onto their surface [15–17]. This process not only enhances the biocompatibility and stability of the material but also increases its surface area and provides a platform for the immobilization of biomolecules. Surface engineering of 2D TMD materials can also be achieved through the introduction of defects, such as vacancies or doping, which alter their electronic properties and improve their sensitivity towards biomolecules. Overall, surface engineering of 2D TMD materials holds great promise for the development of advanced wearable biosensors for real-time monitoring of biomolecules in various biological fluids [16, 17]. Here, Fig. 1 depicts the various 2D materials systems, and their uses in various potential applications with important key properties.

Surface engineering of 2D TMD materials involves modifying their surface chemistry to achieve desired properties, such as enhanced biocompatibility, improved stability, and higher sensitivity toward biomolecules. Therefore, for surface engineering, several approaches have been widely adopted which are also outlined below:

- ***Chemical functionalization***: In this approach the surface modification of 2D TMD materials by covalently attaching chemical groups to the surface. For example, amino or carboxyl functional groups can be attached to the 2D TMD materials surface to improve their biocompatibility and active/enable the immobilization of biomolecules. Chemical functionalization can be achieved using various methods, including diazonium chemistry, thiol-ene click chemistry, and silane coupling chemistry [18].
- ***Deposition of a thin layer of another material***: This approach involves depositing a thin layer of another material, such as gold, graphene, or metal oxides, onto the

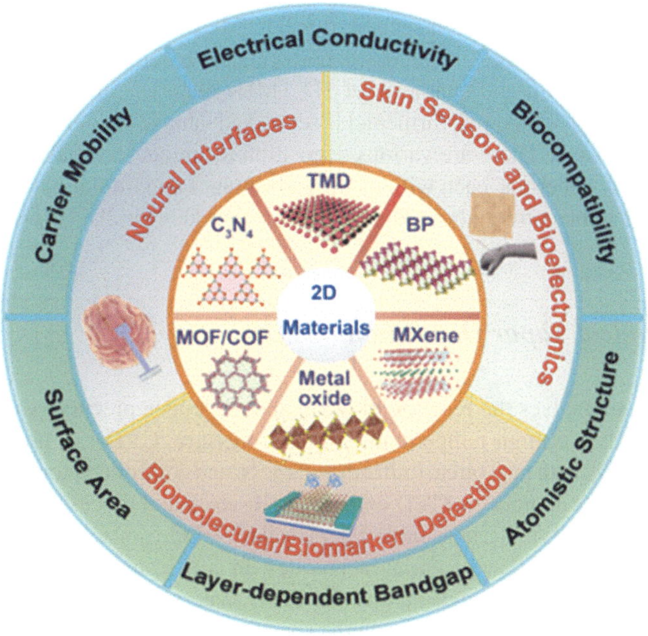

Fig. 1 Various two-dimensional (2D) materials systems and their uses in various potential applications

surface of 2D TMD materials [19–22]. This process can be improved the stability and biocompatibility of 2D TMD materials, as well as increase their surface area for the immobilization of biomolecules.

- ***Defect engineering***: In this approach, introducing defects, such as vacancies or doping, into the crystal structure of 2D TMD materials to alter their electronic properties and improve their sensitivity towards biomolecules [23–26]. For example, introducing sulfur vacancies in MoS_2 can improve its catalytic activity towards biomolecules.
- ***Plasma treatment***: Herein, treating the surface of 2D TMD materials with plasma to modify their surface chemistry. Plasma treatment can create or introduce various functional element groups, such as hydroxyl, carboxyl, and amino groups, over the surface of 2D TMD materials and improve their biocompatibility and sensitivity with respect to the biomolecules [27–29].

Therefore, the surface engineering of 2D TMD materials is a promising approach for developing advanced biosensors with improved sensitivity, selectivity, and stability.

2 Deposition Techniques for 2D TMDs Materials

As discussed in the introduction section, the 2D TMDs materials have shown great potential in the designing of enhanced performance-based wearable biosensors. Therefore, the growth, and synthesis of 2D TMDs materials are critical task and finding cost-effective methods further adds several advantages in designing low-cost wearable biosensors. There are various deposition techniques available for synthesizing 2D TMD materials, each with its own advantages and limitations. Herein, we have outlined some commonly used deposition techniques for 2D TMDs:

2.1 Chemical Vapor Deposition (CVD)

This technique involves the reaction of metal and chalcogen precursors in the presence of a catalyst at high temperature and low pressure. CVD can produce high-quality 2D TMDs with large area, uniform thickness, and controlled morphology [30–34]. Herein, Fig. 2 depicts the CVD experimental set-up and process demonstration to grow the 2D TMDs materials. Table 1 depicts the CVD advantages, disadvantages, and comparison with other potential techniques. In conclusion, CVD is a versatile technique for synthesizing high-quality 2D TMD materials, with some limitations that should be considered depending on the specific application and requirements of the materials.

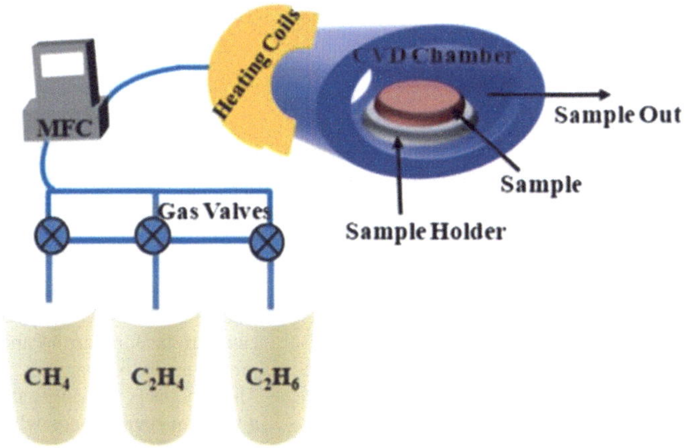

Fig. 2 CVD setup and process demonstration

2.2 *Physical Vapor Deposition (PVD)*

In this process the physical evaporation or sputtering of metal and chalcogen precursors in a vacuum chamber, followed by their condensation on a substrate. PVD can produce 2D TMDs with controlled thickness, morphology, and crystal quality [35–38]. Figure 3 shows the PVD experimental set-up and materials growth process to deposit the 2D materials. Table 1 depicts the PVD benefits, drawbacks, and its comparison with other potential techniques. In conclusion, PVD is a powerful technique for synthesizing high-quality 2D TMD materials with controlled thickness, morphology, and crystal quality. However, the technique is relatively complex and may not be suitable for large-scale production, depending on the specific requirements of the application.

2.3 *Molecular Beam Epitaxy (MBE)*

In this technique the deposition of metal and chalcogen precursors in an ultra-high vacuum environment using a molecular beam source. MBE can be suitable for high-quality 2D TMDs with precise control over their composition, thickness, and crystal orientation [39–45]. Figure 4 shows the MBE illustration with multiple target guns having different materials for the deposition. Table 1 depicts the MBE sakes, snags, and its comparison with other potential techniques. As a result, the MBE is a powerful technique for synthesizing high-quality 2D TMD materials with precise control over their composition, thickness, and crystal orientation. However, the technique is relatively complex and may not be suitable for large-scale production, depending on the specific requirements of the application.

Table 1 Advantages and disadvantages of various deposition techniques for 2D materials widely used in wearable biosensors

Technique	Advantages	Disadvantages
CVD	1. High-quality 2D TMD materials can be produced with large area, uniform thickness, and controlled morphology 2. CVD can be used to synthesize a wide range of 2D TMD materials, including MoS_2, $MoSe_2$, WS_2, WSe_2, and others 3. The process is highly scalable, making it suitable for large-scale production of 2D TMD materials 4. The deposition process is relatively simple, and the growth parameters can be easily optimized to manipulate and control the properties of the 2D TMDs	1. The process requires high-temperature and low-pressure conditions, which can be challenging to achieve and maintain, particularly for large-area deposition 2. The use of a catalyst, such as a transition metal, is required for the growth of 2D TMD materials, which can introduce impurities and affect the properties of the materials 3. The process is typically carried out in a vacuum environment, which limits the utilization of specific types of substrates that may affect the process cost 4. The deposition rate of CVD is relatively slow compared to other techniques such as PVD
PVD	1. PVD can produce 2D TMD materials with controlled thickness, morphology, and crystal quality 2. The process does not require a catalyst, which eliminates the risk of impurity incorporation 3. PVD can be carried out in a variety of vacuum environments, allowing for the use of various substrates and deposition conditions 4. The deposition rate of PVD is relatively high compared to other techniques such as CVD	1. The process requires high vacuum conditions, which can be expensive and limit the size of the deposition chamber 2. PVD may not be appropriate for large-area deposition of 2D TMD materials due to the limited size of the deposition chamber 3. The process is relatively complex and may require multiple deposition steps to achieve the desired properties of the 2D TMD materials 4. PVD may produce 2D TMD materials with lower yields compared to other techniques such as LPE
MBE	1. MBE can grow high-quality 2D TMD materials with precise control over their composition, thickness, and crystal orientation 2. The process does not require a catalyst, which eliminates the risk of impurity incorporation 3. MBE can be carried out in an ultra-high vacuum environment, allowing to produce high-purity 2D TMD materials 4. MBE is highly customizable and can be used to produce complex layered structures	1. MBE requires a complex system of vacuum pumps, which can be expensive and require regular maintenance 2. The deposition rate of MBE is relatively slow compared to other techniques such as PVD 3. The process is relatively complex and may require multiple deposition steps to get the desirable properties of 2D TMD materials 4. MBE may not be the right choice to deposit the large-scale 2D TMD materials because of the limited size of the deposition chamber

(continued)

Table 1 (continued)

Technique	Advantages	Disadvantages
Solvothermal synthesis	1. It is a low-cost and simple process for synthesizing 2D TMD materials in large quantities 2. It can be used to synthesize a vast range of 2D TMD materials with varying thicknesses and morphologies 3. The process is carried out at relatively low temperatures and pressures, which makes it more suitable to produce 2D TMD materials at large-scale 4. It can be used to produce 2D TMD materials with a very high degree of crystallinity, lower roughness with better uniformity	1. Solvothermal synthesis can lead to the incorporation of impurities in the 2D TMD materials due to the use of solvents and the reaction conditions 2. The process may require a long reaction time to achieve the desired properties of the 2D TMD materials 3. The process can be difficult to control and optimize due to the complex interplay between solvent, temperature, pressure, and other reaction parameters 4. The process may produce 2D TMD materials with a limited range of thicknesses and morphologies compared to other techniques such as CVD
LPE	1. LPE is a cost-effective and scalable process for synthesizing 2D TMD materials in large quantities 2. The process can be utilized to synthesize a vast variety of 2D TMD materials with varying thicknesses and morphologies 3. The process does not require high vacuum conditions, making it suitable for large-scale production 4. LPE can be carried out in a variety of solvents, allowing for the production of 2D TMD materials with a high degree of purity	1. The process may result in the production of 2D TMD materials with a wide range of thicknesses and morphologies, leading to poor uniformity 2. The process can be difficult to control and optimize due to the complex interplay between solvent, temperature, and other processing parameters 3. The process can result in the incorporation of impurities in the 2D TMD materials due to the use of solvents 4. LPE may not produce 2D TMD materials with the same level of crystallinity as other methods such as MBE or PVD

2.4 Solvothermal Synthesis

This technique involves the reaction of metal and chalcogen precursors in a high-temperature solvent under high pressure. Solvothermal synthesis can produce 2D TMDs with high crystallinity and controlled morphology [43–45]. Figure 5 shows the Solvothermal synthesis process with its associate components. Table 1 shows the solvothermal synthesis advantages, disadvantages, and its comparison with other potential techniques. In summary, this is a low-cost and simple method for synthesizing 2D TMD materials in large quantities with high crystallinity and uniformity. However, the process may lead to the incorporation of impurities and can be difficult to optimize depending on the specific requirements of the application.

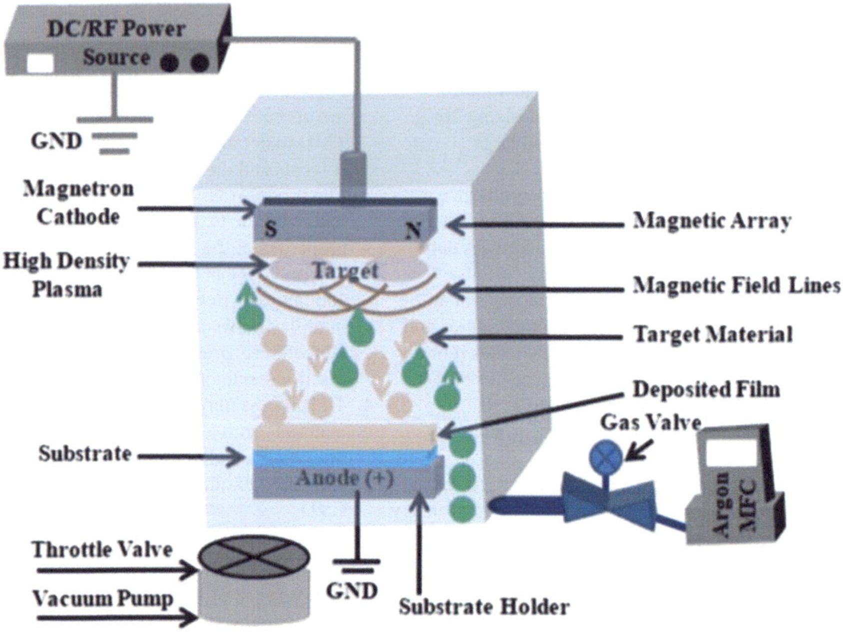

Fig. 3 PVD experimental setup to demonstrate the material deposition process

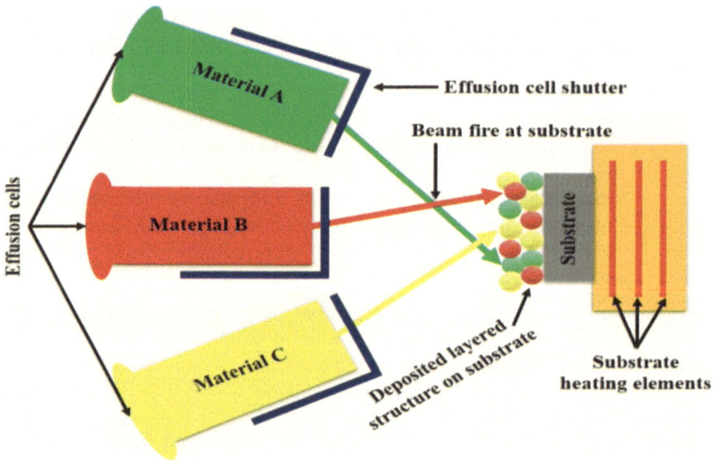

Fig. 4 Illustration of MBE with multiple target guns to deposit layers of different materials system

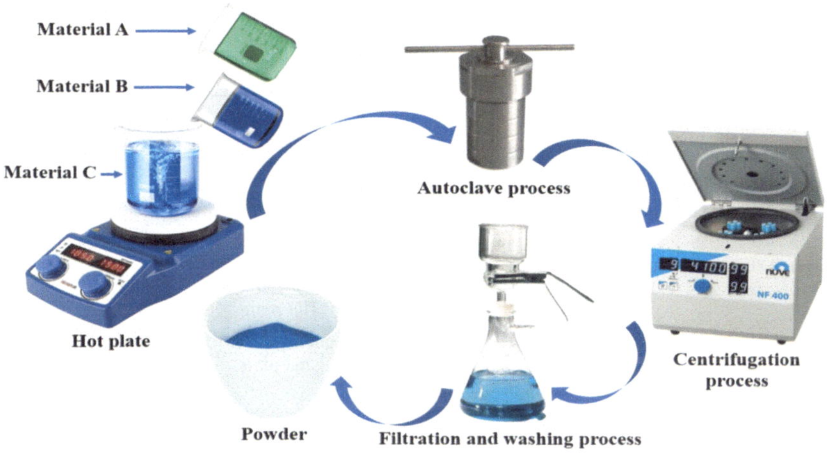

Fig. 5 Schematic illustration of solvothermal synthesis method to prepare the 2D material

2.5 Liquid-Phase Exfoliation (LPE)

In this technique the exfoliation of bulk TMD crystals in a solvent to obtain 2D TMD nanosheets. LPE can produce 2D TMDs with high yield, low cost, and large scale [46–49]. The preference for deposition technique associates with the desirable properties and application of the 2D TMD materials. Herein, Fig. 6 shows the LPE method with their process illustration to prepare the material for device development. Table 1 highlights the LPE advantages, disadvantages, and its comparison with other potential techniques. Therefore, LPE is a cost-effective and scalable process for synthesizing 2D TMD materials in large quantities with a high degree of purity. However, the process may result in poor uniformity and the incorporation of impurities, and may not produce 2D TMD materials with the same level of crystallinity as other methods depending on the specific requirements of the application.

3 Performance Analysis of 2D TMDs in Wearable Biosensors

Performance analysis of fabricated 2D-TMDs-based wearable biosensors refers to the evaluation of the sensor's sensing properties and overall performance in detecting target biomolecules or analytes in a biological sample. The performance analysis of the wearable biosensors depends on several key parameters such as sensitivity, selectivity, stability, response time, and limit of detection (LOD) [50–60].

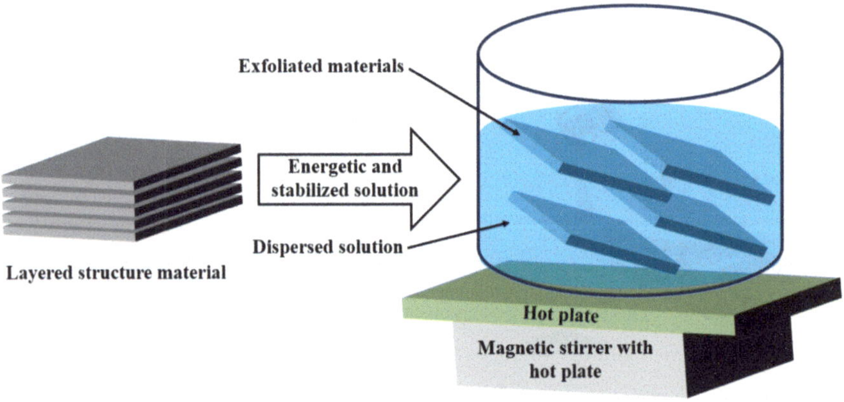

Fig. 6 LPE method with their process illustration to prepare the material for device development

Sensitivity: It is the measure of the sensor's ability to detect low concentrations of the target analyte. It is an important parameter that determines the sensor's accuracy and reliability in monitoring various physiological parameters. The sensitivity value depends on several factors, including the sensor design, the type of biological signal or analyte being measured, and the specific detection technology employed [50–52]. The 2D-TMDs-based biosensor sensitivity can be improved by optimizing the surface functionalization of the material and increasing the contact area between the biomolecules and the sensing surface.

Selectivity: It is the ability of the biosensor to distinguish between the target analyte and other interfering species in the sample. Achieving selectivity is crucial for accurate and reliable measurements in biosensing applications. The following key factors are influenced the selectivity of wearable biosensors such as recognition element, sensor coatings or membranes, signal transduction mechanism, signal processing and analysis, and calibration and validation. Researchers are continuously working on improving the selectivity of wearable biosensors by refining recognition elements, optimizing sensor designs, and implementing advanced signal processing techniques. By enhancing selectivity, biosensors become more reliable and effective tools for monitoring specific analytes or physiological parameters in various applications, such as health monitoring, sports performance tracking, and disease diagnosis [53–55].

Stability: This is the ability of the sensor to maintain its performance over time. The stability of a 2D TMD-based wearable biosensor depends on various factors, including material degradation, device fabrication, device packaging, environmental conditions, substrate choice, biocompatibility, and the specific TMD material used in biosensor development. Stability can be improved by optimizing the fabrication process and storage conditions [56, 57].

Response time: Time required for the sensor to produce a detectable signal in response to the target analyte. The response time of a 2D TMD-based wearable biosensor depends on several factors, including the type of TMD material, device design, sensing mechanism, analyte diffusion, signal processing, and the target analyte being detected. It is important to note that response time is a critical parameter in biosensor performance, especially for real-time applications. A shorter response time is desirable, as it allows for faster and more accurate detection [58–60].

Limit of detection (LOD): It is the lowest concentration of the target analyte that can be detected by the sensor. A lower LOD indicates higher sensitivity and performance of the biosensor device. The LOD of a 2D TMD-based wearable biosensor is affected by several factors, including sensing mechanism, type of TMD material, device design, noise, and signal processing, analyte concentration and diffusion, background interference, and the specific analyte being detected [59, 60].

Therefore, the performance analysis of fabricated 2D-TMDs-based wearable biosensors is important for determining the suitability of the biosensor for specific applications and for identifying areas for optimization and improvement.

4 Effect of Surface Modification on Performance Parameters of Wearable Biosensors

Surface modification of 2D TMD materials can have a significant impact on the sensitivity of wearable biosensors. By modifying the surface of the 2D TMD material, it is possible to enhance the interaction between the material and the biological molecules, thereby improving the sensitivity of the biosensor [61–70]. Herein, some of the ways in which surface modification of 2D TMD materials can affect the sensitivity of wearable biosensors as outlined below:

Increased Surface Area: Surface modification techniques such as chemical functionalization, plasma treatment, and electrochemical methods can effectively enhance the surface area of 2D TMD materials. The enhanced surface area provides the maximum binding sites to the target biomolecules, resulting in increased sensitivity.

Enhanced Biocompatibility: Surface modification can also enhance the biocompatibility of 2D TMD materials, which makes them more suitable for use in wearable biosensors. For example, functionalization with biocompatible polymers or peptides which can be reduced the cytotoxicity of the material and enhance the interaction with biological molecules.

Improved Selectivity: Surface modification can also enhance the selectivity of wearable biosensors by introducing specific functional groups that can selectively bind to target biomolecules. This can improve the accuracy and reliability of biosensors.

Increased Stability: Surface modification can also improve the stability of 2D TMD materials, making them more resistant to degradation and environmental factors. This can improve the longevity and performance of the biosensor.

Overall, surface modification of 2D TMD materials can have a significant impact on the sensitivity, selectivity, and stability of wearable biosensors, making them more effective tools for monitoring and diagnosing various medical conditions. Surface modification of 2D TMD materials can also have a significant impact on the selectivity of wearable biosensors. By modifying the surface of the 2D TMD material, it is possible to introduce specific functional groups that can selectively bind to target biomolecules, improving the selectivity of the biosensor. There are several ways in which surface modification of 2D TMD materials can affect the selectivity of wearable biosensors are:

Introduction of Functional Groups: Surface modification techniques such as chemical functionalization, plasma treatment, and electrochemical methods can introduce specific functional groups over the surface of 2D TMD materials. The induced functional groups can selectively bind to target biomolecules, improving the selectivity of the biosensor.

Molecular Imprinting: Surface modification can also be used to create molecularly imprinted polymers over the surface of 2D TMD materials. MIPs are synthetic receptors that can selectively bind with target biomolecules with high affinity, selectivity, and specificity, which further improve the selectivity of the biosensor.

Aptamer Functionalization: Aptamers are short single-stranded DNA or RNA sequences that can selectively bind to target biomolecules with high affinity and specificity. Surface modification can be used to functionalize 2D TMD materials with aptamers, improving the selectivity of the biosensor.

Selective Coating: Surface modification can also be used to selectively coat the surface of 2D TMD materials with specific materials that can selectively bind to target biomolecules. For example, a selective coating of a specific metal or metal oxide can enhance the selectivity of the biosensor.

Overall, surface modification of 2D TMD materials can improve the selectivity of wearable biosensors by introducing specific functional groups or materials that can selectively bind to target biomolecules, improving the accuracy and reliability of biosensors. Surface modification of 2D TMD materials can also have an effect on the responsivity of wearable biosensors. The responsivity of a biosensor defines its ability to detect and respond to a specific biological or chemical analyte. Specifically, in biosensors, responsivity is a measure of how effectively the device can convert a signal generated by the interaction between the target analyte and the biological recognition element (such as enzymes, antibodies, or nucleic acids) into a measurable output. Some of the ways in which surface modification of 2D TMD materials can affect the responsivity of wearable biosensors as given below:

Enhanced Surface Area: Surface modification techniques such as chemical functionalization, plasma treatment, and electrochemical methods can enhance the surface area of 2D TMD materials. The enhanced surface area can further improve the interaction between the biomolecules and the biosensor, resulting in increased responsivity.

Improved Charge Transfer: Surface modification can also improve the charge transfer mechanism and properties of the 2D TMD materials. For example, functionalization with conductive polymers or graphene can enhance the electron transfer between the biomolecules and the biosensor, resulting in improved responsivity.

Increased Sensitivity: As discussed earlier, surface modification can improve the sensitivity of biosensors. Improved sensitivity can also result in improved responsivity, as the biosensor can detect smaller changes in the input signal.

Reduced Noise: Surface modification can also reduce the noise in the biosensor, resulting in improved responsivity. For example, a selective coating on the surface of the 2D TMD material can reduce non-specific binding and interference, resulting in a cleaner signal.

Therefore, the surface modification of 2D TMD materials can have a significant impact on the responsivity of wearable biosensors, resulting in more accurate and reliable detection of biological signals.

5 Bandgap Tuneability and Functionalization: Challenges and Their Potential Solutions

5.1 Bandgap Tuneability and Functionalization: Challenges

Tuning the bandgap of 2D TMD materials for use in wearable biosensors presents several challenges. Here, we have outlined some of the challenges which are essentially to address them and make a more stable wearable biosensor [67–70]:

Precise Control of Bandgap: The bandgap of 2D TMD materials is sensitive to changes in their thickness, composition, and strain. Therefore, it can be challenging to precisely control the bandgap to achieve the desired properties for a biosensor application.

Integration with Wearable Devices: Integrating 2D TMD materials into a wearable biosensor can be challenging due to the need for a flexible and biocompatible substrate. The choice of substrate and the method of transferring the material onto the substrate can affect the performance and stability of the biosensor. For example, for the growth of the 2D TMD materials, typically silicon carbide or sapphire substrates have been used and after the growth, this substrate must then be transferred to the wearable device, which can be a challenging process. The transfer process for 2D TMD materials is often difficult to pattern and fabricate into devices as these are very thin and fragile. Also, the 2D TMD materials are sensitive to environmental factors, such as moisture and oxygen.

Stability of Bandgap: As discussed above, the bandgap of 2D TMD materials can be sensitive to environmental factors such as temperature, humidity, and exposure to chemicals. Maintaining a stable bandgap over time is critical for the long-term performance of the biosensor.
Large-Scale Synthesis: Scaling up the synthesis of 2D TMD materials while maintaining precise control over their bandgap can be challenging. The synthesis process can also affect the quality and uniformity of the material, which can impact the performance of the biosensor.
Compatibility with Target Biomolecules: The bandgap of 2D TMD materials needs to be tuned to the energy level of the target biomolecules to achieve optimal sensitivity and selectivity. However, the energy levels of biomolecules can vary widely, making it challenging to achieve optimal tuning for a range of target biomolecules.

By addressing these challenges, researchers are exploring various strategies such as chemical doping, alloying, and mechanical strain to precisely control the bandgap of 2D TMD materials. The investigation of the new synthesis methods and transfer techniques to integrate 2D TMD materials into wearable biosensors is still required. Furthermore, computational modeling is being used to optimize the tuning of the bandgap for specific target biomolecules and environmental conditions.

5.2 Bandgap Tuneability and Functionalization: Potential Solutions

The bandgap tuneability and functionalization of 2D TMD materials for wearable biosensors can be addressed by adopting several solutions as listed below [70–75]:

Control of Bandgap: To control the bandgap of 2D TMD materials, researchers can use techniques such as chemical doping, mechanical strain, and alloying. Chemical doping involves introducing foreign atoms into the lattice structure of the material to alter its electronic properties while mechanical strain involves applying stress to the material to alter its bandgap. Alloying involves combining two or more materials to create a new material with a desired bandgap.
Surface Modification: To enhance the sensitivity, selectivity, and responsivity of wearable biosensors based on 2D TMD materials, surface modification is necessary. To minimize defects and maintain the electronic properties of the material, researchers can carefully control the surface modification process [12–16].
Integration with Wearable Devices: Integrating 2D TMD materials with wearable devices can be accomplished by using flexible and biocompatible substrates such as polymer films. Researchers can develop new transfer techniques to preserve the quality of the material during transfer.

Optimization of Sensing Mechanism: The sensing mechanism of 2D TMD-based biosensors can be optimized for specific biomolecules and environmental conditions by using computational modeling to understand the sensing mechanism and optimizing the design of the biosensor for specific applications.
Stability of Bandgap: To maintain a stable bandgap over time, researchers can develop methods to protect the material from environmental factors such as temperature, humidity, and exposure to chemicals.
Large-Scale Synthesis: To scale up the synthesis of 2D TMD materials, researchers can investigate new synthesis methods that maintain precise control over the bandgap of the material. However, several other parameters such as controlling the thickness of the films, achieving uniform films, reducing defects, and increasing the yield are also associated with the large-scale synthesis of 2D TMDs materials.

Overall, the solutions for bandgap tuneability and functionalization of 2D TMD materials for wearable biosensors involve careful control of the synthesis and surface modification processes, optimization of the sensing mechanism, and integration with flexible and biocompatible substrates. By addressing these solutions, researchers can overcome the challenges and achieve the desired properties for a range of biosensor applications.

6 Potential Applications of the Wearable Biosensors

In the recent era, wearable biosensors have gained significant attention in recent years due to their potential to revolutionize healthcare, wellness monitoring, and various other fields. Among many applications, here we have discussed a few of them are some trending applications of wearable biosensors [76–102]:

1. ***Health and Fitness Monitoring***: Wearable biosensors are widely used for monitoring vital signs and physical activity. One promising application for TMD-based wearable biosensors is for health and fitness monitoring. TMDs can be used to detect a variety of bio-markers, including heart rate, blood glucose levels, and body temperature [76–80]. This information can be used to track a person's health and fitness, and to provide early warning signs of potential health problems. In addition to health and fitness monitoring, 2D TMD-based wearable biosensors can also be used for other applications, such as environmental monitoring and military applications. For example, TMDs can be used to detect pollutants in the air or water, or to monitor the movement of troops on the battlefield. However, the development of TMD-based wearable biosensors is still in its early stages, but the potential applications of these devices are vast. TMDs offer a number of advantages over traditional wearable biosensors, including their flexibility, lightweight, and excellent sensing properties. As the technology continues to develop, TMD-based wearable biosensors are likely to become increasingly common in a variety of applications.

2. ***Chronic Disease Management***: Wearable biosensors offer great potential for managing chronic diseases such as diabetes, cardiovascular conditions, and respiratory disorders. They can continuously monitor glucose levels, heart rhythm irregularities, and pulmonary function, allowing individuals to track their health status in real-time and make informed decisions about their treatment and lifestyle [81, 82]. For example, a TMD-based wearable biosensor could be used to monitor a person's blood glucose levels. This information could be used to help people with diabetes manage their condition and prevent serious complications, such as heart disease and stroke. TMD-based wearable biosensors could also be used to monitor other chronic diseases, such as heart disease, hypertension, and asthma. By providing early warning signs of potential problems, these biosensor devices would be helpful to lift-up the quality of life for people with chronic diseases and to reduce the risk of serious complications.
3. ***Remote Patient Monitoring***: Wearable biosensors enable remote patient monitoring (RPM), especially for patients with chronic conditions or those recovering from surgeries or hospital stays. In RPM, 2D TMD wearable biosensors can be used to collect continuous data on a patient's health status. This data can be transmitted to a remote healthcare provider, who can then monitor the patient's condition and provide care as needed. This can help to improve patient outcomes by reducing the risk of complications and readmission to the hospital.
4. ***Mental Health Monitoring***: Wearable biosensors are being explored for monitoring mental health conditions such as stress, anxiety, and depression. These biosensors can measure physiological indicators such as heart rate variability, skin conductance, and sleep patterns to provide objective data on an individual's mental well-being [80]. This information can help in early intervention, personalized treatment plans, and overall mental wellness management.
5. ***Performance Optimization in Sports***: Athletes and sports enthusiasts use wearable biosensors to monitor their performance metrics, including heart rate, oxygen levels, and activity patterns. 2D TMD-based wearable biosensors can be used to collect continuous data on an athlete's performance. This data can be used to track progress, identify areas for improvement, and prevent injuries. For example, heart rate data can be used to monitor exertion levels and ensure that athletes are not overtraining. Respiratory rate data can be used to track recovery and identify signs of fatigue. Blood oxygen saturation data can be used to monitor oxygen levels and ensure that athletes are getting enough oxygen. These data can be analyzed to optimize training programs, prevent injuries, and enhance athletic performance. Wearable biosensors also enable real-time monitoring during competitions, offering valuable insights for coaches and athletes [82].
6. ***Elderly Care***: Wearable biosensors play a crucial role in elderly care by monitoring their health, falls, and daily activities. These biosensors can provide alerts in case of emergencies, such as sudden falls or abnormal vital signs. They can also help track medication adherence and provide valuable insights for caregivers and healthcare providers. The remarkable characteristics of 2D TMDs materials propose them as an ideal candidate for plenty of biosensor applications, including

the detection of several biomolecules including DNA, cells, and proteins [94]. In elderly care, 2D TMD-based biosensors could be used to monitor a variety of health conditions, including:

i. Falls: 2D TMD-based biosensors could be used to detect changes in gait or balance that could indicate an increased risk of falls.
ii. Urinary tract infections (UTIs): 2D TMD-based biosensors could be used to detect the presence of bacteria in urine, which could help to diagnose UTIs at an early stage.
iii. Dehydration: 2D TMD-based biosensors could be used to measure the amount of water in the body, which could help to prevent dehydration, a common problem in the elderly.
iv. Heart conditions: 2D TMD-based biosensors could be used to monitor heart rate and rhythm, which could help to detect and manage heart conditions.

7. ***Environmental Monitoring***: Wearable biosensors are being utilized for environmental monitoring applications. 2D TMDs are a class of materials with unique electronic and optical properties that make them well-suited for environmental monitoring applications. TMDs are highly sensitive to a variety of environmental pollutants, including gases, liquids, and particles. They can be used to detect these pollutants at very low concentrations, making them ideal for early warning systems. One of the most promising applications of TMDs for environmental monitoring is in the detection of toxic gases. TMDs can be used to create sensors that can detect gases such as CO, NO_2, and O_3. These gases are all harmful to human health and can cause respiratory problems, heart disease, and cancer. TMD sensors can be used to monitor the levels of these gases in the air, providing early warning of potential health hazards. Moreover, TMDs can also be used to detect liquids and particles in the environment. For example, TMD sensors can be used to detect water contamination, oil spills, and airborne particles such as dust and pollen. These sensors can be used to monitor the quality of the environment and to protect public health. In addition to their sensitivity to environmental pollutants, TMDs are also very stable and can be used in harsh environments. This makes them ideal for applications such as air quality monitoring in urban areas and water quality monitoring in industrial settings.
8. ***Human–Computer Interaction (HCI)***: Wearable biosensors can facilitate natural and intuitive interactions between humans and computers. They enable gesture recognition, eye-tracking, and brain-computer interfaces, opening up new possibilities for enhanced user experiences in gaming, virtual reality, and augmented reality applications. TMDs are highly transparent and can be used to create displays that are incredibly thin and lightweight [85–90]. They are also very flexible, making them ideal for wearable devices. One of the most promising applications of TMDs for HCI is in the development of transparent displays. TMD-based displays can be used to create heads-up displays (HUDs) that project information onto the user's field of view. HUDs can be used for a variety of applications, such as navigation, gaming, and augmented reality. TMDs can also be used to create flexible displays. Flexible displays can be used to create wearable devices such

as smart watches and fitness trackers. They can also be used to create large-area displays that can be rolled up or folded. In addition to their transparency and flexibility, TMDs are also very efficient at generating and detecting light. This makes them ideal for applications such as touch screens and light sensors. Touch screens based on TMDs can be used to create devices that are more responsive and easier to use [84–102]. Light sensors based on TMDs can be used to create devices that are more energy-efficient and can detect light in a wider range of wavelengths.

7 Present State-Of-The-Art Work of Various 2D Materials in Wearable Biosensors

As discussed, the 2D materials have enormous potential in wearable biosensors technology by adopting various sensing mechanisms and targeting specific groups of chemicals or human tasks. Here, Table 2 shows the various 2D materials, their sensing mechanism, sensing targets, and performance parameters of biosensors including the limit of detection and sensitivity.

8 Conclusion

Herein, we have summarized the different aspects of the 2D TMDs materials which have been playing a very importance role to design and fabricate the 2D TMDs-based wearable biosensors. Among many synthesis processes, CVD is the most commonly used method to synthesize the 2D TMDs for wearable biosensors because of its high yield, better controllability, user-friendly, large-area growth, and reproducibility. The performance parameters of the 2D TMD-based wearable biosensors are affected and dependent on a number of factors, including the type of TMD, the synthesis method, the surface modification, and the bandgap tunability. The type of TMD has a significant impact on the sensitivity and selectivity of the biosensor. As discussed, MoS_2 is a direct bandgap material and can be utilized to detect analytes that absorb light in the visible or near-infrared region. Moreover, surface modification and functionalization can be used to improve the performance of 2D TMDs-based wearable biosensors by enhancing the surface area, increasing or improving the biocompatibility, and preventing oxidation. On the other hand, bandgap tunability and functionalization are two important strategies to improve the performance of 2D TMD-based wearable biosensors. Bandgap tunability can be achieved by doping or by applying an electric field while functionalization can be achieved by covalently or non-covalently binding biomolecules to the surface of the TMDs materials. Currently, several benchmarking potential applications have been proposed by the researchers on 2D TMD-based wearable biosensors, including point-of-care (POC) diagnostics, chronic disease management, remote patient monitoring, mental health

Table 2 Current state-of-the-art work based on 2D materials for wearable biosensors technology

Materials	Mechanism	Target	LOD	Sensitivity
MXene/Prussian blue ($Ti_3C_2T_x$) [103]	Electrochemical	Glucose and lactate	0.33×10^{-6} M 0.67×10^{-6} M	35.3 μA mm^{-1} cm^{-2} 11.4 μA mm^{-1} cm^{-2}
MXene/PVA hydrogel ($Ti_3C_2T_x$) [104]	Electrochemical	Facial expression	NA	25 (GF)
MXene/PDMS (Ti_3C_2) [105]	Piezoelectric	Human motion	NA	7400 (GF)
MXene ($Ti_3C_2T_x$) [106]	Electrical	VOC	50 ppb	<ppm
MXene/PVA ($Ti_3C_2T_x$) [107]	Electrical	VOC	50 ppb	0.10 ppm^{-1}
AuNPs/MXene-PDDA ($Ti_3C_2T_x$) [108]	Electrochemical	Nitrite	0.059 μM	250 μA mM^{-1} cm^{-2}
MXene (Ti_3C_2) [109]	Optical	Hg^{2+}	42.5 pM	0.05–50 mM
MXene ($Ti_3C_2T_x$) [110]	Optical	Ag	0.615 μM	NA
Graphene-pyrene [111]	Electrical	Interferon interleukin tumor necrosis factor	476×10^{-15} M 608×10^{-15} M 611×10^{-15} M	NA
Graphene [112]	Electrical	Brain stimulation	116.07–174.10 μC cm^{-2}	NA
Graphene-MoS_2 [113]	Piezoelectric	Human motion	NA	6.06 kPa^{-1}
rGO [114]	Piezoelectric	Human motion	NA	25.1 kPa^{-1}
rGO/cellulose fibers [115]	Piezoelectric	Human motion	8 μm	−8.9 (GF)
Carboxyphenyl-G [116]	Electrochemical	Ovalbumin	0.9 pg mL^{-1}	NA
rGO-aryl fluoride [117]	Electrochemical	Ammonia	20 ppm	63%
rGO-ferrocenylaniline [118]	Electrochemical	Tumor necrosis factor	0.1 pg mL^{-1}	NA
Pt-loaded/h-BN [119]	Electrical	Propane	2%	20,000 ppm
h-BN [120]	Electrical	NO_2	20 ppb	1.645 ppm^{-1}
$MoTe_2$ [121]	Electrical	Ketone	0.2 ppm	100 ppm
WS_2 [122]	Electrical	NH_3	1 ppm	10 ppm
MoS_2 [123]	Electrical	Pressure and temperature	NA	51kΩ °C^{-1}
MoS_2/hBN [124]	Electrical	NO_X	6 ppb	6 ppb
MoS_2 [125]	Electrical	NO_2	0.15 ppb	1 ppb

(continued)

Table 2 (continued)

Materials	Mechanism	Target	LOD	Sensitivity
MoS_2 [126]	Optical	Methanol	2.7 ppm	NA
MoS_2/Au [127]	Optical	Antigen	0.5 pg/mL	NA
MoS_2/rGO [128]	Electrochemical	H_2O_2	0.19 μM	101.70 μA mM^{-1} cm^{-2}
MoS_2/rGO [129]	Electrochemical	NO_2	0.17 μM	0.46 μA μM^{-1} cm^{-2}
BP/PEI-PEG [130]	Flexible	CO_2	200 ppm	NA
BP [131]	Electrical	NO_2	20 ppb	20% for 20 ppb
BP [132]	Electrical	NO_2	5 ppb	5 ppb
BP [133]	Electrical	NH_3	80 ppb	10 ppm
BP-AuNPs [134]	Electrochemical	Patulin	0.03×10^{-3} M	NA

NA: Not available; rGO: reduced graphene oxide; BP: Black Phosphorous; VOC: Volatile organic compound; PVA: Polyvinyl alcohol; PDMS: Polydimethylsiloxane; PEI: Polyethylene imine; PEG: Polyethylene glycol; ppb: Parts-per billion; ppm: Parts-per million

monitoring, environmental monitoring, human–computer interaction, and personalized medicine. POC diagnostics refers to the ability to perform diagnostic tests at the point of care, such as in a doctor's office or in the home. Remote patient monitoring refers to the ability to continuously monitor a patient's health parameters, such as blood glucose levels or heart rate. Therefore, it is concluded that the 2D TMDs materials have a wide area of interest in the development of wearable biosensors with greater performance and stability and can be useful to implement the real-time applications in various technological domains.

Conflict of Interest The authors have declared there is no conflict of interest in this chapter.

References

1. Silvestri, A., Wetzl, C., Alegret, N., Cardo, L., Hou, H.L., Criado, A., Prato, M.: The era of nano-bionic: 2D materials for wearable and implantable body sensors. Adv. Drug Delivery Rev. **186**, 114315 (2022). https://doi.org/10.1016/j.addr.2022.114315
2. Mia, A.K., Meyyappan, M., Giri, P.K.: Two-dimensional transition metal dichalcogenide based biosensors: from fundamentals to healthcare applications. Biosensors **13**, 169 (2023). https://doi.org/10.3390/bios13020169
3. Xia, F., Mueller, T., Lin, Y., Valdes-Garcia, A., Avouris, P.: Ultrafast graphene photodetector. Nat. Nanotechnol. **4**, 839–843 (2009). https://doi.org/10.1038/nnano.2009.292
4. Lin, Y.M., Jenkins, K.A., Valdes-Garcia, A., Small, J.P., Farmer, D.B., Avouris, P.: Operation of graphene transistors at gigahertz frequencies. Nano Lett. **9**, 422–426 (2009). https://doi.org/10.1021/nl803316h

5. Mittendorff, M., Winnerl, S., Kamann, J., Eroms, J., Weiss, D., Schneider, H., Helm, M.: Ultrafast graphene-based broadband THz detector. Appl. Phys. Lett. **103**, 021113 (2013). https://doi.org/10.1063/1.4813621
6. Mueller, T., Xia, F., Avouris, P.: Graphene photodetectors for high-speed optical communications. Nat. Photon. **4**, 297–301 (2010). https://doi.org/10.1038/nphoton.2010.40
7. Zhang, B.Y., Liu, T., Meng, B., Li, X., Liang, G., Hu, X., Wang, Q. J.: Broadband high photoresponse from pure monolayer graphene photodetector. Nat. Commun. **4**, 1811 (2013). https://doi.org/10.1038/ncomms2830
8. Yoon, H.J., Jun, D.H., Yang, J.H., Zhou, Z., Yang, S.S., Cheng, M.M.C.: Carbon dioxide gas sensor using a graphene sheet. Sens. Actuators B Chem. **157**, 310–313 (2011). https://doi.org/10.1016/j.snb.2011.03.035
9. Tian, W., Liu, X., Yu, W.: Research progress of gas sensor based on graphene and its derivatives: a review. Appl. Sci. **8**, 1118 (2018). https://doi.org/10.3390/app8071118
10. Justino, C.I.L., Gomes, A.R., Freitas, A.C., Duarte, A.C., Rocha-Santos, T.A.P.: Graphene based sensors and biosensors. TrAC Trends Anal. Chem. **91**, 53–66 (2017). https://doi.org/10.1016/j.trac.2017.04.003
11. De, S., Coleman, J.N.: Are there fundamental limitations on the sheet resistance and transmittance of thin graphene films? ACS Nano **4**, 2713–2720 (2010). https://doi.org/10.1021/nn100343f
12. Niu, T., Zhang, J., Chen, W.: Surface engineering of two-dimensional materials. ChemNanoMat **5**, 6–23 (2019). https://doi.org/10.1002/cnma.201800181
13. Azadmanjiri, J., Kumar, P., Srivastava, V.K., Sofer, Z.: Surface functionalization of 2D transition metal oxides and dichalcogenides via covalent and non-covalent bonding for sustainable energy and biomedical applications. ACS Appl. Nano Mater. **3**, 3116–3143 (2020). https://doi.org/10.1021/acsanm.0c00120
14. Guo, Y., Xu, K., Wu, C., Zhao, J., Xie, Y.: Surface chemical modification for engineering the intrinsic physical properties of inorganic two-dimensional nanomaterials. Chem. Soc. Rev. **44**(3), 637–646 (2015). https://doi.org/10.1039/C4CS00302K
15. Yin, F., Gu, B., Lin, Y., Panwar, N., Tjin, S.C., Qu, J., Lau, S.P., Yong, K.-T.: Functionalized 2D nanomaterials for gene delivery applications. Coord. Chem. Rev. **347**, 77–97 (2017). https://doi.org/10.1016/j.ccr.2017.06.024
16. Zhang, H., Arutchelvan, G., Meersschaut, J., Gaur, A., Conard, T., Bender, H., Lin, D., Asselberghs, I., Heyns, M., Radu, I., Vandervorst, W., Delabie, A.: MoS_2 functionalization with a sub-nm thin SiO_2 layer for atomic layer deposition of high-κ dielectrics. Chem. Mater. **29**(16), 6772–6780 (2017). https://doi.org/10.1021/acs.chemmater.7b01695
17. Liu, M., Zhu, H., Wang, Y., Sevencan, C., Li, B.L.: Functionalized MoS_2-based nanomaterials for cancer phototherapy and other biomedical applications. ACS Mater. Lett. **3**(5), 462–496 (2021). https://doi.org/10.1021/acsmaterialslett.1c00073
18. Martin, N., Tagmatarchis, N., Wang, Q.H., Zhang, X.: Chemical functionalization of 2D materials. Chem. Eur. J. **26**, 6292–6295 (2020). https://doi.org/10.1002/chem.202001304
19. Nam, T., Seo, S., Kim, H.: Atomic layer deposition of a uniform thin film on two-dimensional transition metal dichalcogenides. J. Vac. Sci. Technol. A **38**, 030803 (2020). https://doi.org/10.1116/6.0000068
20. Zhu, C., Xia, X., Liu, J., Fan, Z., Chao, D., Zhang, H., Fan, H.J.: TiO_2 nanotube @ SnO_2 nanoflake core–branch arrays for lithium-ion battery anode. Nano Energy **4**, 105 (2014). https://doi.org/10.1016/j.nanoen.2013.12.018
21. Ansari, M.Z., Parveen, N., Nandi, D.K., Ramesh, R., Ansari, S.A., Cheon, T., Kim, S.H.: Enhanced activity of highly conformal and layered tin sulfide (SnS_x) prepared by atomic layer deposition (ALD) on 3D metal scaffold towards high performance supercapacitor electrode. Sci. Rep. **9**, 10225 (2019). https://doi.org/10.1038/s41598-019-46679-7
22. Cremers, V., Puurunen, R.L., Dendooven, J.: Conformality in atomic layer deposition: Current status overview of analysis and modelling. Appl. Phys. Rev. **6**, 021302 (2019). https://doi.org/10.1063/1.5060967

23. Liu, F., Fan, Z.: Defect engineering of two-dimensional materials for advanced energy conversion and storage. Chem. Soc. Rev. **52**, 1723 (2023). https://doi.org/10.1039/d2cs00931e
24. Novoselov, K.S., Geim, A.K., Morozov, S.V., Jiang, D.E., Zhang, Y., Dubonos, S.V., Grigorieva, I.V., Firsov, A.A.: Electric field effect in atomically thin carbon films. Science **306**, 666–669 (2004). https://doi.org/10.1126/science.1102896
25. Pertsinidis, A., Ling, X.: Diffusion of point defects in two-dimensional colloidal crystals. Nature **413**, 147–150 (2001). https://doi.org/10.1038/35093077
26. Setvin, M., Wagner, M., Schmid, M., Parkinson, G.S., Diebold, U.: (2017), Surface point defects on bulk oxides: atomically-resolved scanning probe microscopy. Chem. Soc. Rev. **46**, 1772–1784 (2017). https://doi.org/10.1039/C7CS00076F
27. Nan, H., Zhou, R., Gu, X., Xiao, S., Ostrikov, K.: Recent advances in plasma modification of 2D transition metal dichalcogenides. Nanoscale **11**, 19202 (2019). https://doi.org/10.1039/c9nr05522c
28. Jelil, R.A.: A review of low-temperature plasma treatment of textile materials. J. Mater. Sci. **50**, 5913–5943 (2015). https://doi.org/10.1007/s10853-015-9152-4
29. Ostrikov, K., Neyts, E.C., Meyyappan, M.: Plasma nanoscience: from nano-solids in plasmas to nano-plasmas in solids. Adv. Phys. **62**, 113–224 (2013). https://doi.org/10.1080/00018732.2013.808047
30. Bhowmik, S., Rajan, A.G.: Chemical vapor deposition of 2D materials: a review of modeling, simulation, and machine learning studies. iScience **25**, 103832 (2022). https://doi.org/10.1016/j.isci.2022.103832
31. Zhang, Y., Yao, Y., Sendeku, M.G., Yin, L., Zhan, X., Wang, F., Wang, Z., He, J.: Recent progress in CVD growth of 2D transition metal dichalcogenides and related heterostructures. Adv. Mater. **31**, 1901694 (2019). https://doi.org/10.1002/adma.201901694
32. Ahn, C., Park, Y., Shin, S., Ahn, J.-G., Song, I., An, Y., Jung, J., Kim, C.S., Kim, J.H., Bang, J., et al.: Growth of monolayer and multilayer MoS_2 films by selection of growth mode: two pathways via chemisorption and physisorption of an inorganic molecular precursor. ACS Appl. Mater. Interfaces **13**(5), 6805–6812 (2021). https://doi.org/10.1021/acsami.0c19591
33. Lee, Y., Zhang, X., Zhang, W., Chang, M., Lin, C., Chang, K., Yu, Y., Wang, J.T., Chang, C., Li, L.: Synthesis of large-area MoS_2 atomic layers with chemical vapor deposition. Adv. Mater. **24**(17), 2320–2025 (2012). https://doi.org/10.1002/adma.201104798
34. Wei, Z., Li, B., Xia, C., Cui, Y., He, J., Xia, J.-B., Li, J.: Various structures of 2D transition-metal dichalcogenides and their applications. Small Methods **2**, 1800094 (2018). https://doi.org/10.1002/smtd.201800094
35. Muratorea, C., Voevodinb, A.A., Glavinc, N.R.: Physical vapor deposition of 2D Van der Waals materials: a review. Thin Solid Films **688**, 137500 (2019). https://doi.org/10.1016/j.tsf.2019.137500
36. Greene, J.E.: Tracing the recorded history of thin-film sputter deposition: from the 1800s to 2017. J. Vac. Sci. Technol. A **35**, 05C204 (2017). https://doi.org/10.1116/1.4998940
37. Window, B., Savvides, N.: Charged particle fluxes from planar magnetron sputtering sources. J. Vac. Sci. Technol. **A4**, 196–202 (1986). https://doi.org/10.1116/1.573470
38. Petrov, I., Barna, P.B., Hultman, L., Greene, J.E.: Microstructural evolution during film growth. J. Vac. Sci. Tech. A **21**, S117–S128 (2003). https://doi.org/10.1116/1.1601610
39. Cheng, P., Zhang, W., Zhang, L., Gou, J., Wong, P.K.J., Chen, L.: Molecular beam epitaxy fabrication of two-dimensional materials. Semicond. Mater. Devices **2D**, 103–134 (2020). https://doi.org/10.1016/B978-0-12-816187-6.00004-2
40. Cho, A.Y., Arthur, J.R.: Molecular beam epitaxy. Prog. Solid State Chem. **10**, 3, 157–191 (1975). https://doi.org/10.1016/0079-6786(75)90005-9.
41. Franchi, S.: Molecular beam epitaxy: fundamentals, historical background and future prospects. pp. 1–46 (2013). https://doi.org/10.1016/B978-0-12-387839-7.00001-4
42. Dimoulas, A.: Perspectives for the growth of epitaxial 2D van der Waals Layers with an Emphasis on ferromagnetic metals for spintronics. Adv. Mater. Interfaces **9**, 2201469. https://doi.org/10.1002/admi.202201469

43. Alam, S., Chowdhury, M.A., Shahid, A., Alam, R., Rahim, A.: Synthesis of emerging two-dimensional (2D) materials-advances, challenges and prospects. FlatChem **30**, 100305 (2021). https://doi.org/10.1016/j.flatc.2021.100305
44. Xie, H., Li, Z., Cheng, L., Haidry, A. A., Tao, J., Xu, Y., Xu, K., Ou, J.Z.: Recent advances in the fabrication of 2D metal oxides. iScience **25**, 103598 (2022). https://doi.org/10.1016/j.isci.2021.103598.
45. Shu, Y., Guo, J., Fan, T., Xu, Y., Guo, P., Wang, Z., Wu, L., Ge, Y., Lin, Z., Ma, D., Wei, S., Li, J., Zhang, H., Chen, W.: Ultrafast and programmable shape memory hydrogel of gelatin soaked in tannic acid solution. ACS Appl. Mater. Interfaces **12**, 46509 (2020). https://doi.org/10.1021/acsami.0c13531
46. Alzakia, F.I., Tan, S.C.: Liquid-exfoliated 2D materials for optoelectronic applications. Adv. Sci. **8**, 2003864 (2021). https://doi.org/10.1002/advs.202003864
47. Choi, S.H., Yun, S.J., Won, Y.S., Oh, C.S., Kim, S.M., Kim, K.K., Lee, Y.H.: Large-scale synthesis of graphene and other 2D materials towards industrialization. Nat. Commun. **13**, 1484 (2022). https://doi.org/10.1038/s41467-022-29182-y
48. Nicolosi, V., Chhowall, M., Kanatzidis, M.G., Strano, M.S., Coleman, J.N.: Liquid exfoliation of layered materials. Science **340**(6139), 1226419 (2013). https://doi.org/10.1126/science.12264
49. Niu, L., Coleman, J.N., Zhang, H., Shin, H., Chhowalla, M., Zheng, Z.: Production of two-dimensional nanomaterials via liquid-based direct exfoliation. Small **12**(3), 272–293 (2016). https://doi.org/10.1002/smll.201502207
50. Lopez, A.T., Pasadas, F., Marin, E.G., Rull, A.M., Medina, J.M.G., Ruiz, F.G., Jimenez, D., Godoy, A.: Multi-scale analysis of radio-frequency performance of 2D-material based field-effect transistors. Nanoscale Adv. **3**, 2377 (2021). https://doi.org/10.1039/d0na00953a
51. Wang, G., Hou, H., Yan, Y., Jagatramka, R., Shirsalimian, A., Wang, Y., Li, B., Daly, M., Cao, C.: Recent advances in the mechanics of 2D Materials. Int. J. Extrem. Manuf. **5**, 032002 (56pp) (2023). https://doi.org/10.1088/2631-7990/accda2
52. Mathew, M., Radhakrishnan, S., Vaidyanathan, A., Chakraborty, B., Rout, C.S.: Flexible and wearable electrochemical biosensors based on two-dimensional materials: Recent developments. Anal. Bioanal. Chem. **413**(3), 727–762 (2021). https://doi.org/10.1007/s00216-020-03002-y
53. Wang, Y., Li, T., Li, Y., Yang, R., Zhang, G.: 2D-materials-basedwearable biosensor systems. Biosensors **12**, 936 (2022). https://doi.org/10.3390/bios12110936
54. Vaghasiya, J. V., Carmen C., Martinez, M., Pumera, M.: Wearable sensors for telehealth based on emerging materials and nanoarchitectonics. NPJ Flexible Electron **7**, 26 (2023). https://doi.org/10.1038/s41528-023-00261-4
55. Bolotsky, A., Butler, D., Gerace, C.K., Glavin, N.R., Muratore, C., Robinson, J.A., Ebrahimi, A.: Two-dimensional materials in biosensing and healthcare: from in vitro diagnostics to optogenetics and beyond. ACS Nano **13**(9), 9781–9810 (2019). https://doi.org/10.1021/acsnano.9b03632
56. Wang, Z., Jing, X., Duan, S., Liu, C., Kang, D., Xu, X., Chen, J., Xia, Y., Chang, B., Zhao, C., Zhu, B., Xu, T., Lin, H., Lu, W., Ren, Y., Sun, L., Wu, J., Tao, L.: 2D $PtSe_2$ enabled wireless wearable gas monitoring circuits with distinctive strain-enhanced performance. ACS Nano (2023). https://doi.org/10.1021/acsnano.3c01582
57. Radhakrishnan, S., Mathew, M., Rout, C. S.: Microfluidic sensors based on two-dimensional materials for chemical and biological assessments. Mater. Adv. **3**, 1874 (2022). https://doi.org/10.1039/d1ma00929j
58. Gao, L.: Flexible device applications of 2D semiconductors. Small **13**(35), 1603994 (2017). https://doi.org/10.1002/smll.201603994
59. Chen, F., Tang, Q., Ma, T., Zhu, B., Wang, L., He, C., Luo, X., Cao, S., Ma, L., Cheng, C.: Structures, properties, and challenges of emerging 2D materials in bioelectronics and biosensors. InfoMat. **4**, e12299 (2022). https://doi.org/10.1002/inf2.12299
60. Chen, Y., Tan, C., Zhang, H., Wang, L.: Two-dimensional graphene analogues for biomedical applications. Chem. Soc. Rev. **44**(9), 2681–2701 (2015). https://doi.org/10.1039/C4CS00300D

61. Chu, D., Kim, E.K.: Hole conduction of tungsten diselenide crystalline transistors by niobium dopant. Adv. Electron. Mater. **5**(2), 1800695 (2019). https://doi.org/10.1002/aelm.201800695
62. Ko, K.Y., Song, J.G., Kim, Y., Choi, T., Shin, S., Lee, C.W., Lee, K., Koo, J., Lee, H., Kim, J., Lee, T., Park, J., Kim, H.: Improvement of gas-sensing performance of large-area tungsten disulfide nanosheets by surface functionalization. ACS Nano **10**(10), 9287–9296 (2016). https://doi.org/10.1021/acsnano.6b03631
63. Chaves, A., Azadani, J. G., Alsalman, H., da Costa. D., R., Frisenda, R., Chaves, A. J., Song, S. H., Kim, Y. D., He, D., Zhou, J., Castellanos-Gomez, A., Peeters, F. M., Liu, Z., Hinkle. C. L., Oh, S. H., Ye, P. D., Koester, S. J., Lee, Y. H., Avouris, P., Wang, X., Low, T.: NPJ 2D Mater. Appl. **4**, 29 (2020). https://doi.org/10.1038/s41699-020-00162-4
64. Kappera, R., Voiry, D., Yalcin, S.E., Branch, B., Gupta, G., Mohite, A.D., Chhowalla, M.: Phase-engineered low-resistance contacts for ultrathin MoS_2 transistors. Nat. Mater. **13**, 1128 (2014). https://doi.org/10.1038/nmat4080
65. Fei, Z., Palomaki, T., Wu, S., Zhao, W., Cai, X., Sun, B., Nguyen, P., Finney, J., Xu, X., Cobden, D.H.: Edge conduction in monolayer WTe_2. Nat. Phys. **13**, 677 (2017). https://doi.org/10.1038/nphys4091
66. Li, L., Chen, Z., Hu, Y., Wang, X., Zhang, T., Chen, W., Wang, Q.: Single-layer single-crystalline SnSe nanosheets. J. Am. Chem. Soc. **135**, 1213–1216 (2013). https://doi.org/10.1021/ja3108017
67. Anasori, B., Lukatskaya, M.R., Gogotsi, Y.: 2D metal carbides and nitrides (MXenes) for energy storage. Nat. Rev. Mater. **2**, 16098 (2017). https://doi.org/10.1038/natrevmats.2016.98
68. Balushi, Z.Y.A., Wang, K., Ghosh, R.K., Vila, R.A., Eichfeld, S.M., Caldwell, J.D., Qin, X., Lin, Y.C., DeSario, P.A., Stone, G., Subramanian, S., Paul, D.F., Wallace, R.M., Datta, S., Redwing, J.M., Robinson, J.A.: Two-dimensional gallium nitride realized via graphene encapsulation. Nat. Mater. **15**, 1166 (2016). https://doi.org/10.1038/nmat4742
69. Zhao, W., Ghorannevis, Z., Chu, L., Toh, M., Kloc, C., Tan, P.H., Eda, G.: Evolution of electronic structure in atomically thin sheets of WS_2 and WSe_2. ACS Nano **7**, 791–797 (2013). https://doi.org/10.1021/nn305275h
70. Zhang, G., Amani, M., Chaturvedi, A., Tan, C., Bullock, J., Song, X., Kim, H., Lien, D.H., Scott, M.C., Zhang, H., Javey, A.: Optical and electrical properties of two-dimensional palladium diselenide. Appl. Phys. Lett. **114**, 253102 (2019). https://doi.org/10.1063/1.5097825
71. Hong, X., Kim, J., Shi, S.F., Zhang, Y., Jin, C., Sun, Y., Tongay, S., Wu, J., Zhang, Y., Wang, F.: Ultrafast charge transfer in atomically thin MoS_2 /WS_2 heterostructures. Nat. Nanotechnol. **9**, 682 (2014). https://doi.org/10.1038/nnano.2014.167
72. Lee, H., Choi, T.K., Lee, Y.B., Cho, H.R., Ghaffari, R., Wang, L., Choi, H.J., Chung, T.D., Lu, N., Hyeon, T., Choi, S.H., Kim, D.H.: A graphene-based electrochemical device with thermoresponsive microneedles for diabetes monitoring and therapy. Nat. Nanotechnol. **11**, 566–572 (2016). https://doi.org/10.1038/nnano.2016.38
73. Wood, A., O'Neal, D., Furler, J., Ekinci, E.I.: Continuous glucose monitoring: a review of the evidence, opportunities for future use and ongoing challenges. Intern. Med. J. **48**, 499–508 (2018). https://doi.org/10.1111/imj.13770
74. Pu, Z., Zou, C., Wang, R., Lai, X., Yu, H., Xu, K., Li, D.: A continuous glucose monitoring device by graphene modified electrochemical sensor in microfluidic system. Biomicrofluidics **10**(1), 011910 (2016). https://doi.org/10.1063/1.4942437
75. Lee, H., Hong, Y.J., Baik, S., Hyeon, T., Kim, D.H.: Enzyme-based glucose sensor: from invasive to wearable device. Adv. Healthcare Mater. **7**(8), 1701150 (2018). https://doi.org/10.1002/adhm.201701150
76. Facchinetti, A., Sparacino, G., Guerra, S., Luijf, Y.M., De Vries, J.H., Mader, J.K., Ellmerer, M., Benesch, C., Heinemann, L., Bruttomesso, D., Avogaro, A., Cobelli, C.: Real-time improvement of continuous glucose monitoring accuracy: the smart sensor concept. Diabetes Care **36**, 793–800 (2013). https://doi.org/10.2337/dc12-0736
77. Ameri, K.S., Ho, R., Jang, H., Tao, L., Wang, Y., Wang, L., Schnyer, D.M., Akinwande, D., Lu, N.: Graphene electronic tattoo sensors. ACS Nano **11**, 7634–7641 (2017). https://doi.org/10.1021/acsnano.7b02182

78. Tao, L.Q., Zhang, K.N., Tian, H., Liu, Y., Wang, D.Y., Chen, Y.Q., Yang, Y., Ren, T.Y.: Graphene-paper pressure sensor for detecting human motions. ACS Nano **11**(9), 8790–8795 (2017). https://doi.org/10.1021/acsnano.7b0282610.1021/acsnano.7b02826.s001
79. Chen, X., Luo, F., Yuan, M., Xie, D., Shen, L., Zheng, K., Wang, Z., Li, X., Tao, L.Q.: A dual-functional graphene-based self-alarm health-monitoring e-skin. Adv. Funct. Mater. **29**(51), 1904706 (2019). https://doi.org/10.1002/adfm.v29.5110.1002/adfm.201904706
80. He, J., Xiao, P., Shi, J., Liang, Y., Lu, W., Chen, Y., Wang, W., Theato, P., Kuo, S.W., Chen, T.: High performance humidity fluctuation sensor for wearable devices via a bioinspired atomic-precise tunable graphene-polymer heterogeneous sensing junction. Chem. Mater. **30**(13), 4343–4354 (2018). https://doi.org/10.1021/acs.chemmater.8b01587
81. Torrente-Rodriguez, R.M., Tu, J., Yang, Y., Min, J., Wang, M., Song, Y., Yu, Y., Xu, C., Ye, C., IsHak, W.W., Gao, W.: Investigation of cortisol dynamics in human sweat using a graphene-based wireless mhealth system. Matter. **2**(4), 921–937 (2020). https://doi.org/10.1016/j.matt.2020.01.021
82. Gholizadeh, A., Voiry, D., Weisel, C., Gow, A., Laumbach, R., Kipen, H., Chhowalla, M., Javanmard, M.: Toward point-of-care management of chronic respiratory conditions: Electrochemical sensing of nitrite content in exhaled breath condensate using reduced graphene oxide. Microsyst. Nanoeng. **3**, 17022 (2017). https://doi.org/10.1038/micronano.2017.22
83. Hwang, B.U., Lee, J.H., Trung, T.Q., Roh, E., Kim, D.I., Kim, S.W., Lee, N.E.: Transparent stretchable self-powered patchable sensor platform with ultrasensitive recognition of human activities. ACS Nano **9**(9), 8801–8810 (2018). https://doi.org/10.1021/acsnano.5b01835
84. Blaschke, B. M., Colet, N. T., Guimera-Brunet, A., Weinert, J., Rousseau, L., Heimann, A., Drieschner, S., Kempski, O., Villa, R., Sanchez-Vives, M. V., Garrido, J. A.: Mapping brain activity with flexible graphene micro-transistors. 2D Mater. 4, 2, 025040 (2017). https://doi.org/10.1088/2053-1583/aa5eff
85. Masvidal-Codina, E., Illa, X., Dasilva, M., Calia, A.B., Dragojevic, T., Vidal- Rosas, E.E., Prats-Alfonso, E., Martínez-Aguilar, J., De la Cruz, J.M., Garcia- Cortadella, R., Godignon, P., Rius, G., Camassa, A., Corro, E.D., Bousquet, J., Hebert, C., Durduran, T., Villa, R., Sanchez-Vives, M.V., Garrido, J.A., Guimera- Brunet, A.: High-resolution mapping of infraslow cortical brain activity enabled by graphene microtransistors. Nat. Mater. **18**(3), 280–288 (2019). https://doi.org/10.1038/s41563-018-0249-4
86. Park, D.W., Schendel, D.A., Mikael, S., Brodnick, S.K., Richner, T.J., Ness, J.P., Hayat, M.R., Atry, F., Frye, S.T., Pashaie, R., Thongpang, S., Ma, Z., Williams, J.C.: Graphene-based carbon-layered electrode array technology for neural imaging and optogenetic applications. Nat. Commun. **5**, 5258 (2014). https://doi.org/10.1038/ncomms6258
87. Pang, Y., Yang, Z., Yang, Y., Ren, T.L.: Wearable electronics based on 2D materials for human physiological information detection. Small **16**, 1901124 (2020). https://doi.org/10.1002/smll.201901124
88. Kuzum, D., Takano, Shim, E., Reed, J. C., Juul, H., Richardson, A. G., de Vries, J., Bink, H., Dichter, M. A., Lucas, T. H., Coulter, D. A., Cubukcu, E., Litt, B.: Transparent and flexible low noise graphene electrodes for simultaneous electrophysiology and neuroimaging. Nat. Commun. **5**, 5259 (2014). https://doi.org/10.1038/ncomms6259
89. Faisal, S.N., Iacopi, F.: Thin-film electrodes based on two-dimensional nanomaterials for neural interfaces. ACS Appl. Nano Mater. **5**(8), 10137–10150 (2022). https://doi.org/10.1021/acsanm.2c03056
90. Han, N., Yao, X., Wang, Y., Huang, W., Niu, M., Zhu, P., Mao, Y.: Recent progress of biomaterials-based epidermal electronics for healthcare monitoring and human–machine interaction. Biosensors **13**(3), 393 (2023). https://doi.org/10.3390/bios13030393
91. Lv, Z., Zhou, Y., Han, S.T., Roy, V.A.L.: From biomaterial-based data storage to bio-inspired artificial synapse. Mater. Today **21**, 537–552 (2018). https://doi.org/10.1016/j.mattod.2017.12.001
92. Pradhan, S., Brooks, A.K., Yadavalli, V.K.: Nature-derived materials for the fabrication of functional biodevices. Mater. Today Bio. **7**, 100065 (2020). https://doi.org/10.1016/j.mtbio.2020.100065

93. Hang, Y., Zhang, Y., Jin, Y., Shao, H., Hu, X.: Preparation of regenerated silk fibroin/silk sericin fibers by coaxial electrospinning. Int. J. Biol. Macromol. **51**, 980–986 (2012). https://doi.org/10.1016/j.ijbiomac.2012.08.010
94. Gogurla, N., Kim, Y., Cho, S., Kim, J., Kim, S.: Multifunctional and ultrathin electronic tattoo for on-skin diagnostic and therapeutic applications. Adv. Mater. **33**, 2008308 (2021). https://doi.org/10.1002/adma.202008308
95. Xu, W., Gracias, D.H.: Soft three-dimensional robots with hard two-dimensional materials. ACS Nano **13**(5), 4883–4892 (2019). https://doi.org/10.1021/acsnano.9b03051
96. Tajik, S., Dourandish, Z., Nejad, F.G., Beitollahi, H., Jahani, P.M., Bartolomeo, A.D.: Transition metal dichalcogenides: synthesis and use in the development of electrochemical sensors and biosensors. Biosens. Bioelectron. **216**, 14674 (2022). https://doi.org/10.1016/j.bios.2022.114674
97. Derakhshi, M., Daemi, S., Shahini, P., Habibzadeh, A., Mostafavi, E., Ashkarran, A.A.: Two-dimensional nanomaterials beyond graphene for biomedical applications. J. Funct. Biomater. **13**, 27 (2022). https://doi.org/10.3390/jfb13010027
98. Singh, E., Singh, P., Kim, K.S., Yeom, G.Y., Nalwa, H.S.: Flexible molybdenum disulfide (MoS_2) atomic layers for wearable electronics and optoelectronics. ACS Appl. Mater. Interfaces **11**(12), 11061–11105 (2019). https://doi.org/10.1021/acsami.8b19859
99. Chimene, D., Alge, D.L., Gaharwar, A.K.: Two-dimensional nanomaterials for biomedical applications: Emerging trends and future prospects. Adv. Mater. **27**, 7261–7284 (2015). https://doi.org/10.1002/adma.201502422
100. Kwon, Y.T., Kim, Y.S., Kwon, S., Mahmood, M., Lim, H.R., Park, S.W., Kang, S.O., Choi, J.J., Herbert, R., Jang, Y.C., Choa, Y.H., Yeo, W.H.: All-printed nanomembrane wireless bioelectronics using a biocompatible solderable graphene for multimodal human-machine interfaces. Nat. Commun. **11**, 1–11 (2020). https://doi.org/10.1038/s41467-020-17288-0
101. Driscoll, N., Richardson, A.G., Maleski, K., Anasori, B., Adewole, O., Lelyukh, P., Escobedo, L., Cullen, D.K., Lucas, T.H., Gogotsi, Y., Vitale, F.: Two-dimensional Ti_3C_2 MXene for high-resolution neural interfaces. ACS Nano **12**, 10419–10429 (2018). https://doi.org/10.1021/acsnano.8b06014
102. Guo, W., Qiu, J., Liu, J., Liu, H.: Graphene microfiber as a scaffold for regulation of neural stem cells differentiation. Sci. Rep. **7**, 1–8 (2017). https://doi.org/10.1038/s41598-017-06051-z
103. Lei, Y., Zhao, W., Zhang, Y., Jiang, Q., He, J.H., Baeumner, A.J., Wolfbeis, O.S., Wang, Z.L., Salama, K.N., Alshareef, H.N.: A MXene-based wearable biosensor system for high-performance in vitro perspiration analysis. Small **15**(19), 1901190 (2019). https://doi.org/10.1002/smll.201901190
104. Zhang, Y.Z., Lee, K.H., Anjum, D.H., Sougrat, R., Jiang, Q., Kim, H., Alshareef, H.N.: (2022) MXenes stretch hydrogel sensor performance to new limits, Sci. Adv. 4, 6, eaat0098. https://doi.org/10.1126/sciadv.aat0098
105. Kedambaimoole, V., Kumar, N., Shirhatti, V., Nuthalapati, S., Sen, P., Nayak, M.M., Rajanna, K., Kumar, S.: Laser-induced direct patterning of free-standing Ti3C2–MXene films for skin conformal tattoo sensors. ACS Sens. **5**(7), 2086–2095 (2020). https://doi.org/10.1021/acssensors.0c00647
106. Kim, S.J., Koh, H.J., Ren, C.E., Kwon, O., Maleski, K., S.Y., Cho, Anasori, B., Kim, C.K., Choi, Y.K., Kim, J., Gogotsi, Y., Jung, H.T.: Metallic $Ti_3C_2T_x$ MXene gas sensors with ultrahigh signal-to-noise ratio. ACS Nano **12**, 2, 986–993 (2018). https://doi.org/10.1021/acsnano.7b07460
107. Yuan, W., Yang, K., Peng, H., Li, F., Yin, F.: A flexible VOCs sensor based on a 3D MXene framework with a high sensing performance. J. Mater. Chem. A **6**(37), 18116–18124 (2018). https://doi.org/10.1039/C8TA06928J
108. Wang, Y., Zeng, Z., Qiao, J., Dong, S., Liang, Q., Shao, S.: Ultrasensitive determination of nitrite based on electrochemical platform of AuNPs deposited on PDDA modified MXene nanosheets. Talanta **221**, 121605 (2021). https://doi.org/ https://doi.org/10.1016/j.talanta.2020.121605.

109. Lu, L., Han, X., Lin, J., Zhang, Y., Qiu, M., Chen, Y., Li, M., Tang, D.: Ultrasensitive fluorometric biosensor based on Ti_3C_2 MXenes with Hg^{2+}-triggered exonuclease III-assisted recycling amplification. Analyst 146, 8, 2664–2669 (2021). https://doi.org/10.1039/D1AN00178G
110. Wang, Y., Wang, S., Dong, N., Kang, W., Li, K., Nie, Z.: Titanium carbide MXenes mediated in situ reduction allows label-free and visualized nanoplasmonic sensing of silver ions. Anal. Chem. **92**, 6, 4623–4629 (2020). https://doi.org/10.1021/acs.analchem.0c00164
111. Hao, Z., Luo, Y., Huang, C., Wang, Z., Song, G., Pan, Y., Zhao, X., Liu, S.: An intelligent graphene-based biosensing device for cytokine storm syndrome biomarkers detection in human biofluids. Small **17**, 29, 2101508 (2021). https://doi.org/10.1002/smll.202101508
112. Park, D. W., Ness, J., P., Brodnick, S., K., Esquibel, C., Novello, J., Atry, F., Baek, D. H., Kim, H., Bong, J., Swanson, K. I., Suminski, A. J., Otto, K. J., Pashaie, R., Williams, J. C., Ma, Z.: Electrical neural stimulation and simultaneous in vivo monitoring with transparent graphene electrode arrays implanted in $GCaMP_6f$ mice. ACS Nano **12**, 1, 148–157 (2018). https://doi.org/10.1021/acsnano.7b04321
113. Kim, S.J., Mondal, S., Min, B.K., Choi, C.G.: Highly sensitive and flexible strain–pressure sensors with cracked paddy-shaped MoS_2/Graphene Foam/ Ecoflex hybrid nanostructures. ACS Appl. Mater. Interfaces **10**(42), 36377–36384 (2018). https://doi.org/10.1021/acsami.8b11233
114. Pang, Y., Zhang, K., Yang, Z., Jiang, S., Ju, Z., Li, Y., Wang, X., Wang, D., Jian, M., Zhang, Y., Liang, R., Tian, H., Yang, Y., Ren, T.L.: Epidermis microstructure inspired graphene pressure sensor with random distributed spinosum for high sensitivity and large linearity. ACS Nano **12**, 3, 2346–2354 (2018). https://doi.org/10.1021/acsnano.7b07613
115. Yin, F., Li, X., Peng, H., Li, F., Yang, K., Yuan, W.: A highly sensitive, multifunctional, and wearable mechanical sensor based on RGO/synergetic fiber bundles for monitoring human actions and physiological signals. Sens. Actuators B Chem. **285**, 179–185 (2019). https://doi.org/10.1016/j.snb.2019.01.063
116. Eissa, S., Jimenez, G. C., Mahvash, F., Guermoune, A., Tlili, C., Szkopek, T., Zourob, M., Siaj, M.: (2015) Functionalized CVD monolayer graphene for label-free impedimetric biosensing, Nano Res. 8, 5, 1698–1709. https://doi.org/10.1007/s12274-014-0671-0
117. Khurshid, F., Jeyavelan, M., Takahashi, K., Leo Hudson, M. S., Nagarajan, S.: Aryl fluoride functionalized graphene oxides for excellent room temperature ammonia sensitivity/selectivity. RSC Adv. **8**, 36, 20440–20449 (2018). https://doi.org/10.1039/C8RA01818A
118. Qi, M., Zhang, Y., Cao, C., Zhang, M., Liu, S., Liu, G.: Decoration of reduced graphene oxide nanosheets with aryldiazonium salts and gold nanoparticles toward a label free amperometric immunosensor for detecting cytokine tumor necrosis factor-α in live cells. Anal. Chem. **88**(19), 9614–9621 (2016). https://doi.org/10.1021/acs.analchem.6b02353
119. Trochimczyk, A.H., Pham, T., Chang, J., Chen, E., Worsley, M.A., Zettl, A., Mickelson, W., Maboudian, R.: Platinum nanoparticle loading of boron nitride aerogel and its use as a novel material for low-power catalytic gas sensing. Adv. Funct. Mater. **26**(3), 433–439 (2016). https://doi.org/10.1002/adfm.201503605
120. He, Y., Li, D., Gao, W., Yin, H., Chen, F., Sun, Y.: High-performance NO_2 sensors based on spontaneously functionalized hexagonal boron nitride nanosheets via chemical exfoliation. Nanoscale **11**, 45, 21909–21916 (2019). https://doi.org/10.1039/C9NR07153A
121. Wu, E., Xie, Y., Yuan, B., Hao, D., An, C., Zhang, H., Wu, S., Hu, X., Liu, J., Zhang, D.: Specific and highly sensitive detection of ketone compounds based on p-Type MoTe2 under ultraviolet illumination. ACS Appl. Mater. Interfaces **10**(41), 35664–35669 (2018). https://doi.org/10.1021/acsami.8b14142
122. Li, X., Li, X., Li, Z., Wang, J., Zhang, J.: WS_2 nanoflakes based selective ammonia sensors at room temperature. Sens. Actuators B Chem. **240**(2017), 273–277 (2017). https://doi.org/10.1016/j.snb.2016.08.163
123. Chen, X., Park, Y.J., Kang, M., Kang, S.K., Koo, J., Shinde, S.M., Shin, J., Jeon, S., Park, G., Yan, Y., MacEwan, M.R., Ray, W.Z., Lee, K.M., Rogers, J.A., Ahn, J.H.: CVD grown monolayer MoS_2 in bioabsorbable electronics and biosensors. Nat. Commun. **9**(1), 1690 (2018). https://doi.org/10.1038/s41467-018-03956-9

124. Ali, A., Koybasi, O., Xing, W., Wright, D.N., Varandani, D., Taniguchi, T., Watanabe, K., Mehta, B.R., Belle, B.D.: Single digit parts-per-billion NO_x detection using MoS2/hBN transistors. Sens. Actuators A Phys. **315**, 112247 (2020). https://doi.org/10.1016/j.sna.2020.112247
125. Tabata, H., Matsuyama, H., Goto, T., Kubo, O., Katayama, M.: Visible-light-activated response originating from carrier-mobility modulation of NO_2 gas sensors based on MoS_2 monolayers. ACS Nano **15**, 2, 2542–2553 (2021). https://doi.org/10.1021/acsnano.0c06996
126. Zhao, C., Gan, X., Yuan, Q., Hu, S., Fang, L., Zhao, J.: High-performance volatile organic compounds microsensor based on few-layer MoS_2-Coated photonic crystal cavity. Adv. Optic. Materials, **6**, 6, 1700882 (2018). https://doi.org/10.1002/adom.201700882
127. Su, S., Li, J., Yao, Y., Sun, Q., Zhao, Q., Wang, F., Li, Q., Liu, X., Wang, L.: Colorimetric analysis of carcinoembryonic antigen using highly catalytic gold nanoparticles decorated MoS_2 nanocomposites. ACS Appl. Bio Mat. **2**(1), 292–298 (2019). https://doi.org/10.1021/acsabm.8b00598
128. Yang, H., Zhou, J., Bao, J., Ma, Y., Zhou, J., Shen, C., Luo, H., Yang, M., Hou, C., Huo, D.: A simple hydrothermal one-step synthesis of 3D-MoS_2/rGO for the construction of sensitive enzyme-free hydrogen peroxide sensor. Microchem. J. **162**, 105746 (2021). https://doi.org/10.1016/j.microc.2020.105746
129. Hu, J., Zhang, J., Zhao, Z., Liu, J., Shi, J., Li, G., Li, P., Zhang, W., Lian, K., Zhuiykov, S.: Synthesis and electrochemical properties of rGO-MoS_2 heterostructures for highly sensitive nitrite detection. Ionics, **24**, 2, 577–587. https://doi.org/10.1007/s11581-017-2202-y
130. Liu, L., Sang, S., Han, D., Liu, Z., Han, X., Li, D., Chen, Y., Liu, D., Liu, X., Yang, K., Cheng, Y.: PEI/PEG functionalized Black Phosphorus prepared by a One-Pot method for a wide detection range CO_2 gas sensor. Sensors Actuators B Chem. **369**, 132303 (2022). https://doi.org/10.1016/j.snb.2022.132303
131. Cui, S., Pu, H., Wells, S.A., Wen, Z., Mao, S., Chang, J., Hersam, M.C., Chen, J.: Ultrahigh sensitivity and layer-dependent sensing performance of phosphorene-based gas sensors. Nat. Commun. **6**(1), 8632 (2015). https://doi.org/10.1038/ncomms9632
132. Abbas, A.N., Liu, B., Chen, L., Ma, Y., Cong, S., Aroonyadet, N., Kopf, M., Nilges, T., Zhou, C.: Black phosphorus gas sensors. ACS Nano **9**(5), 5618–5624 (2015). https://doi.org/10.1021/acsnano.5b01961
133. Cho, S.Y., Lee, Y., Koh, H.J., Jung, H., Kim, J.S., Yoo, H.W., Kim, J., Jung, H.T.: Superior chemical sensing performance of black phosphorus: comparison with MoS_2 and graphene. Adv. Mater. **28**, 32, 7020–7028 (2016). https://doi.org/10.1002/adma.201601167
134. Xu, J., Qiao, X., Wang, Y., Sheng, Q., Yue, T., Zheng, J., Zhou, M.: Electrostatic assembly of gold nanoparticles on black phosphorus nanosheets for electrochemical aptasensing of patulin. Microchim. Acta **186**(4), 238 (2019). https://doi.org/10.1007/s00604-019-3339-3

Recent Developments in Various 2D Material-Based Gas Sensors: Diagnostic Perspective of Human Exhalation

Levna Chacko and Zdeněk Sofer

Abstract A crucial goal in modern medical technology is to develop effective early-stage disease detection methods that can quickly identify the underlying causes or health conditions, all while minimizing the invasiveness of diagnostic procedures. Conventional approach for diagnosing various medical illnesses and pathological conditions has primarily relied on chemical analysis of human samples and invasive medical procedures that are expensive, time-intensive, and necessitate complex chemical analysis equipment. The composition of chemicals in human breath is highly intricate, consisting of volatile organic compounds (VOCs) emitted due to metabolic processes within the human body. These compounds provide insights into various biochemical and physiological processes occurring in the body. Importantly, disease processes can lead to the production of VOC metabolites, which result from alterations in the normal physiological and metabolic pathways. In light of these considerations, it is imperative to develop a novel diagnostic strategy employing efficient nanomaterial-based electronic nose (e-nose) sensor devices for analyzing exhaled breath enabling early disease screenings with the focused identification of established disease biomarkers. Such a device delivers cost-effective, non-invasive, user-friendly, and reliable diagnostic outcomes. Two-dimensional materials, known for their unique properties, have shown significant potential as biomarkers for disease diagnosis. In this context, we will review and explore new possibilities in the synthesis, fabrication of sensors, and disease diagnosis using two-dimensional materials.

L. Chacko (✉) · Z. Sofer
Department of Inorganic Chemistry, University of Chemistry and Technology Prague, Technická 5, 166 28 Prague 6, Czech Republic
e-mail: chackoe@vscht.cz

Z. Sofer
e-mail: zdenek.sofer@vscht.cz

A. Katoch et al. (eds.), *2D Materials*, Engineering Materials,
https://doi.org/10.1007/978-981-97-6258-3_4

1 Introduction

The unprecedented reality and causalities engendered by the COVID-19 pandemic has already made its imprints on various human aspects and ideologies. Unremarkably, medical technology worked assiduously and took a rapid, essential step forward to save the life of humans and it will undoubtedly usher in a new era of global governance. Appreciating the diligent support from the medical field, the future perspective must rely on the advancement and improvement representing a new frontier in medical framework such as non-invasive, reliable, portable, disease diagnostic techniques competent of prompt disease diagnosis that is easily available to common people.

Among the current sensing technologies, the electronic nose (e-nose) analysis approach stands out with significant benefits, due to its high sensitivity, fast response time, real-time monitoring capability, non-invasive nature, and portability. In the realm of clinical medicine, the utilization of e-nose technology primarily involves early diagnosis of a range of medical conditions. This includes the detection of various cancers, lung diseases, diabetes, etc. and can be extended to various domains such as chemical, industrial and pharmaceutical sectors, agriculture, environmental protection, food and beverage, military, transportation, and more. When normal human physiological processes fail or are altered by disease (pathogenesis) or metabolic disorders, the mixture of gases released by the lungs in the breath changes as a result of the altered chemical pathways resulting from the abnormal metabolic changes caused by these various illnesses. Hence, it is imperative to adopt a preventative health strategy to analyze the volatile organic compounds (VOCs) present in the exhaled breath, assess the effectiveness of breath analysis for the diagnosis of diseases through the efficient development of e-nose sensing technique, facilitating the current clinical technologies which still rely on to invasive, long-time, and costly diagnosis techniques such as endoscopic investigations, blood test, various scanning techniques, and conventional culture methods. VOC components can be used as a potential biomarker for a number of disorders, including oxidative stress, inflammation, carcinogenesis, and others. Development of e-nose sensors, mimicking the olfactory system, by sensing volatile compounds employing efficient sensing materials is highly critical for early qualitative diagnosis. Similar to the functioning of the human olfactory system toward various compounds employing neural networks, e-noses consist of sensor arrays to respond to volatile compounds through redox reactions and other series of processes resulting in a set of sensor signals (Fig. 1). For example, pulmonary disorders, pancreatic abnormalities, and kidney malfunctions produce certain specific volatile compounds such as increased concentrations of nitric oxide, hydrogen sulfide, carbon monoxide, and isoprene in the exhaled breath profile that can be easily detected using e-nose devices.

So far, a wide array of sensors has prospered for e-nose systems which is a smart system that integrates a cross-sensitive chemical sensor array with a powerful suite of pattern recognition algorithms. With the advances in nanotechnology such as the possibility to synthesize various nanomaterials, nanostructures, and heterostructures,

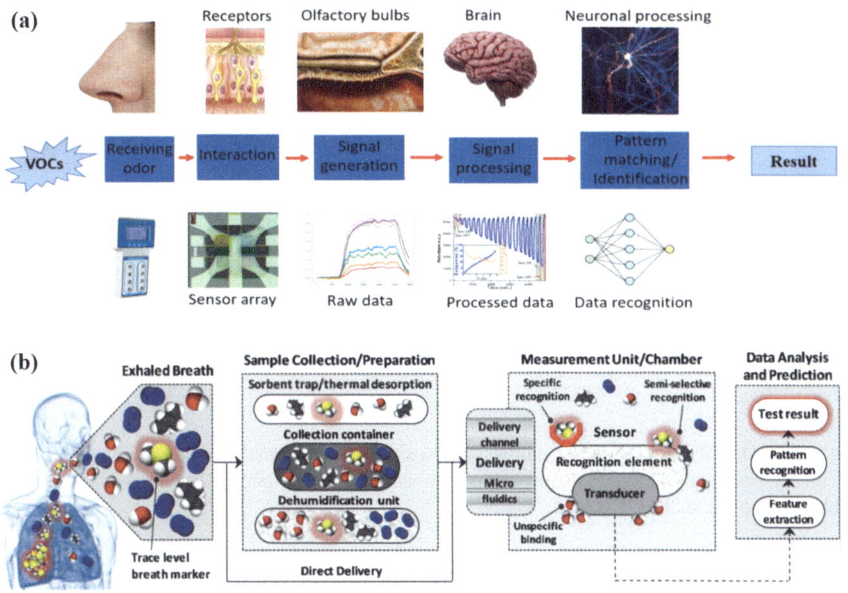

Fig. 1 **a** Comparison of biological and artificial olfactory systems and **b** Outline of various processes during exhaled breath analysis [1], Reprinted with permission from Konvalina et al. [1]. Copyright 2014, the American Chemical Society

it has become feasible to design and implement gas sensors. Gady Konvalina and Hossam Haick [1], for instance, have utilized gas sensors made from nanomaterials in the context of disease breath analysis. In another study, Ulrike Tisch and Hossam Haick instigated a sensor array comprising multiple sensors composed of monolayer-capped metallic nanoparticles (MCNPs) as an efficient device for breath screening for disease detection [2]. Markedly, the significant triumphs of e-nose technology have been in the medical field, with a particular emphasis on the non-invasive, early screening diagnostic technology. In fact, every disease has certain volatile biomarkers that travel through the patient's blood, urine, and breath due to the phenomena of metastasis. Low amounts of specific alkanes and aromatic chemicals can be found in the breath produced by a tumor and are circulated from the patient's blood to the lungs.

In the breath of healthy individuals, you can typically find significant VOCs such as acetone (ranging from 1.2 to 900 ppb), ethanol (ranging from 13 to 1,000 ppb), methanol (ranging from 160 to 2,000 ppb), isoprene (ranging from 12 to 580 ppb), as well as ammonia, and various other analytes like pentane, higher alcohols, aldehydes, and ketones. These VOC levels can vary from person to person, influenced by various factors like their internal health status, environmental conditions, dietary habits, lifestyle factors such as smoking or drinking alcohol and more. Cigarette smoking, for example, results in elevated levels of acetonitrile and furans. Likewise, the presence of an acetone-like odor can be a sign of severe diabetes, while

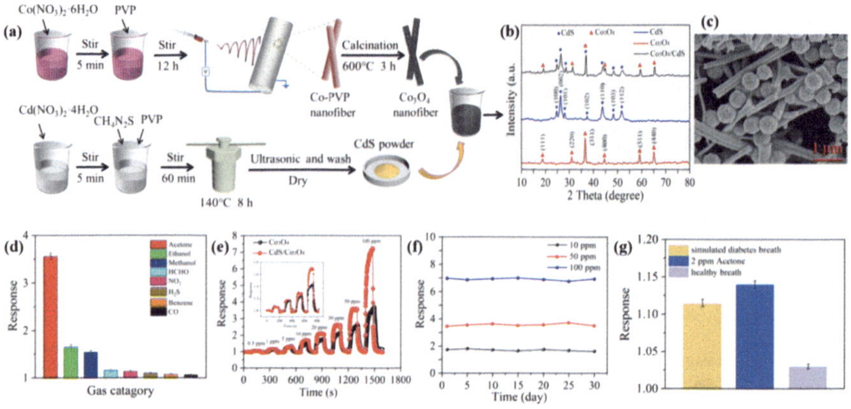

Fig. 2 Co_3O_4-CdS hybrid nanostructures synthesized by hydrothermal method exhibiting excellent selectivity, sensitivity, and stability toward acetone sensing [3], Reprinted with permission from Guo et al. [3]. Copyright 2021, Elsevier

a strong, stinging smell may be an indicator of diabetes and hepatic disorder. A hybrid nanostructure of cobalt oxide (Co_3O_4) nanofibers and cadmium sulfide (CdS) nanospheres-based acetone sensor synthesized by hydrothermal method showed enhanced response with excellent selectivity, sensitivity, and stability toward acetone demonstrating the potential to identify biomarkers (Fig. 2) [3]. In a similar vein, elevated levels of pentane and ethane serve as indicators of various persistent respiratory and pulmonary ailments, including malignant pleural mesothelioma (MPM), cystic fibrosis, and bronchial asthma. According to Mazzatenta et al., their research findings indicate that the standard concentration of VOCs tends to decrease in individuals with Alzheimer's disease [4]. When compared to healthy controls, it seemed that the condition affected brain metabolism due to cell death and pathology. Recently, phosphorene/hexagonal boron nitride (hBN) sensor device demonstrated remarkable selectivity toward biomolecules correlated with influenza, pneumonia, SARS-CoV-2 and molybdenum disulphide/hBN system enabled the identification of cyclohexanone, which is distinctive to tuberculosis [5]. Detecting specific VOCs from the breath is difficult since there is always the risk of getting incorrect results due to the presence of foreign chemicals.

2 Prominence of 2D Materials for Sensing Applications

The atomically thin structure of 2D materials provides an enormous surface-to-volume ratio, varied chemical structures, versatile surface functionalization, excellent communion with device integration, tunable bandgaps and electronic structure attributes, great absorption coefficient, increased accessibility of active centers suitable for redox mechanism, increased charge transport properties, and diverse conducting characteristics spanning from metallic, semi-metallic to semiconducting can be observed in 2D structures like graphene, black phosphorous, transition-metal dichalcogenides (TMDCs), transition metal oxides (TMOs), and their hybrid structures are being investigated extensively for sensing (Fig. 3). Moreover, the above-mentioned 2D materials hold the potential of achieving energy efficiency, accurate gas detection at room-temperature (RT) with varying surface framework for different gases, implying good selectivity contrary to the traditional metal-oxide-based gas sensors portraying dynamic sensing behaviors with substandard cross-sensitivity, high operation temperatures around greater than 200 °C, high power consumption, and poor stability. Also, the primary sensing mechanism observed in metal oxide materials involves the surface adsorption of oxygen ions such as O_2^-. O^- and O^{2-}, creating a depletion layer. On exposure of n-type metal-oxide material to reducing gas, the resistance increases, whereas exposure to oxidizing gas increases the resistance. Conversely, for p-type materials, the opposite effect is observed. In gas sensors based on 2D materials, the sensing mechanism primarily relies on the charge-transfer process between the 2D materials and adsorbed gas molecules and does not require oxygen adsorption.

Notably, together with the above-mentioned advantages of 2D nanomaterials and related hybrid heterostructure, allows for the modification of interface properties, electronic structure, surface chemistry, layer-dependent sizable band gap with proper lattice achieving enhanced responsiveness, lower power utilization, and enhanced cross-sensitivity for gas sensing at room temperature. Being one of the supreme dominant, ultra-large surface-to-volume ratio of 2D materials compared to other nanostructures have a significant contribution toward the increase in specific surface area

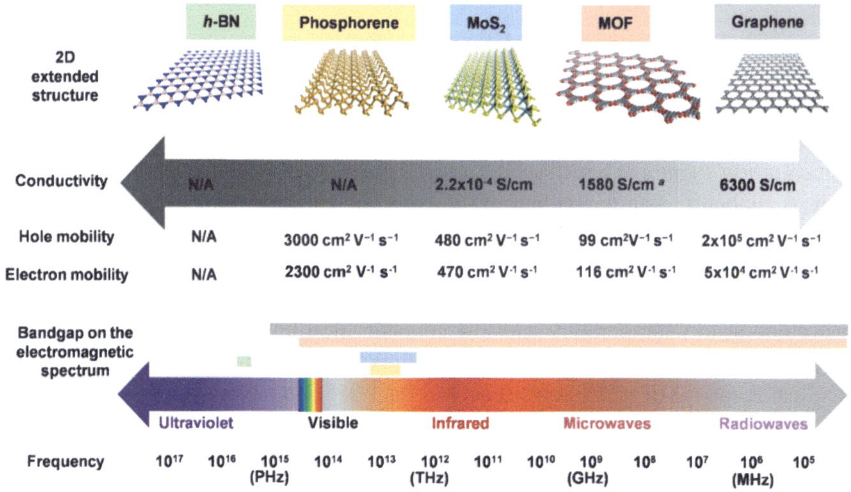

Fig. 3 2D materials exhibiting a vast variety of chemical structures, conductivities, carrier mobility, and bandgaps, [16] Reprinted with permission from Meng et al. [16]. Copyright 2019, the American Chemical Society

and capability of providing excessive surface active sites for redox reactions, thereby contributing to the sensor parameters such as resulting in high sensitivity, low detection limits, and fast response time. Simultaneously, the tunable electronic structure provides an efficient pathway for combining materials overcoming lattice mismatch issues, and possessing increased carrier mobility. The synergy between morphology, enhanced surface area, tunable electronic structure together with enhanced carrier mobility, further serves as an efficient platform for the efficient adsorption–desorption mechanism on exposure toward various gaseous species through effective carrier charge-transport process. For example, reports have demonstrated the increased carrier mobility of graphene around 10^2–10^4 cm^2/Vs [6] with a specific surface area of 2630 m^2/g [7], MoS_2 with 200–500 cm^2/Vs at room temperature [8,9] and a specific surface area of ~ 27.82 m^2/g [10], MoS_2-WS_2 composite with a surface area of 142.6 m^2/g [11], an ultra-large specific surface area of 200.8 m^2/g for nitrogen-doped $Ti_3C_2T_x$ aerogel [12], 2D Ti_2CO_2 with electron–hole mobility in the range of 10^2–10^4 cm^2/Vs [13], Sc_2CF_2 MXene with mobility of 5.03×10^3 and 1.07×10^3 cm^2/Vs in the zigzag and armchair directions [14], black phosphorous with mobility in the range of 1.0–2.6×10^4 cm^2/Vs [15]. All these distinct chemical, physical, structural, and electronic properties of 2D materials make them promising substitutes for traditionally used metal oxide-based materials.

3 Advancements in 2D Material-Based Biomarkers

3.1 Acetone Sensing

Human exhalation, comprising of several gases essentially nitrogen (N_2—78%), oxygen (O_2—17%), carbon dioxide (CO_2—4%), and 1% other, affirm to include diagnostic indication to pathophysiological processes. As a byproduct of metabolic processes in the human body, a variety of gases are present in the exhaled breath, the concentration change of which serves as a diagnostic tool to identify several diseases such as diabetes, asthma, kidney diseases, pancreatic disorders, and pulmonary malfunctions. To state, for a normal healthy person the concentration of acetone in the exhaled breath ranges around 0.3–0.9 ppm, whereas that for a diabetic patient is greater than 1.76 ppm [17,18]. Unlike several works which lack a selective sensing of acetone at low temperature and low concentration levels, a report on Fe-doped reduced graphene oxide (rGO) decorated with WO_3 nanocomposite demonstrated an excellent response of ~78% toward acetone at 130 °C at a concentration of 10 ppm (Fig. 4a).

The cross-sensitivity measurement demonstrated the effectiveness of the nanocomposite-based sensor at selectively detecting acetone in the presence of other interfering gases such as ethanol, ammonia, formaldehyde, carbon monoxide and methane at a 10 ppm level (Fig. 4b). Similarly, sensitivity measurements displayed an increase in the sensing response with an increase in the concentration of acetone

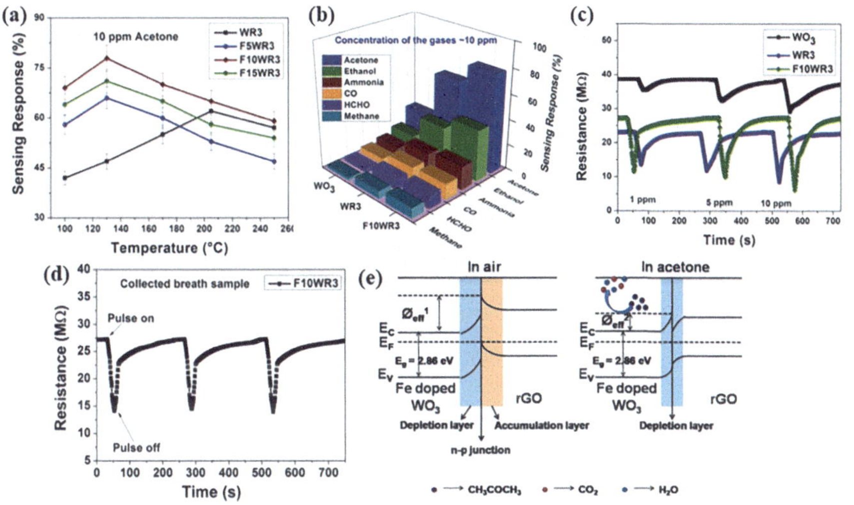

Fig. 4 **a** Gas sensing response of 3 wt% rGO decorated WO_3 (WR3) and Fe (5 wt%, 10 wt%, 15 wt%) doped rGO decorated WO_3 nanocomposites with respect to various temperatures, **b** Selectivity measurement of WO_3, WR3, 10 wt% Fe doped rGO decorated WO_3 nanocomposites (F10WR3), **c** Sensitivity measurement of WO_3, WR3 and F10WR3, **d** Response of the F10WR3 nanocomposite toward collected human breath sample and **e** Sensing response mechanism of F10WR3 toward air and acetone [18], Reprinted with permission from Sen et al. [18]. Copyright 2022, Elsevier

from 1 to 5 ppm to 10 ppm with responses being 56%, 64%, and 78%, respectively, for Fe doped rGO-WO_3 nanocomposite (Fig. 4c) attaining a saturation limit around 25 ppm. Further, the sensor also exhibited excellent dynamic response-recovery time of 20–75 s with extended long-term stability over three months span. More interestingly, the exhaled breath of a diabetic patient (~0.7 ppm) was collected and reflected a sensing response of ~45% which is in compliance with a 1 ppm level of acetone showing 56% response by Fe-doped rGO-WO_3 nanocomposite (Fig. 4d). The enhanced surface area leading to increased adsorption sites and excellent conductivity provided by rGO, increased oxygen vacancies and defects due to incorporation of Fe into rGO-WO_3 nanocomposite and modified electronic structure properties facilitated efficient charge-transfer mechanism between the gaseous species and the nanomaterial (Fig. 4e) [18].

Yet another effort led to the fabrication of SnO_2-MoS_2-based fiber optic sensor displaying a sensing response of ~23.5% with an excellent response-recovery time of 14–17 s respectively [17]. On the other hand, a solid-state FET sensor device employing MoS_2 showed the lowest detection limit of 0.4 ppm at room temperature [19]. Development of novel sensing material by designing a 1D-2D nanocomposite made of 1D KWO ($K_2W_7O_{22}$) nanorods/2D $Ti_3C_2T_x$ nanosheets (Fig. 5a–c), wherein T_x stands for OH^-, –O, and F^{-1} surface terminated groups, showed promising selectivity and sensitivity toward acetone detection. The distinctive multi-layered structure of $Ti_3C_2T_x$ decorated with KWO nanorods, provided an incredibly broad surface and

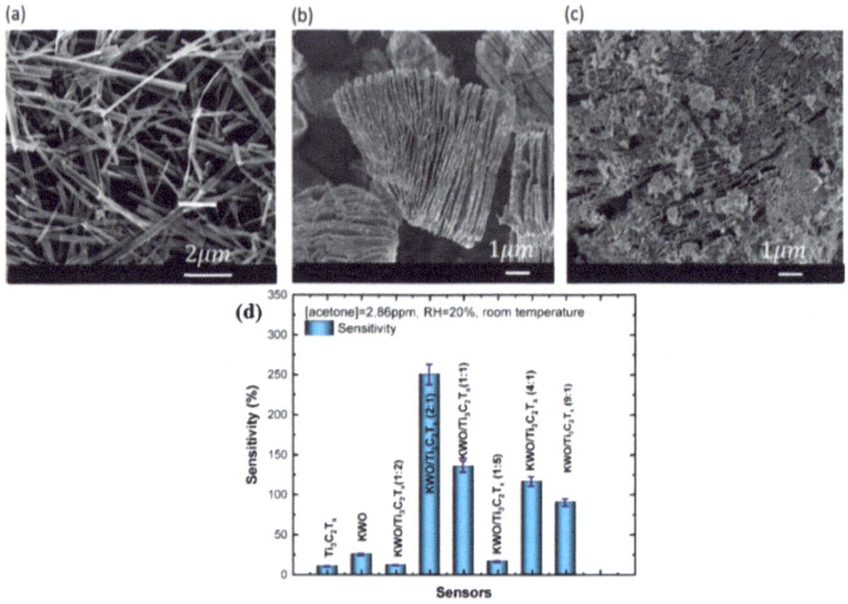

Fig. 5 Scanning electron microscopic images of **a** KWO nanorods, **b** $Ti_3C_2T_x$, **c** KWO/$Ti_3C_2T_x$ (2:1) nanocomposite and **d** Acetone sensing response curve of various sensors [20]. Reprinted with permission from Ama et al. [20]. Copyright 2020, MDPI

interface area, enhanced the electric properties and flexible surface functioning due to T_x groups facilitating the charge transfer mechanism together with the ferroelectric properties of KWO contributing to enhanced chemiresistive response toward acetone. KWO/$Ti_3C_2T_x$ (2:1) nanocomposite displayed an increased sensing response toward 2.86 ppm acetone at room temperature (Fig. 5d) [20].

Key characteristics of MXenes such as their significant metallic and semi-conductivity, unique 2D morphology, large surface area, and great flexibility were further exploited by Sanjit et al. to develop room temperature (23 °C) acetone sensor with the lowest detection limit in sub-ppm level. The accordion-like $Ti_3C_2T_x$ MXene multilayers were screen-printed on alumina (Al_2O_3) substrate-based Pt-interdigitated electrodes. The acetone sensing response transients of $Ti_3C_2T_x$ demonstrated a p-type chemiresistive behavior with a sensitivity of 45.7% at 15 ppm level of acetone and 17.3% at the lowest detection limit of 0.25 ppm. Moreover, $Ti_3C_2T_x$ well-addressed the crucial selectivity test of the sensor toward various gases such as acetone, H_2, CO, H_2S, and CO_2 at a concentration level of 100 ppm, reflecting around 99% response toward acetone in contrast to other gaseous species. The contact performance of $Ti_3C_2T_x$ multilayer MXene is affected by the large molecular size of acetone relative to other gaseous species resulting in a noticeable change in the resistance when acetone molecules enter the multilayer sheets of MXene, yielding excellent selectivity toward acetone. The sensor also manifested a response time of 53 s and a recovery time of 120 s on exposure to 100 ppm acetone with long-term durability [21].

An investigation of heterogeneous composite made of MXene ($Ti_3C_2T_x$ nanosheet) and metal oxide semiconductor (α-Fe_2O_3 nanocube) amalgam (Fig. 6a) illustrated ameliorated selectivity toward acetone sensing. The sensor presented a response of 16.6% for 5 ppm level of acetone at room temperature with response-recovery time of 5 s each, excellent sensitivity, and abiding stability over 28 days (Fig. 6b–e). The MXene-metal oxide semiconductor composite hindered the aggregation of materials and led to enhanced charge transfer reaction, further promoting the sensing behavior. Akin to the other n-type oxide semiconducting materials, the sensing process of $Ti_3C_2T_x$/α-Fe_2O_3 composite toward acetone includes the reaction of acetone to adsorbed oxygen species and transfers the collected electrons to the conduction band of the composite. This narrows down the electron depletion region, thereby decreasing the resistance of the composite. The heterogeneous composite structure also provided the greater active surface area with the adsorption energy of acetone molecules on the surface of the composite being—6.7 eV, enhancing adsorption-diffusion process and further driving toward enhanced sensing performance [22].

Being a new class of 2D transition metal carbides, $V_4C_3T_x$ MXene was synthesized by employing Al and Co fluxes and catalysts forming intermediate V_4AlC_3 and then subjected to selective etching by hydrofluoric acid treatment. The new MXene material showed pronounced semiconducting behavior with the increased response toward acetone sensing at 25 °C exhibiting the lowest detection limit of

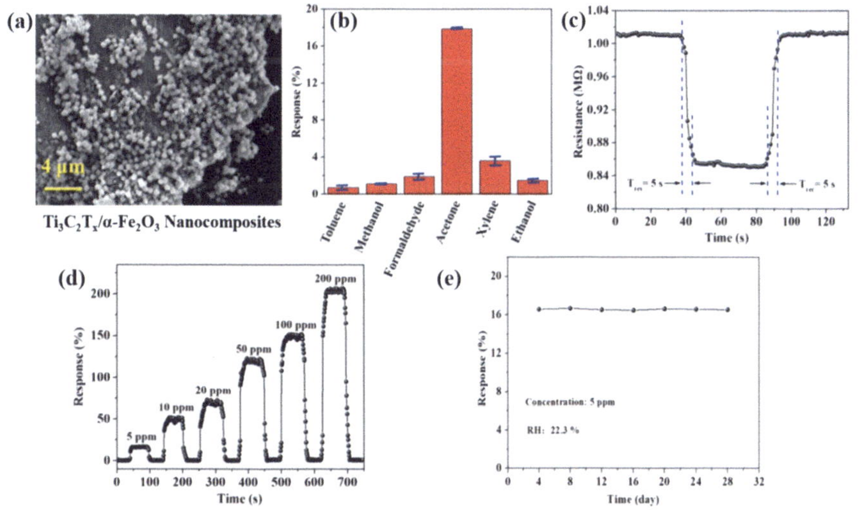

Fig. 6 **a** Scanning electron microscopic image, **b** Cross-sensitivity test toward various analytes, **c** Resistance-time plot depicting response-recovery time toward acetone, **d** Sensitivity test on exposure toward various concentrations of acetone, and **e** Stability test of $Ti_3C_2T_x$/α-Fe_2O_3 nanocomposite [22], Reprinted with permission from Liu et al. [22]. Copyright 2021, Elsevier

1 ppm which is less than the diabetes diagnosis threshold. The protonic current measurement also revealed the increased sensing response of almost >104 times toward acetone compared to water vapor of the same concentration [23]. Recently, the nanoengineering of WO_3/WS_2-based 2D/2D heterostructure-based sensor [24] reflected a remarkable response toward acetone sensing on exposure to a concentration range of 1–1000 ppm. The 2D/2D heterostructure demonstrated a sensor response of 17.00 when exposed to 1 ppm acetone and escalated to 132.59 when the acetone concentration was increased to 1000 ppm. This increase in sensor response with rising acetone concentration correlates directly with the adsorption of acetone on the surface of 2D/2D heterostructure comprising of WO_3/WS_2 nanoflakes. Additionally, the momentary sensor traits for exhaled breath detection were executed by collecting the exhaled breath in Tedlar bags. The introduction of collected exhaled breath displayed excellent sensor response that peaked at 271.56 for a concentration of 99 vol% and at 98.74 for a concentration of 8 vol%. The research of Gabriel et al. [25] conducted a numerical study on a sensor embodying graphene, chitosan, PANI designed for examining acetone levels in exhaled breath and exhaled breath condensate (EBC) by leveraging the surface plasmon resonance (SPR) principle. Chitosan demonstrated the highest sensitivities, measuring 246°/RIU, 216°/RIU, and 150°/RIU when paired with gold, copper, and silver, respectively. Graphene yielded sensitivities of 210°/RIU for gold and 150°/RIU for silver. The reports have proved the efficiency of acetone detection using several materials, which could be used as a potential biomarker to diagnose type I and type II diabetics from the exhaled breath of humans.

3.2 Ammonia (NH_3) Sensing

Increased traces of ammonia ranging from 0.82 to 14.70 ppm in human exhaled breath produced as a result of biological metabolism serve as exemplary biomarkers identifying kidney, liver disorders and peptic ulcers. However, for a normal person, the concertation of NH_3 in the exhaled breath remains in the range of 0.43 to 1.80 ppm. 2D materials with their increased specific surface area compared to other dimensionalities, other unique surface chemistries offering a large number of surface-active sites, and excellent charge transport properties stemming from versatile and tunable electronic structures prove their promising application in the field of sensors. In a report Wang et al. [26] fabricated a MXene-polymer (Nb_2CT_x nanosheets and PANI nanofibers as shown in Fig. 7b) based NH_3 sensor directed by triboelectric nanogenerator (TENG) (Fig. 7a) with a response varying from 9.33 to 301% on exposing to NH_3 ranging from 1 to 100 ppm with a linear response and long-term stability over 35 days (Fig. 7c). The results also presented the diminished impact of humidity (72.2–87.1%) on the sensing characteristics emanating from the formation of intermolecular hydrogen bonds between MXene and polymer (Fig. 7d). The effect of temperature ranging from 25 to 50 °C on the sensing performance of Nb_2CT_x/PANI sensor blazed a decrease in the response with an increase in temperature, besides

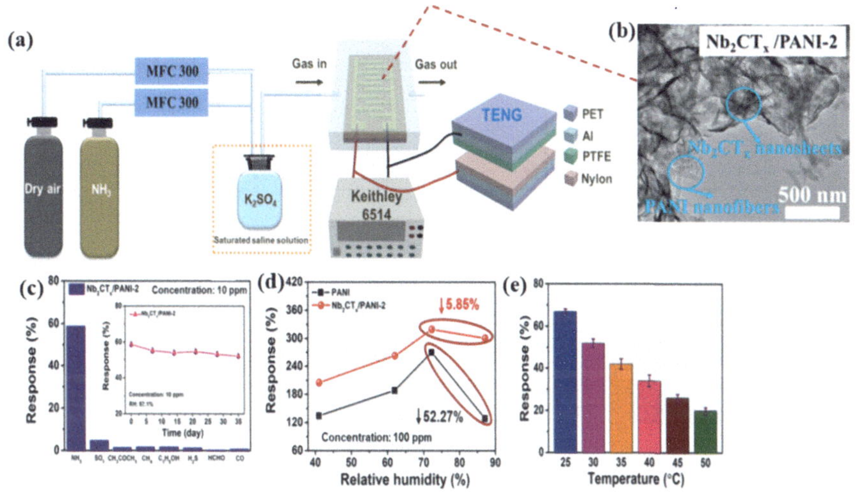

Fig. 7 **a** Schematic representation of the gas sensing system employing triboelectric nanogenerator (TENG), **b** SEM image, **c** Selectivity test with stability measurement in the inset, **d** Sensing response curve of NH_3 toward various percentage of humidity, **d** Influence of various temperatures on the sensing response of Nb_2CT_x/PANI-based sensor on exposure toward 10 ppm NH_3 with relative humidity of 87.1% and **e** Gas sensing response of Nb_2CT_x/PANI-2 sensor at 25–50 °C under 87.1% RH toward 10 ppm NH_3 [26], Reprinted with permission from Wang et al. [26]. Copyright 2020, Elsevier

exhibiting a better sensitivity of 20% at 50 °C which makes it feasible for real-time applications like NH_3 detection from exhaled breath (34–37 °C) (Fig. 7e).

Persuaded by the versatile features and unprecedented reactivities toward ammonia sensing, 2D $Ti_3C_2T_x$ based MXene/MoS_2 composite polymerized by conductive polymer, polypyrrole (PPy) based sensor [27] was conscientiously investigated with ameliorated selective NH_3 responsiveness at 26 °C exhibiting sensitivity of 21.1% (100 ppm), fast response-recovery time of 43.6/40.8 s for 100 ppm NH_3 concentration, low detection limits of 10 ppm with a response of 2.31% in the presence relative humidity level of 45%. Further, a heterojunction of 2D $Ti_3C_2T_x$ MXene with V_2O_5/$CuWO_4$-based sensor showcased an amended selective sensing response toward NH_3 over 15 other gaseous species (Fig. 8a). Impressively, the sensor displayed linear response over a wide concentration range of 1–160 ppm (Fig. 8b), with detection limit reaching as low as 0.3 ppm, ultra-fast response-recovery time of 10/5 s for 1 ppm and 1.6/4 s for 51 ppm of NH_3 (Fig. 8c) and long-term stability over 70 days prospering the propriety suitable for diagnosis of traces of ammonia in the exhaled human breath [28]. The charge transfer mechanism clearly depicts a decrease in the depletion layer on exposure to ammonia compared to air exposure, due to the significant heterojunction formation between metallic $Ti_3C_2T_x$ MXene with V_2O_5/$CuWO_4$ expediting response toward ammonia (Fig. 8d).

Reports have suggested that the functionalization of MXenes ($M_{n+1}X_nT_x$) enforces a significant role in stemming tunable electronic, optical, catalytic, and

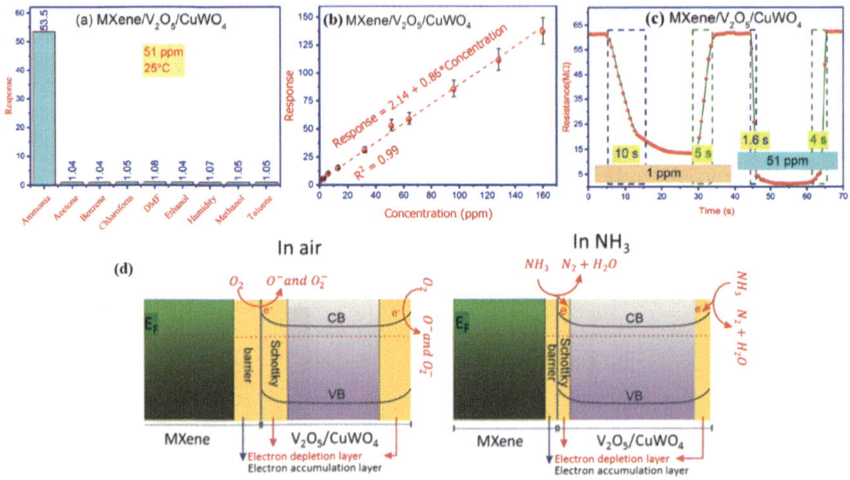

Fig. 8 $Ti_3C_2T_x$ MXene/V_2O_5/$CuWO_4$ depicting **a** ultra-high response toward ammonia in the presence of various analytes against 51 ppm concentration, **b** linear-response curve with respect to variation in ammonia concentration from 1 to 160 ppm, **c** ultra-fast response-recovery time against 1 and 51 ppm of ammonia and **d** the charge-transfer sensing mechanism on exposure toward air and ammonia [28], Reprinted with permission from Ranjbar et al. [28]. Copyright 20,221, Elsevier

thermal properties. For example, studies have reported that Ti_2CO_2, Zr_2CO_2, Hf_2CO_2, Sc_2CT_2 with T = O, F, OH have a semiconductor property, while Ti_2C, Ti_2CF_2, $Ti_2C(OH)_2$, etc. have a metallic characteristic [28]. Also, these terminated MXenes are found to exhibit superior sensing performances, increased stability, and durability on exposure to various analytes. Theoretical studies demonstrating the gas adsorption energy (E_{ad}) of bare and O-functionalized MXenes toward gaseous species are shown in Fig. 9, manifesting the high E_{ad} value of O-terminated MXenes compared to bare ones. It is clear that Ti_2CO_2 and Nb_2CO_2 presented E_{ad} of −0.37 and −0.50 eV toward NH_3 while Mo_2CO_2 and V_2CO_2 exhibited E_{ad} of −0.80 and −0.73 eV toward NO gas, confirming the strong binding interaction and excellent selectivity of these O-terminated MXenes to particular gases compared to other gases. The very low E_{ad} value of Ti_2CO_2 (0.05 eV), Nb_2CO_2 (0.06 eV), Mo_2CO_2 (0.07 eV), and V_2CO_2 (0.09 eV) MXenes toward hydrogen gas represented the weak interaction and least selectivity with H_2 [29]. Yet another theoretical study revealed the interaction of NH_3 molecules on pristine and O-functionalized Cr_2C- and Fe_2C-based MXenes layers [30], with bare MXene displaying strong NH_3 interaction. NH_3 molecule showed strong interaction with Cr_2C with an adsorption energy of −1.40 eV and charge transfer of +1.257, whereas Fe_2C exhibited an adsorption energy of −1.33 eV with charge transfer of +1.2307 which is a much larger than their O-terminated counterparts with adsorption energies of −0.29 and −0.22 eV for Cr_2CO_2 and Fe_2CO_2 respectively. It is observed that among the pristine and O-terminated MXenes, Cr_2C and Cr_2CO_2-based MXenes present strong adsorption

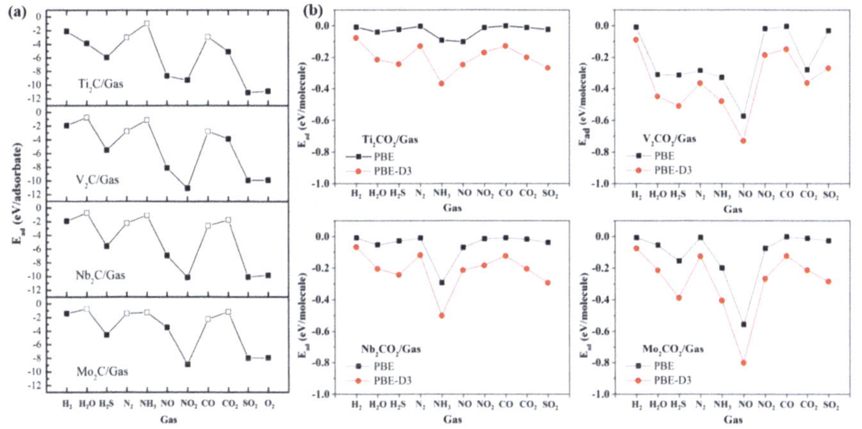

Fig. 9 Observation of adsorption energy values of **a** bare and **b** functionalized MXenes toward various gaseous species [29], Reprinted with permission from Junkaew et al. [29]. Copyright 2018, the Royal Society of Chemistry

energies compared to pristine and O-terminated Fe-based MXenes with NH_3 adsorption energy on Cr and O in Cr_2C and Cr_2CO_2 around 70 meV higher than that at Fe-site and O-site of Fe_2C and Fe_2CO_2, attributed to the magnetic moment changes observed in the MXenes. The magnetic moment of the Cr atom was found to increase by 0.029 μB in the course of NH_3 adsorption on the surface of Cr_2C MXenes, accompanied by a loss of magnetic moment (0.024 μB) from C-atom. A similar trend is observed in the case of Cr_2CO_2, with a change in the localized magnetic moment between Cr and O atoms. Experiments with ultra-low NH_3 concentration detection capability of Ti-S coordinated MXenes, increased adsorption energies, charge transfer, and strong interaction between the NH_3 analytes and MXenes make them potential candidates to be exploited for advanced gas sensing applications [31].

3.3 *Nitrogen Oxide Sensing*

Another signaling molecule, NO, in the exhaled breath of humans has been used as a biomarker to identify pulmonary and respiratory infections and inflammations above a concentration of 50 ppb in the exhaled breath. NO is being produced by various cells such as neutrophils, red blood cells, endothelial cells, and nerve cells as a result of the nitric oxide synthase process. Also, it is important to note that because NO is readily oxidized to NO_2, detecting NO_2 often provides an indirect reflection of NO concentration. Two-dimensional materials with superior surface chemistries and transport properties are widely exploited to identify the presence of increased concentrations of various signaling molecules from human exhalation. In an attempt to improve the sensing characteristics of 2D materials, Levna et al. [32]

have fabricated a functionalized MoS_2 sensor device $Si/SiO_2/Pd\text{-}MoS_2$, displaying excellent sensitivity toward NO gas with strong sensitivity of 700% (2 ppm NO) and minimum detection limit of 0.1 ppm. First principle calculations proved the efficiency of using Sc_2CO_2 monolayer with an O-vacancy-based resistive sensor for NO sensing with an adsorption energy of 3.19 eV. On exposure to NO gas, O-vacancy site adsorbs the N atom from the NO molecule forming bonds with Sc and C atoms. Also, with a dominant charge transfer of 2.051 e from O-vacancy containing Sc_2CO_2 monolayer, the NO molecule functions as an acceptor [33]. The exceptional conductivity, increased surface chemistries and mechanical strength of rGO were manipulated in association with the high electroactivity of Sb^{3+}/Sb^{4+} from antimony tetroxide (Sb_2O_4) nanoflowers to form Sb_2O_4/rGO nanocomposite, characterizing exquisite real-time NO response performances produced from tumor cells [34]. When using electrochemical sensing with a glassy carbon electrode coated with the Sb_2O_4/rGO nanocomposite (as shown in Fig. 10a), an Ag/AgCl reference electrode, and a platinum wire counter electrode, a prominent oxidation peak at 0.84 V was observed in cyclic voltammetry experiments. This occurred in the presence of 3.57 μM NO in a 0.01 M phosphate buffer solution (PBS) with a pH of 7.4 (as depicted in Fig. 10b). The composite presented a linear sensitivity for a range of 3.98 nM–0.772 μM with LOD 3.98 nM, sensitivity of 74.59 μAcm^{-2} μM^{-1}, and long-term stability over 25 days with 9.19% decay. The amperometric response employing Sb_2O_4/rGO nanocomposite bestowed the real-time in-situ monitoring of increased NO release from A549 tumor cells compared to that from HUVECs normal cells. Through the injection of 2- and 5-mM acetylcholine (Ach), a prototype medication that releases NO, into tumor and normal cells, around 56.67 (2 mM Ach) and 93.33 nA (5 mM Ach) of NO were emitted from A549 tumor cells. In contrast, about 11.25 (2 mM Ach) and 18.87 nA (5 mM Ach) of NO were generated by HUVECs cells (Fig. 10c). Notably, these experimental findings indicate that the amount of NO produced by A549 is relatively five times greater than that released by HUVECs. This suggests that NO synthase is more active in tumor cells than in normal cells, potentially contributing to the further growth and metastasis of tumor cells.

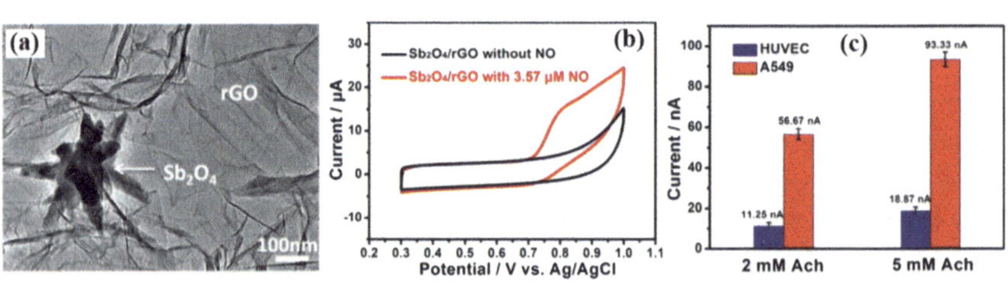

Fig. 10 **a** SEM representing the formation of Sb_2O_4/rGO nanocomposite, **b** Cyclic voltammetry measurement of Sb_2O_4/rGO nanocomposite-based electrode at a potential range of 0.3–1.0 V and scan rate of 50 mV/s and **c** current response characteristics on release of NO from tumor and normal cells [34], Reprinted with permission from Deng et al. [34]. Copyright 2020, Elsevier

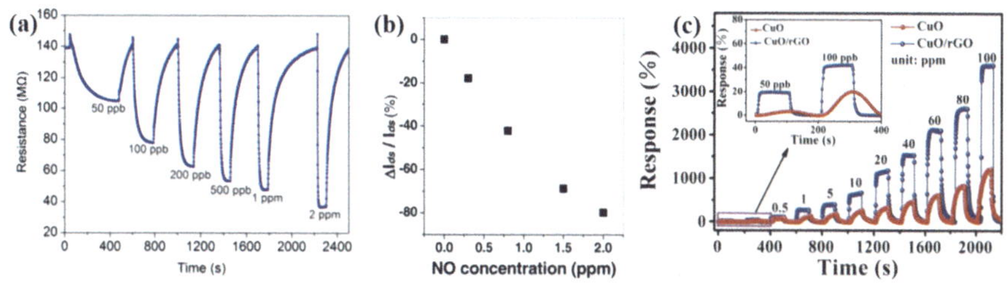

Fig. 11 Dynamic response characteristics of **a** Pd-SnO_2-rGO toward NO_2 [35], Reprinted with permission from Wang et al. [35]. [36][36][37][37] Copyright 2017, Elsevier. **b** MoS_2 toward NO, Reprinted with permission from Li et al.. Copyright 2012, WILEY-VCH Verlag GmbH & Co. KGaA, Weinheim and **c** CuO/rGO toward NO_2 gases representing the lowest detection limit of sensing NO/NO_2 gases, Reprinted with permission from Bai et al.. Copyright 2021, Elsevier

Since a low level of NO around 50 ppb itself indicates airway inflammation, it is highly crucial to design sensors that can detect exhaled NO at a very low detection limit. Wang et al. fabricated NO_2 gas sensors based on palladium and SnO_2 nanoparticles decorated on reduced graphene oxide forming Pd-SnO_2-rGO hybrid structures. The hybrid sensor was found to display an excellent sensitivity of 3.92 higher than that of SnO_2-rGO composite and pure rGO with a sensitivity of 2.13 and 1.10, respectively, on exposure toward 1 ppm NO_2 gas with a lowest detection limit of 50 ppb (Fig. 11a). Even though results demonstrated the impact of humidity on Pd-SnO_2-rGO-based sensor, still the sensor showed excellent sensing dynamics with a response of 2.36 toward 1 ppm NO_2 gas in the presence of 80% relative humidity [35]. The introduction of Pd nanoparticles into p-type rGO and n-type SnO_2 was found to create a nano-Schottky contact, further improving the surface chemistries on exposure toward gaseous NO_2. A field effect transistors (FETs) based prototype characterized by bilayer, trilayer, and quadrilayer MoS_2 developed to detect NO, demonstrated excellent sensitivity toward NO with the lowest detection limit of 0.8 ppm (Fig. 11b) [36].

In a work, the synergetic effect of CuO porous nanoflakes modified with rGO nanosheets based NO_2 gas sensor [37] displayed remarkable room temperature (23 °C) sensing performances with improved selectivity, ultrahigh response of 400.8% on exposure toward 5 ppm NO_2 gas with a LOD of 50 ppb with a responsivity of 20.6% (Fig. 11c) and long term stability over 30 days. The dynamic response curves also illustrated accelerated response/recovery time of 31.8/60.6 s and 6.8/55.1 s on exposure toward 0.05 and 5 ppm concentration of NO_2 gas. The mesoporous design, coupled with an elevated specific surface area of 76.07 m^2/g in CuO-rGO, not only facilitated enhanced gas diffusion but also provided abundant surface active sites for gas, thereby leading to an improvement in gas response kinetics and redox reaction on exposure toward NO_2. Besides, it's obvious that on increasing the concentration of NO_2 from 0.1 to 100 ppm, the responsivity also suffered a hysteric improvement from 42.3 to 3612.8%.

3.4 Hydrogen Sulfide Sensing

Hydrogen sulfide (H_2S), an endogenous gasotransmitter produced by several enzymatic reactions, serves as a biomarker to identify a variety of diseases such as Alzheimer's, chronic pulmonary and hepatic disorders. Being an extremely sensitive material toward H_2S, CuO nanosheets exhibited a low detection limit of 0.2 ppm with a response/recovery time of 20/120 s, at an operating temperature of 250 °C [38]. Yet another experiment on Ni-MoS_2-based gas sensor [32] exhibited a response reaching around 80% on exposure to 2 ppm H_2S gas with a detection limit of 0.3 ppm. In another report, cellulose nanofiber template CuO decorated with WS_2 nanosheets presented a responsivity of 37 at 0.5 ppm with response/recovery time of 37.2/33.9 s at a reduced operating temperature of 100.1 °C, however limiting the practical applicability of the device due to high operating temperatures [39]. Most of the conventional H_2S sensing materials are found to suffer from this limitation of high operation temperature. A room temperature (25 °C) H_2S sensor was fabricated employing $SnSe_2$ nanosheet/WO_3 nanoparticle-based nanocomposite [40] with a sensitivity of 33.80% on exposure toward 10 ppm H_2S gas at 82% RH with a response–recovery time of 136/459 s. Further, on increasing the concentration of H_2S from 10 to 30 ppm, the nanocomposite displayed an increase in the response to 56.72% at 30 ppm H_2S. The sensor also presented a theoretical low detection limit down to 28 ppb, opening up a larger range of application possibilities. In another report, 1.0 at% Indium doped Co_3O_4 porous nanosheets [41] demonstrated an excellent response of 6.81 at 50 ppm concentration of H_2S and a response of 1.4 on exposure to 1 ppm H_2S with long-term stability over 30 days. On exposure against H_2S gaseous species, In doped Co_3O_4 suffered a large change in the resistance compared to that of undoped Co_3O_4 due to a reaction between chemisorbed oxygen species and dH2S releasing electrons which further associates with holes in the valence band resulting in decrease of hole concentration (Fig. 12a). It was observed that indium doping improved the dynamic response characteristics of the sensor by virtue of increased charge carrier concentration, oxygen vacancy regions, and chemisorbed oxygen together with the favorable gas adsorption provided by the porous nanosheet structure with increased surface reactive sites. Taking the advantage of strong chemical affinity of p-type CuO semiconductor against acidic H_2S and high stability of n-type In_2S_3, a p-CuO nanotube/n-In_2S_3 nanosheet-based heterostructure [42] was formed exhibiting a remarkable response toward H_2S at room temperature with a detection limit of 1 ppm. However, the sensor exposes a sharp increase in the response at a concentration of 50 ppm, attributed to the conversion of CuO/In_2S_3 heterostructure to CuS/In_2S_3 Schottky junction (Fig. 12b). Whereas, at low H_2S concentrations, the gas sensing mechanism is promoted by the oxidation reaction between the H_2S analyte and the oxygen species on the surface of the heterostructure without any transformation of CuO to CuS (Fig. 12b). By virtue of the unique and versatile electronic properties, chemical stability, inherent and tunable conductivity, synchronizing surface characteristics, etc. when compared to their oxide counterparts, make them potential materials for sensing applications. Table 1 summarizes various 2D materials/hybrid structures

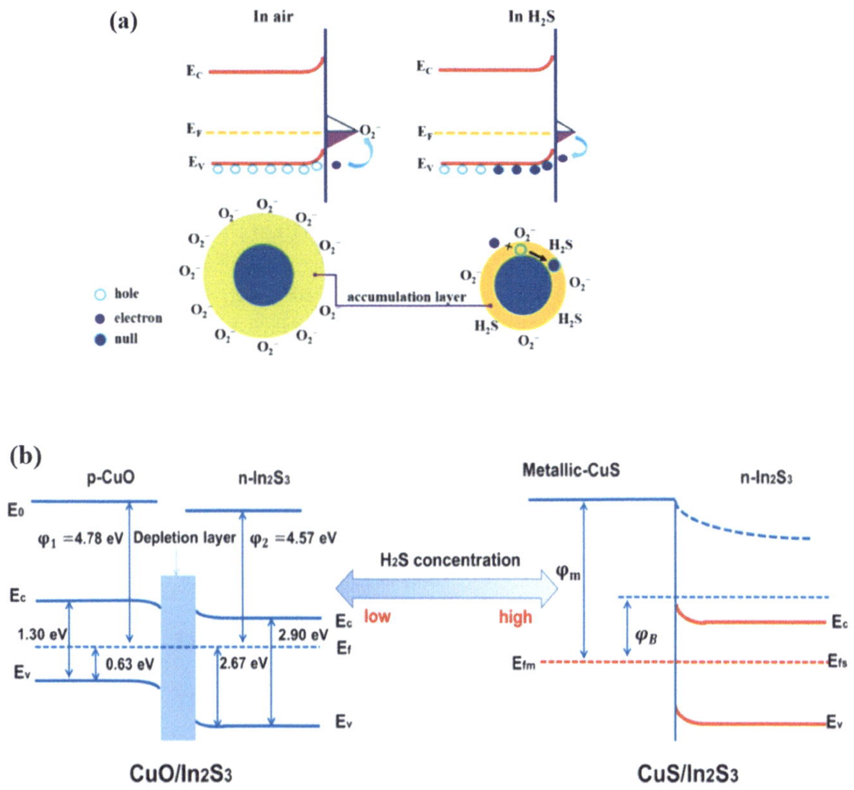

Fig. 12 Energy band diagram depicting the gas sensing mechanism in **a** In doped Co_3O_4 on exposure against air and H_2S gas [41], Reprinted with permission from Wu et al. [41]. [42][42] Copyright 2019, the Royal Society of Chemistry. **b** CuO/In_2S_3 nanostructures on exposure toward low and high concentrations of H_2S, Reprinted with permission from Zhang et al.. Copyright 2021, the American Chemical Society

with promising sensing characteristics explicitly toward acetone, NO, NH_3 and H_2S gases.

4 Summary and Future Perspectives

In this chapter, we have summarized the sensing response characteristics of various 2D materials toward specific gases, especially acetone, ammonia, nitric oxide and hydrogen sulphide revealing a prospective impact in e-nose-based sensing technology. The e-nose prototype-based technology will be extremely beneficial for the non-invasive diagnosis of several diseases from human exhalation, further revolutionizing point-of-care diagnostics and health care. The development of diverse

Table 1 Dynamic gas sensing response characteristics of various 2D materials toward acetone, ammonia, NO, H_2S

Materials	Gas	Lowest detection limit (LOD) (ppm)	Response (%)	T_{res}/T_{rec}	Temperature (°C)	Ref
Accordion like-Ti_3C_2 MXene	Acetone	0.25	17.30	53/120 s	23	[21]
Fe doped rGO-WO_3	Acetone	1	78 at 10 ppm	20/75 s at 10 ppm	130	[18]
$Ti_3C_2T_x$ MXene/ α-Fe_2O_3	Acetone	5	16.6	5/5 s	RT	[22]
V_2CT_x/V_2O_5	Acetone	0.25	11.9 at 15 ppm	115/180 s at 15 ppm	RT	[43]
$V_4C_3T_x$	Acetone	1	~ 5	40/- s	25	[23]
Nb_2CT_x nanosheets and PANI nanofibers	NH_3	1	9.33	105/143 @ 100 ppm	RT at 87%RH	26
MXene/V_2O_5/ $CuWO_4$	NH_3	0.3	53.5 at 51 ppm	10/5 s @ 1 ppm	26	[28]
$Ti_3C_2T_x$-S QCM	NH_3	0.01	49 Hz at 100 ppb	33/58 s at 100 ppb	RT	44
40 wt%-MXene/ In_2O_3	NH_3	1	100.7 at 20 ppm	60/300	RT	[45]
$Ti_3C_2T_x/WO_3$	NH_3	1	22.3	119/228 s	RT	[46]
Pd-SnO_2-rGO	NO_2	0.05	3.92 at 1 ppm	13/105 s at 1 ppm	RT	[35]
CuO-rGO	NO_2	0.05	20.6	6.8/55.1 s	23	37
Co_3O_4@PEI/ $Ti_3C_2T_x$ MXene	NOx	0.03	~ 27.9 at 100 ppm	1.6/73.1 s at 100 ppm	RT	[47]
SnO_2/MXene	NO_2	0.03	231	146/102 s	RT	[48]
MoS_2/ZnO	NO_2	0.0002	91 at 0.005 ppm	132/110 s at 0.1 ppm	RT	[49]
ssDNA functionalized graphene	H_2S	0.0053	98.5 at 2 ppm	-	RT/80% of RH	[50]
α-Fe_2O_3 nanosheets	H_2S	1	5.2 at 5 ppm	1/10 s at 5 ppm	135	[51]
PSS doped PANI/ graphene	H_2S	1	~ 20	90/150 s	RT	[52]
γ-Fe_2O_3/rGO	H_2S	97	520.73	-	RT	[53]
In-doped Co_3O_4 porous nanosheets	H_2S	1	6.83 at 50 ppm	6.81/9 s at 50 ppm	RT	[41]
p-CuO Nanotube/ n-In_2S_3 Nanosheet	H_2S	1	3511 at 50 ppm	30/1200 s at 50 ppm	RT	[42]

2D sensing materials with enhanced surface chemistries was found to enhance the sensing pattern recognition, however still faces some challenges.

Even though most of the materials displayed excellent selectivity, sensitivity, stability and fast response/recovery time on exposure to various analytes, it is crucial to fabricate and test e-nose sensors with respect to the exhaled human breath with exceptional response features providing consistent and ongoing real-time monitoring for breath markers. Also, the fabrication of sensor arrays can provide better recognition and responsive behavior with long-term stability.

While reviewing the sensing behavior of various 2D materials toward H_2S gas, the optimum response was observed at higher operating temperatures limiting their practical applications. On that account, creating materials that can operate at room temperature and with little power consumption is another major issue in order to be relevant to diverse applications, such as compact and portable devices. Also, these research findings are still largely limited to laboratory experiments only and it is essential to advance to real-time applications in industries and medical care.

Additionally, owing to the possibility of low working temperatures and increased mobility, many 2D material gas sensors excel in detecting gases, even at parts per billion (ppb) levels. Howbeit, reports still face the issue of sensing at ppb level, which is crucial for early-stage diagnosis of exhaled breath and large-scale applications because the percentage of these gases is in very small concentration in the human exhalation. One significant barrier includes the decrease in the response due to air humidity and it is not practicable to introduce a relatively inert atmosphere for real-time applications. Further, most of the research works display long-term stability for several days, which is another consequential barricade to be overpowered for practical monitoring. As a result, more research is required to overcome these challenges and produce a highly sensitive, long-term stable sensor capable of ppb level response characteristics.

Regardless of the theoretical studies administering the adsorption energies, band structures, electronic features, etc. of materials, and more research has to be done to conceptualize the chemical interaction between each of the materials and chemical species involved in the gaseous state, blazing the increased selectivity of particular materials toward specific gaseous phases.

Taking into account the benefits of 2D materials and an eye toward use in a real-world situation, we hope that more constructive research work will be produced in the field of e-nose system, facilitating a non-invasive, early-stage disease diagnosis from human exhalation and taken forward from a cradle-like laboratory system to a real world, real-time application for pervasive breath detection.

References

1. Konvalina, G., Haick, H.: Sensors for breath testing: from nanomaterials to comprehensive disease detection. Acc. Chem. Res. **47**, 66–76 (2014). https://doi.org/10.1021/ar400070m

2. Tisch, U., Haick, H.: Arrays of chemisensitive monolayer-capped metallic nanoparticles for diagnostic breath testing. **26**, 171–179 (2010). https://doi.org/10.1515/revce.2010.009
3. Guo, J., Zhang, D., Li, T., Zhang, J., Yu, L.: Green light-driven acetone gas sensor based on electrospinned CdS nanospheres/Co_3O_4 nanofibers hybrid for the detection of exhaled diabetes biomarker. J. Colloid Interface Sci. **606**, 261–271 (2022). https://doi.org:https://doi.org/10.1016/j.jcis.2021.08.022
4. Mazzatenta, A., Pokorski, M., Sartucci, F., Domenici, L., Di Giulio, C.: Volatile organic compounds (VOCs) fingerprint of Alzheimer's disease. *Respiratory Physiology & Neurobiology* **209**, 81–84 (2015). https://doi.org/10.1016/j.resp.2014.10.001
5. Allosh, A., Calin-Andrei P.-S., Nicolae, F., Amanda, T.P., George, N., Ghitiu, L., Dragos-Victor, A., Mihnea, A.D., George, A.N.: Tuning phosphorene and MoS_2 2D materials for detecting volatile organic compounds associated with respiratory diseases. RSC Adv. **14**, 1803–1812 (2024). https://doi.org/10.1039/D3RA07685G
6. Gosling, J. H., Makarovsky, O., Wang, F., Cottam, N. D., Greenaway, M. T., Amalia, P., Ricky, D. W., Christopher, J. T., Lyudmila, T., Fromhold, T. M.: Universal mobility characteristics of graphene originating from charge scattering by ionised impurities. Commun. Phys. **4**, 30 (2021). https://doi.org/10.1038/s42005-021-00518-2
7. Zhang, S., Wang, H., Liu, J., Bao, C.: Measuring the specific surface area of monolayer graphene oxide in water. Mater. Lett. **261**, 127098 (2020). https://doi.org/10.1016/j.matlet.2019.127098
8. Radisavljevic, B., Radenovic, A., Brivio, J., Giacometti, V., Kis, A.: Single-layer MoS_2 transistors. Nat. Nanotechnol. **6**, 147–150 (2011). https://doi.org/10.1038/nnano.2010.279
9. Eda, G., Hisato, Y., Damien, V., Takeshi, F., Mingwei, Chen., Manish, C.: Photoluminescence from Chemically Exfoliated MoS_2. Nano Lett. **11**, 5111–5116 (2011). https://doi.org/10.1021/nl201874w
10. Li, Y., Xiang, F., Lou, W., Zhang, X.: MoS_2 with structure tuned photocatalytic ability for degradation of methylene blue. IOP Conf. Ser. Earth Environ. Sci. **300**, 052021 (2019). https://doi.org/10.1088/1755-1315/300/5/052021
11. Ikram, M., Lujia, L., Yang, L., Laifeng, M., He, L., Mohib, U., Lang, H., Hongyuan, W., Ruihong, W., Keying, S.: Fabrication and characterization of a high-surface area $MoS_2@WS_2$ heterojunction for the ultra-sensitive NO_2 detection at room temperature. J. Mater. Chem. A **7**, 14602–14612 (2019). https://doi.org/10.1039/C9TA03452H
12. Liu, X., Liu, Y., Dong, S., Zhang, X., Hou, S.: Synthesis of ultra-high specific surface area aerogels with nitrogen-enriched $Ti_3C_2T_x$ nanosheets as high-performance supercapacitor electrodes. J. Mater. Chem. C **10**, 14929–14938 (2022). https://doi.org/10.1039/D2TC01987F
13. Zhang, X., Zhao, X., Wu, D., Jing, Y., Zhou, Z.: High and anisotropic carrier mobility in experimentally possible Ti_2CO_2 (MXene) monolayers and nanoribbons. Nanoscale **7**, 16020–16025 (2015). https://doi.org/10.1039/C5NR04717J
14. Zha, X.-H., Jie, Z., Yuhong, Z., Qing, H., Jian, He., Joseph, S. F., Kan, L., Shiyu, D.: Promising electron mobility and high thermal conductivity in Sc_2CT_2 (T = F, OH) MXenes. Nanoscale **8**, 6110–6117 (2016). https://doi.org/10.1039/C5NR08639F
15. Qiao, J., Kong, X., Hu, Z.-X., Yang, F., Ji, W.: High-mobility transport anisotropy and linear dichroism in few-layer black phosphorus. *Nature Communications* **5**, 4475 (2014). https://doi.org/10.1038/ncomms5475
16. Meng, Z., Stolz, R. M., Mendecki, L., Mirica, K. A.: Electrically-transduced chemical sensors based on two-dimensional nanomaterials. Chem. Rev. **119**, 478–598 (2019). https://doi.org/10.1021/acs.chemrev.8b00311
17. Prasanth, A., Selamawit, G., Tseganesh, S., Velumani, M., Meher, S.R., Alex, Z.C.: A bilayer SnO_2/MoS_2-coated evanescent wave fiber optic sensor for acetone detection-an experimental study. Biosensors **12**, 734 (2022)
18. Sen, S., Maity, S., Kundu, S.: Fabrication of Fe doped reduced graphene oxide (rGO) decorated WO_3 based low temperature ppm level acetone sensor: Unveiling sensing mechanism by impedance spectroscopy. Sens. Actuators B: Chem. **361**, 131706 (2022). https://doi.org/10.1016/j.snb.2022.131706

19. Sarkar, S., Viegas, A. E., Raghavan, S. R.: MoS_2 Field Effect Transistors based sensors for low concentration acetone detection. arXiv: Appl. Phys. (2019)
20. Ama, O., Sadiq, M., Johnson, M., Zhang, Q., Wang, D.: Novel 1D/2D KWO/$Ti_3C_2T_x$ nanocomposite-based acetone sensor for diabetes prevention and monitoring. Chemosensors **8**, 102 (2020)
21. Majhi, S.M., Ashraf, A., Yaser, E.G., Hesham, F.El-M., Naser, N.Q., Abdul, R.H., Saleh, T.M.: Accordion-like-Ti_3C_2 MXene-based gas sensors with sub-ppm level detection of acetone at room temperature. ACS Appl. Electron. Mater. **4**, 4094–4103 (2022). https://doi.org/10.1021/acsaelm.2c00717
22. Liu, M., Ji, J., Song, P., Liu, M., Wang, Q.: α-Fe_2O_3 nanocubes/$Ti_3C_2T_x$ MXene composites for improvement of acetone sensing performance at room temperature. Sens. Actuators B Chem. **349**, 130782 (2021). https://doi.org/10.1016/j.snb.2021.130782
23. Arpit, V., Bal, C.Y.: 2D/2D nanostructured system based on wo_3/ws_2 for acetone sensor and breath analyzer. ACS Appl. Nano Mater. **6**, 5493 (2023). https://doi.org/10.1021/acsanm.2c05596
24. Monteiro Fernandes, G.B., Nascimento, H., Santa Cruz, R.M., et al.: Investigation of a plasmonic optical sensor for acetone detection in exhaled breath and exhaled breath condensate. Plasmonics (2024). https://doi.org/10.1007/s11468-023-02190-4
25. Zhao, W.-N., Yun, N., Dai, Z.-H., Li, Y.-F.: A high-performance trace level acetone sensor using an indispensable $V_4C_3T_x$ MXene. RSC Adv. **10**, 1261–1270 (2020). https://doi.org/10.1039/C9RA09069J
26. Wang, S., Bohao, L., Zaihua, D., Qiuni, Z., Yajie, Z., Guangzhong, X., Yadong, J., Shaorong, L., Huiling, T.: PANI nanofibers-supported Nb_2CT_x nanosheets-enabled selective NH_3 detection driven by TENG at room temperature. Sens. Actuators B: Chem. **327**, 128923 (2021). https://doi.org/10.1016/j.snb.2020.128923
27. Lu, L., Mengyuan, L., Qingli, S., Chuhan, Z., Yongjin, Z., Fen, X., Lixian, S., Cuili, X.: MXene/MoS_2 nanosheet/polypyrrole for high-sensitivity detection of ammonia gas at room temperature. Mater. Today Commun. **35**, 106239 (2023). https://doi.org/10.1016/j.mtcomm.2023.106239
28. Ranjbar, F., Hajati, S., Ghaedi, M., Dashtian, K., Naderi, H., Toth, J.: Highly selective MXene/V_2O_5/$CuWO_4$-based ultra-sensitive room temperature ammonia sensor. J. Hazard. Mater. **416**, 126196 (2021). https://doi.org/10.1016/j.jhazmat.2021.126196
29. Junkaew, A., Arróyave, R.: Enhancement of the selectivity of MXenes (M_2C, M = Ti, V, Nb, Mo) via oxygen-functionalization: promising materials for gas-sensing and -separation. Phys. Chem. Chem. Phys. **20**, 6073–6082 (2018). https://doi.org/10.1039/C7CP08622A
30. Khazaei, M., Masao, A., Taizo, S., Chan-Yeup, C., Natarajan, S. V., Mehdi, E., Yoshio, S., Yoshiyuki, K.: Novel electronic and magnetic properties of two-dimensional transition metal carbides and nitrides. Adv. Funct. Mater. **23**, 2185–2192 (2013). https://doi.org/10.1002/adfm.201202502
31. Banu, A.A., Sinthika, S., Premkumar, S., Vigneshwaran, J., Smagul, Z.K., Sujin, P.J.: DFT study of NH_3 adsorption on 2D monolayer MXenes (M_2C, M = Cr, Fe) via oxygen functionalization: Suitable materials for gas sensors. FlatChem **31**, 100329 (2022). https://doi.org/10.1016/j.flatc.2021.100329
32. Levna, C., Ettore, M., Aneesh, P.M.: Enhancement in the selectivity and sensitivity of Ni and Pd functionalized MoS_2 toxic gas sensors. J. Electrochem. Soc. **167**, 106506 (2020). https://doi.org/10.1149/1945-7111/ab992c
33. Khang, P. D., Hai, L. L., Hong, N. T. T., Tuan, V. V.: Study on gas adsorption properties (N_2, H_2, O_2, NO, NO_2, CO, CO_2, SO_2, H_2S, H_2O and NH_3) on the O-vacancy-containing Sc_2CO_2 monolayer. VNU J. Sci. Math.–Phys. **38** (2022). https://doi.org/10.25073/2588-1124/vnumap.4653
34. Deng, X., Zhuo, Z., Yuhuan, Z., Jiechang, G., Taotao, L., Zhisong, L., Chang, M. L.: Synthesis of merit-combined antimony tetroxide nanoflowers/reduced graphene oxide to synergistically boost real-time detection of nitric oxide released from living cells for high sensitivity. J. Colloid Interface Sci. **581**, 465–474 (2021). https://doi.org/10.1016/j.jcis.2020.07.076

35. Wang, Z., Tong, Z., Chen, Z., Tianyi, H., Teng, F., Sen, L., Geyu, L.: Anchoring ultrafine Pd nanoparticles and SnO_2 nanoparticles on reduced graphene oxide for high-performance room temperature NO_2 sensing. J. Colloid Interface Sci. **514**, 599–608 (2018). https://doi.org/10.1016/j.jcis.2017.12.075
36. Li, H., Zongyou, Y., Qiyuan, H., Hong, L., Xiao, H., Gang, L., Derrick, W. H. F., Alfred, I. Y. T., Qing, Z., Hua, Z.: Fabrication of single- and multilayer MoS_2 film-based field-effect transistors for sensing NO at room temperature. Small **8**, 63–67 (2012). https://doi.org/10.1002/smll.201101016
37. Bai, H., Hui, G., Jin, W., Yan, D., Bin, L., Zili, X., Fuqiang, G., Dunjun, C., Rong, Z., Youdou, Z.: A room-temperature NO_2 gas sensor based on CuO nanoflakes modified with rGO nanosheets. Sens. Actuators B: Chem. **337**, 129783 (2021). https://doi.org/10.1016/j.snb.2021.129783
38. Miao, J., Chen, C., Meng, L., Lin, Y.S.: Self-assembled monolayer of metal oxide nanosheet and structure and gas-sensing property relationship. ACS Sensors **4**, 1279–1290 (2019). https://doi.org/10.1021/acssensors.9b00162
39. Li, J., Hongchao, Z., Yanjie, W., Ruijie, Z., Cheng, Z., Yong, Z.: Mesoporous WS_2-decorated cellulose nanofiber-templated CuO heterostructures for high-performance chemiresistive hydrogen sulfide sensors. Anal. Chem. **94**, 16160–16170 (2022). https://doi.org/10.1021/acs.analchem.2c03596
40. Guo, X., Yanqiao, D., Chengyao, L., Bingsheng, Du., Chengjiu, Z., Yiling, T., Yijie, S., Pinglei, Z., Xi, Y., Yong, H.: Humidity-activated H_2S sensor based on $SnSe_2/WO_3$ composite for evaluating the spoilage of eggs at room temperature. Sens. Actuators B: Chem. **357**, 131424 (2022). https://doi.org/10.1016/j.snb.2022.131424
41. Wu, J., Ying, Y., Chengxin, Z., Hui, Y., Licheng, H., Xiangting, D., Jinxian, W., Xinlu, W.: Extremely sensitive and accurate H_2S sensor at room temperature fabricated with In-doped Co_3O_4 porous nanosheets. Dalton Trans **48**, 7720–7727 (2019). https://doi.org/10.1039/C9DT01043B
42. Zhang, W., Xukun, W., Ziye, F., Juan, L., Guo, L., Xueliang, L., Bingsheng, L., Jinyuan, Z., Erqing, X., Zhenxing, Z.: Switching Effect of p-CuO Nanotube/n-In_2S_3 Nanosheet Heterostructures for High-Performance Room-Temperature H_2S Sensing. ACS Appl. Mater. Interfaces **13**, 52938–52949 (2021). https://doi.org/10.1021/acsami.1c15564
43. Majhi, S.M., Ali, A., Greish, Y.E., El-Maghraby, H.F., Mahmoud, S.T.: V_2CT_X MXene-based hybrid sensor with high selectivity and ppb-level detection for acetone at room temperature. Scientific Reports **13**, 3114 (2023). https://doi.org/10.1038/s41598-023-30002-6
44. Zhang, B., Zhiqiong, L., Chong, L., Min, L., Chen, F., Ran, T., Xian-hu, Z., Honglang, L., Jingting, L.: High-sensitive ppb-level ammonia QCM sensor based on sulfur doped $Ti_3C_2T_x$ MXene. Sens. Actuators A Phys. **350**, 114138 (2023). https://doi.org/10.1016/j.sna.2022.114138
45. Liu, Z., He, T., Sun, H., Huang, B., Li, X.: Layered MXene heterostructured with In_2O_3 nanoparticles for ammonia sensors at room temperature. Sens. Actuators B: Chem. **365**, 131918 (2022). https://doi.org/10.1016/j.snb.2022.131918
46. Guo, X., Yanqiao, D., Delin, K., Zhilin, W., Xia, S., Bingsheng, D., Chengyao, L., Yingjie, W., Weijie, Q., Lian, X., Yong, H.: Enhanced ammonia sensing performance based on MXene-$Ti_3C_2T_x$ multilayer nanoflakes functionalized by tungsten trioxide nanoparticles. J. Colloid Interface Sci. **595**, 6–14 (2021). https://doi.org/10.1016/j.jcis.2021.03.115
47. Sun, B., He, L., Zhuo, L., Jue, W., Xue, B., Yang, Z., Junkun, C., Kan, K., Keying, S.: Co_3O_4@PEI/$Ti_3C_2T_x$ MXene nanocomposites for a highly sensitive NO_x gas sensor with a low detection limit. J. Mater. Chem. A **9**, 6335–6344 (2021). https://doi.org/10.1039/D0TA11392A
48. Gasso, S., Sohal, M. K., Mahajan, A.: MXene modulated SnO_2 gas sensor for ultra-responsive room-temperature detection of NO_2. Sens. Actuators B: Chem. **357**, 131427 (2022). https://doi.org/10.1016/j.snb.2022.131427
49. Kumar, R. R., Thangapandian, M., Ajit, D., Cheng-Hsuan, H., Shivam, G., Arumugam, M., Aswin, K. A., Chih-Hao, L., Nyan-Hwa, T., Yu-Lun, C., Heh-Nan, L.: Ultrasensitive and light-activated NO_2 gas sensor based on networked MoS_2/ZnO nanohybrid with adsorption/

desorption kinetics study. Appl. Surface Sci. **536**, 147933 (2021). https://doi.org/10.1016/j.apsusc.2020.147933
50. Jung, Y., Hi, G. M., Chaehyun, L., Keunsu, C., Hyun, S. S., Sukang, B., Soo, M. K., Minah, S., Taikjin, L., Seok, L., Hyung-Ho, P., Seong, C. J., Chong-Yun, K., Chulki, K.: Humidity-Tolerant Single-Stranded DNA-Functionalized Graphene Probe for Medical Applications of Exhaled Breath Analysis. Adv. Funct. Mater. **27**, 1700068 (2017). https://doi.org/10.1002/adfm.201700068
51. Zhang, H.-J., Meng, F.-N., Liu, L.-Z., Chen, Y.-J.: Convenient route for synthesis of alpha-Fe_2O_3 and sensors for H_2S gas. J. Alloys CompD. **774**, 1181–1188 (2019). https://doi.org/10.1016/j.jallcom.2018.09.384
52. Cho, S., Lee, J. S., Jun, J., Kim, S. G., Jang, J.: Fabrication of water-dispersible and highly conductive PSS-doped PANI/graphene nanocomposites using a high-molecular weight PSS dopant and their application in H2S detection. Nanoscale **6**, 15181–15195 (2014). https://doi.org/10.1039/C4NR04413D
53. Zhang, C., Zhang, S., Yang, Y., Yu, H., Dong, X.: Highly sensitive H_2S sensors based on metal-organic framework driven γ-Fe_2O_3 on reduced graphene oxide composites at room temperature. Sens. Actuators B: Chem. **325**, 128804 (2020). https://doi.org/10.1016/j.snb.2020.128804

Electrochemical Sensors Based on 2D Materials (2DMs) and Their Heterostructures

Ranjeet Kaur, Randeep Kaur, Bhupender Singh, Jaiveer Singh, Gurpreet Singh, Shweta Rana, Puneet Kaur, and Deepender Kaushik

Abstract The present book chapter explores the potential of 2DMs and their heterostructures in electrochemical sensing. Electrochemical sensing is an inexpensive and hassle-free approach to the identification of analytes. It is widely employed in the food, environment, and health sectors. We will also outline the various types of 2DMs including graphene (GR), transition metal dichalcogenides/oxides (TMDs/TMOs), transition metal nitrides (TMNs), materials based on black phosphorus (BP), carbon nitrides (CNs), metal–organic frameworks (MOFs), boron nitrides (BNs) being utilized for electrochemical sensing applications. In the present chapter, we have focused on the synthesis and properties of 2DMs. The exceptional properties such as high electronic conductivity, thin-film forming ability, flexibility, and tuneable nature make them apt candidates for electrochemical sensing applications. The latest advancements in the designing and construction of electrochemical sensors based on food quality, health monitoring, environmental monitoring, and biomolecular detection are also discussed. Therefore, through understanding the existing electrochemical sensors constructed using 2D materials, this chapter will facilitate the advancement of state-of-the-art electrochemical sensors and devices.

R. Kaur (✉)
University Centre for Research & Development (UCRD), Department of Chemistry, Chandigarh University, Mohali, Gharuan, Punjab 140413, India
e-mail: ranjeet.e14885@cumail.in

R. Kaur · B. Singh · J. Singh · S. Rana
Department of Chemistry, Panjab University, Chandigarh 160014, India

G. Singh
University College of Dhilwan, Barnala, Punjab 148107, India

P. Kaur
Department of Physics, National Cheng Kung University, Tainan 70101, Taiwan ROC

D. Kaushik
College of Pharmacy, University of Nebraska Medical Center, Omaha, NE 68198-6125, USA

A. Katoch et al. (eds.), *2D Materials*, Engineering Materials,
https://doi.org/10.1007/978-981-97-6258-3_5

1 Introduction

As the new era of research and industrialization dawns, the demand for food quality, health, and environmental monitoring is growing due to emerging food safety challenges, globalization, varying consumer expectations, regulatory requirements, and rising awareness about health and environmental concerns. Hence, implementing robust monitoring systems ensures the quality, integrity, safety of food, health, and environment, benefitting consumers and the industry as a whole. Among different methods, electrochemical sensors play a crucial role in the progress of inexpensive and highly sensitive tools for the quantification of harmful analytes and biomolecular detection [1, 2]. It is helpful in determining the amount of analyte by relating to the current signal corresponding to the analyte and offers applications in food and health monitoring, environment monitoring, and industrial applications with high accuracy and reliability.

A variety of electrochemical analysis methods are accessible, each suitable for specific applications and target analytes. These techniques encompass cyclic voltammetry (CV), chronoamperometry (CA), normal pulse voltammetry (NPV), differential pulse voltammetry (DPV), square wave voltammetry (SWV), impedance spectroscopy, and chronocoulometry (CC). Among them, DPV and SWV are pulse techniques that provide picomolar-level detection of analytes within a fraction of seconds due to lower background currents associated with these techniques. A range of pulse techniques is available which differ in the potential wave forms (waveform is a variation of potential with time), type of electrode used or number of sampled points. The waveform is of different type basically linear waveform and pulse waveform.

The type of electrode, electrolyte, sensor design, and method development can be tailored for enhanced sensitivity and selectivity of electrochemical sensors. Nonetheless, specific factors have the potential to enhance sensor sensitivity. These encompass augmenting electrode surface area, managing electrode surface properties, and utilizing techniques for amplifying signals. Over the years, electrochemical sensing has witnessed advancements by developing novel materials and innovative sensor technologies which provide control over the electrode surface properties [3–8].

In recent years, the augmentation of electrode properties through the integration of nanomaterials or carbon-based substances has proven instrumental in elevating the sensitivity of electrochemical sensors. The utilization of two-dimensional materials (2DMs) has captured the attention of researchers due to their remarkable electronic, optical and chemical attributes, thereby ushering in novel avenues for sensor advancement and conception. These 2DMs manifest as extremely thin atomic layers, ranging from monolayers to a few nanometers in thickness. Intrinsically, covalent bonds maintain atomic cohesion within a layer, while interlayer interactions are governed by weak van der Waals forces. Notably, graphene (GR), transition metal dichalcogenides (TMDs), transition metal oxides (TMOs), transition metal nitrides (TMNs), materials derived from black phosphorus (BP), carbon nitrides (CNs), metal–organic frameworks (MOFs), and boron nitrides (BNs) all occur naturally in bulk configurations [9–12].

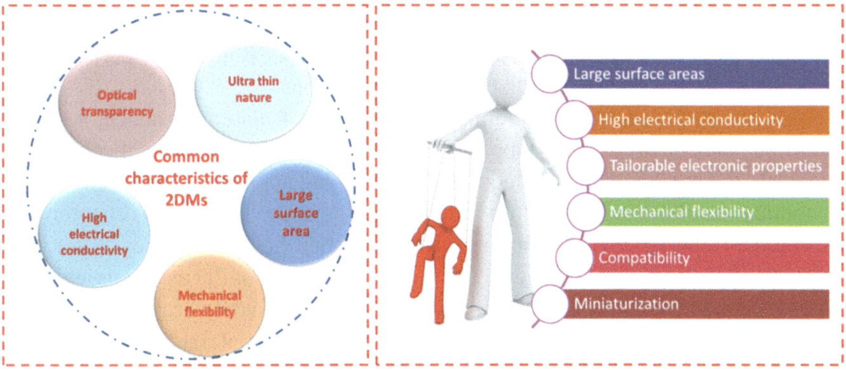

Fig. 1 **a** Common characteristics of different 2DMs. **b** Important aspects of 2DMs in electrochemical sensing

These 2DMs can be most commonly synthesized via exfoliation of bulk materials and is a top-down approach [13, 14]. However, there are several other top-down or bottom-up approaches available for synthesis. The electronic properties of these exfoliated materials are different from those of bulk materials due to the rupture of forces between the adjacent layers.

These substances showcase distinct attributes such as optical translucency, expansive surface coverage, heightened electrical conductivity, and exceptional mechanical suppleness [15] (depicted in Fig. 1a). In addition to these shared traits, two-dimensional materials (2DMs) display an elevated surface-to-volume ratio, outstanding electrical conductance, the capacity to manipulate electronic characteristics, straightforward functionalization, compatibility, and the potential for miniaturization. These aspects collectively contribute to their substantial promise in the realm of electrochemical sensing. (Fig. 1b).

Although 2DMs can provide excellent properties for their use in electrochemical sensing due to their remarkable conductivity and large surface areas, heterostructures of 2DMs offer additional benefits like improved catalytic activity, selectivity, or charge transfer kinetics compared to standalone materials [16, 17]. These heterostructures have attracted a lot of attention in recent years as stacking would result in unique properties that are otherwise absent in single-layer 2DMs. This will result in new electronic, optical, and structural features different than that of individual materials. The preparation of heterostructures allows maneuvering electrochemical properties, interface structures, and compositions which are otherwise difficult to obtain. The desired specific properties can be tailored by controlling layer stacking order and orientation and composition of composite materials. Thus, the resultant heterostructures offer potential in a variety of fields including electrochemical sensing. Depending on the analyte to be analyzed, we can carefully select heterostructures to create novel material with tailored electronic states that will help to improve the detection of an analyte. Using the appropriate synthetic approach, surface functionalities, material composition, etc. we can engineer the wide detection range, selectivity, sensitivity, etc. By following these guidelines for synthesis and design criteria, researchers can leverage the unique properties of heterostructures to develop robust and high-performance electrochemical sensors for a wide range of applications.

2DMs showed tremendous potential in the fields of material science, sensing/biosensing, and environmental applications. The exceptional physical, chemical, biological, and mechanical traits led to their promising applications in electrochemical applications like health and food monitoring devices. The present chapter will focus on the importance of developing 2DMs and their heterostructures. After that, the focus will be laid on the synthesis and properties of various 2D materials and their heterostructures. We will also outline the various 2DMs including graphene (GR) [9], transition metal dichalcogenides (TMDs) [18], transition metal oxides [19] (TMOs), transition metal nitrides [20] (TMNs), materials based on black phosphorus (BP) [21], carbon nitrides (CNs) [22], metal–organic frameworks (MOFs) [23], boron nitrides (BNs) [17, 24] being utilized for electrochemical sensing applications. Apart from this, the chapter will also deal with recent progress in the fabrication of electrochemical sensors based on 2D materials for food, health, environment monitoring, and biomolecular detection. Thus, by gaining insight into already developed electrochemical sensors based on 2D materials, the present chapter will pave the way towards the development of cutting-edge electrochemical sensors and devices.

2 Synthesis and Characteristics of 2DMs

A large number of 2DMs have gained considerable attention and have been extensively investigated in electrocatalysis, energy storage, and electrochemical sensing applications as they exhibit exceptional physical and chemical characteristics relative to their bulk counterparts. The commonly employed 2DMs in electrochemical sensing include graphene (GR) [9], transition metal dichalcogenides (TMDs) [18], transition metal oxides [19] (TMOs), transition metal nitrides [20] (TMNs), materials based on black phosphorus (BP) [21], carbon nitrides (CNs) [22], metal–organic frameworks (MOFs) [23], boron nitrides (BNs) [17, 24], etc. 2DMs exhibit thin sheet with layered structure having infinite planar lengths contributing to large surface areas and tunable physical and chemical properties. The most common synthetic route for 2DMs includes the exfoliation of bulk materials, commonly called the Top-down approach [13, 14]. There are frail interplanar van der walls interactions existing between the layers which can be broken down to obtain a 2D open structure and their physical and chemical features can be maneuvered for specific applications in the field of electrochemistry. There are various Top-down methods available to date for the synthesis of 2DMs from their bulk counterparts which are mechanical exfoliation, liquid exfoliation, and ion/molecules assisted exfoliation [25–28].

Various researchers studied the criterion of feasibility for the exfoliation without fractures and its effect on interfacial properties. The most common and effective method utilized for obtaining few-layer sheets of 2DMs from stacked layers is mechanical exfoliation. The mechanical exfoliation can be carried out in various ways (like ball milling, grinding in mortar, and scotch tape method). The scotch tape method was used originally to isolate GR (Scotch tape method), and was the most popular method for the synthesis of GR. Exfoliation of bulk materials with scotch tape into 2D layers of different precursors; the transfer of these 2DMs onto suitable substrates is in high demand for various applications. Since this method involves simple peeling off without the use of laborious chemical methods, the developed 2DMs will resemble bulk material and retain their pristine structure. However, the scotch tape method is associated with low scalability, inefficiency, and contamination leading to its limited use. Apart from the scotch tape method, grinding in mortar is a famous method to attain 2DMs as an application of shear mixing allows sheets to slip over and create 2DMs. Apart from this, the ball milling approach allows uniform formation of 2DMs with nano or micro dimensions. This also provides good dispersion stability to these materials for film fabrication in electrochemical sensing. The mechanical exfoliation has played a crucial role in the discovery and early research of 2DMs, its limitations have led researchers to develop alternative exfoliation techniques. In the liquid exfoliation method, the bulk layered materials are exfoliated to form few- or single-layered 2D materials which consist of various methodologies like ultrasonication or electrochemical exfoliation or high shear mixing [25–28]. However, the substrate employed for electrochemical exfoliation may interfere with synthetic protocol leading to impurities. Figure 2 depicting the commonly employed mechanical and liquid exfoliation technique for mixing is shown above.

Apart from above-mentioned methods for the synthesis of 2DMs, bottom-up approach (chemical vapour deposition (CVD) and physical vapour deposition (PVD)) is a promising way to attain control over the chemical structure and physical properties but the yields are too low [29]. CVD approach for the synthesis of GR and CNTs requires the use of catalysts while 2D chalcogenides-based materials operate without the use of catalysts. This method can also be extended to develop other listed 2DMs to attain uniform and larger surface areas and hence can be exploited for device applications. The CVD approach was found to be the most efficient method for producing large monolayer GR sheets onto metal substrates like commercially available Cu-Ni foils. GR sheets up to several meters in width can now be produced using the CVD approach, making it ideal for roll-to-roll production processes and easy integration into flexible electronics applications [15, 18]. CVD is a rapidly evolving technique with the synthesis duration of 10–20 min, for the bottom-up synthesis of other 2DMs like transition metal dichalcogenides (TMDs). The synthesis using CVD depends on maintaining high-temperature (700–900 °C) within a CVD furnace's active zone, between the precursors required for required TMDs. In the process of formation of 2DMs, e.g., TMDs, control over the environment (generally Ar or N_2 atmosphere is preferred) is a necessary condition for desired properties and morphology of 2DMs. This whole process can be carried out in a two-zone furnace setup, which allows control over the temperatures of precursors of 2DMs. For example, vacuum thermal evaporation of chalcogenides is carried out to reach the substrate with thin layers of metal oxide, resulting in TMDs. General three-furnace CVD device is shown in Fig. 3, which can be employed for bottom-up synthesis of 2DMs using different substrates.

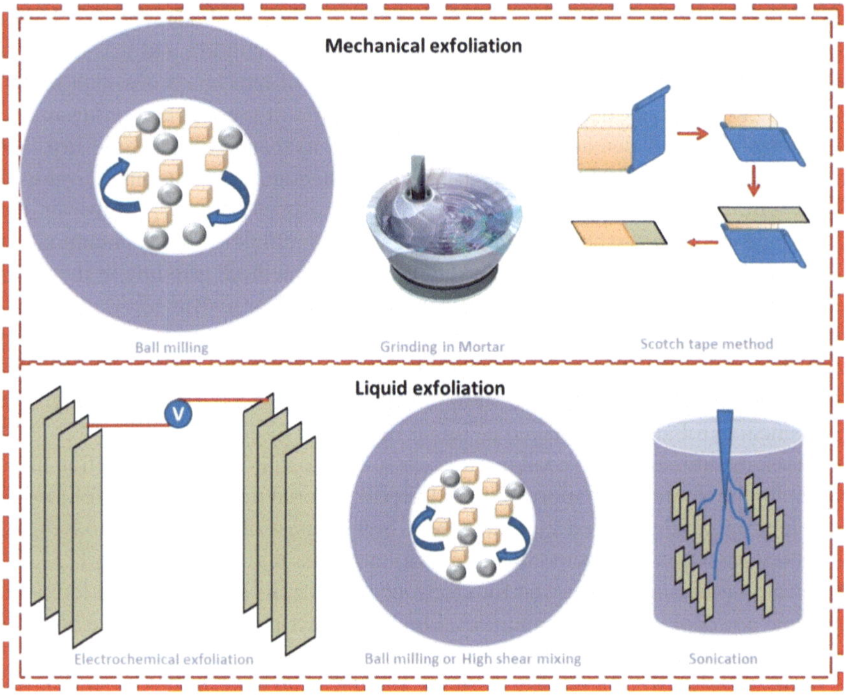

Fig. 2 Methods for the synthesis of 2DMs

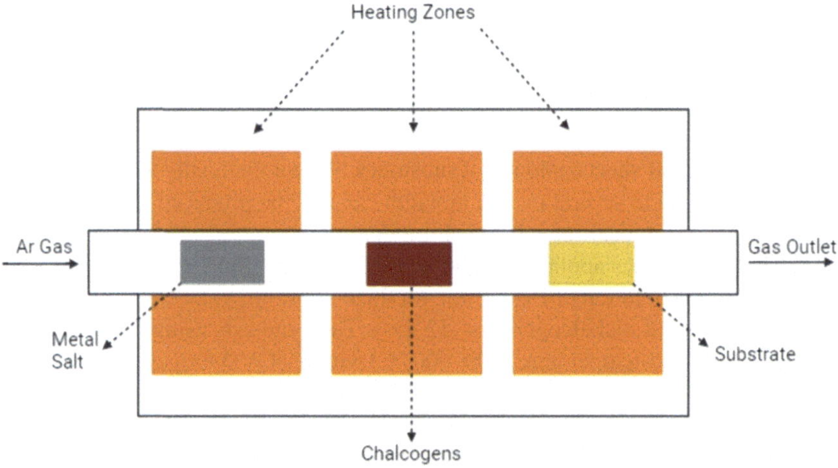

Fig. 3 Three furnace CVD device, metal salt, and chalcogens are placed in quartz and inert atmosphere

However, the control over other parameters such as nucleation rate, concentration of precursor, gas flow, and substrate treatment route can help to achieve high-quality and large-area single crystals of 2DMs usually TMDs. Hence, CVD holds great promise in synthesizing high-quality 2DMs, thus a key method for science and innovation. However, it is a very challenging task to obtain freely suspended TMDs using CVD as they can decompose quickly in contact with air accompanied by defects during their formation leading to changes in response of 2DMs during testing experiments [15, 18, 30, 31].

Another bottom-up approach for the production of 2DMs includes a hydrothermal/solvothermal approach that will result in nanomaterial with specific properties and thus provide the desired application. The solvothermal route employs organic ligands which provide control over the size, dispensability, and morphology of 2DMs. This method is particularly useful for the preparation of TMDs as they prevent the oxidation of metal. In contrast, the hydrothermal route utilizes water as reaction media at elevated temperatures which garnered its broad applicability. To prepare 2DMs using the above routes, the reaction parameters like reactant concentrations, temperature, time, and pH must be controlled to tailor desired applications. Moreover, the morphology and size can also be tailored by choosing solvent and reaction conditions carefully. However, the presence of organic ligands in the vicinity of metal surface hinders the applicability of the solvothermal route in catalysis, biomedical applications [15, 18].

These 2DMs exhibit exceptional properties rendering them appropriate for electrochemical sensing applications. The following are some listed properties: Large surface area: the exceptionally high surface area provides a greater number of active sites resulting in excellent charge transfer and hence electrochemical performance. Good electrical conductivity: 2DMs exhibit excellent electrical conductivity which promotes the electron transfer making the redox process feasible and quick. It also provides larger currents with improved charge transfer kinetics. Thickness of sheets: 2DMs possess ultrathin sheets making them an appropriate candidate for electrode materials. This nature of 2DMs allows enhanced diffusion of ions through the material inducing faster transport of mass. This in turn will improve the sensitivity of the sensor. Mechanical flexibility: 2DMs due to their extreme thickness and lattice structure exhibit good flexibility making them apt candidates for sensing and energy harvesting applications [30]. Tunable chemical properties: 2DMs when maneuvered with functional groups can enhance the electrocatalytic activity or provide a base for the introduction of other materials [31]. This ability of 2DMs provides versatility to design materials with good electrochemical performance. These properties of 2DMs make them apt candidates in the field of electrochemistry.

Figure 4 depicts the applications of 2DMs and their heterostructures in the field of electrochemical sensing. In the present chapter, we have thrown light on the above applications of 2DMs and their heterostructures ahead.

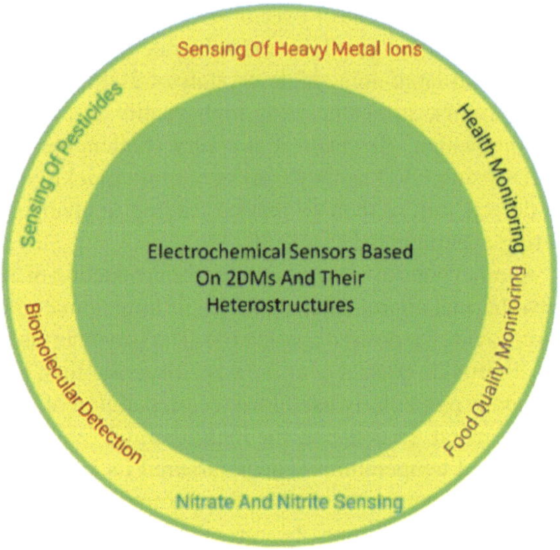

Fig. 4 Applications of 2DMs and their heterostructures towards electrochemical sensing

3 Electrochemical Sensors Based on 2DMs and Their Heterostructures

2DMs have contributed greatly in building electrochemical sensors that monitor the food quality guaranteeing the integrity and safety of food products throughout the supply chain. 2DMs offer a significant amount of surface area, excellent electrical conductivity, adaptable mechanical qualities, chemical stability, conjunction with other sensor components, scalability, and affordable manufacturing [32]. Whereas electrochemical sensors have the ability to offer on-site, real-time analysis, which facilitates prompt decision-making cutting down on the time needed for sample transportation and lab analysis [33]. Electrochemical sensors are also extremely sensitive and selective, enabling accurate detection of certain analytes in intricate food matrices. This feature is especially useful for identifying contaminants that could endanger consumer health, such as heavy metals, pesticides, and microbial toxins [34]. Therefore, the high-value collaboration of 2D materials and electrochemical sensing techniques paved a way for easily accessible control over food quality, environmental monitoring, health monitoring, and biomarkers detection.

3.1 Sensors for Food Quality Monitoring

This part of the chapter discusses the role of 2D materials in the electrochemical detection of various pathogens, heavy metal ions, pesticides, toxins, and physical parameters that need continuous monitoring to prevent any food born health condition. Following are the various applications of 2D material as sensors in food quality control.

3.1.1 Sensing of Pesticides

Pesticides pose a potential risk to human food, and prolonged exposure to these chemicals can have detrimental effects on health. Studies have shown that intake of pesticides through food and water can result in serious health issues such as disruptions in the endocrine and reproductive systems, as well as neurodegenerative diseases such as Parkinson's disease [35]. In this context, the use of 2D materials has shown great promise in the precise identification of hazardous organophosphorus pesticides (OPs). The remarkable surface-to-volume ratio leads to a substantial amount of analyte adsorption delivering high sensitivity to the sensor. Yan et al. developed a sensor by utilizing the hydroxyl (–OH) groups present in graphene oxide (GO) [36]. This approach led to improved electrical and catalytic performance, achieving a rapid response time of 1.8 s and a low detection limit of 0.21 μM with a wide linear range from 0.43 to 218.40 μM. In another study, the graphene-based framework has been functionalized with copper nanoparticles (CuNPs) to develop an electrochemical sensor. Setznagl co-workers have constructed a glyphosate herbicide sensor by incorporating CuNPs in the rGO matrix. The inhibition of Cu^0 oxidation to Cu^{2+} takes place by the addition of such pesticides and hence can be employed for signal generation in the sensor. The sensor exhibits a linear response between 0.1 and 1.1 μM and a detection limit of 0.19 μM. In this report, rGO is providing an extravagant surface area for pesticide adsorption and the act of binding is transduced via CuNPs in the form of a reduced current response at the working electrode [37]. In a separate study, another GR-based heterostructure has been reported. Facure et al. have investigated the impact of incorporating conductive organic polymers (PEDOT:PSS/polypyrrole) and gold nanoparticles (Au NPs) on the sensor's performance [38]. This modification enabled the successful detection of various OPs. Additionally, the sensor achieved a low detection limit of 0.19 μM using 0.2 M phosphate-buffered saline (PBS) at pH 7.0. Jiao and colleagues devised a novel electrode membrane utilizing heterostructures composed of ferrite-decorated graphene oxide ($GO@Fe_3O_4$) for the highly sensitive detection of chlorpyrifos [39]. This $GO@Fe_3O_4$ nanocomposite established a unique sensing layer, leveraging potent synergistic interactions. The graphene oxide endowed a substantial surface area, while the uniform deposition of Fe_3O_4 onto the GO facilitated rapid electron transfer, culminating in an impressively low detection threshold of 94.12 pM. In addition to the above-mentioned reports, Jirasirichote et al. created a portable electrochemical sensor for the detection of carbofuran-phenol. Carbofuran-phenol is the main by-product of carbofuran hydrolysis [40]. The curated screen-printed carbon electrode (SPE) was coated with GO and gold nanoparticles (AuNPs). A low limit of detection of 0.22 μM was obtained for a wide linear range of 1.0–250.0 μM using DPV. The method was also inspected for the detection of carbofuran in the samples of cucumber and rice.

Kaur et al. 2020 designed a highly sensitive electrode material for the detection of methyl parathion using electrochemically reduced graphene oxide-chitosan (ERGO-CS) decorated with redox-active hemoglobin (Hb) bio-element [41]. The increment in the number of redox-active moieties at the surface of the designed electrode was observed leading to an exceptionally low limit of detection of 79.77 nM. The methyl parathion adulteration of various vegetable samples was also investigated using square wave voltammetry (SWV). The electrochemical detection of pesticides using other 2DMs has been employed (Table 1). Ding et al. have synthesized MXene/AuPt/AChE-CS/GCE electrochemical sensor for the real-time detection of chlorpyrifos [42]. The sensor showed an exceptionally linear range of 28.5 pM to 2.85 μM with a limit of detection (LOD) of 4.42×10^{-6} μM. Mi et al. have designed MXenes based heterostructure $CuCo_2S_4/CuS/Ti_3C_2$. By employing acetylcholinesterase as the sensing bioelement, (Fig. 5) the electrode material was utilized for the voltammetric detection of chorphyrifos [43]. The sensor showed high sensitivity with an extremely low limit of detection of 1 pM. In this study of electrochemical sensors, there are other candidates like MCs that show 2D structural characteristics and have been used extensively. Nasir et al. have synthesized a series of TMDs nanosheets (MoS_2, $MoSe_2$, WS_2, WSe_2) using tertiary butyllithium stripping [44]. The heterostructures exhibit good sensing ability towards OP insecticide fenitrothion. Nevertheless, WS_2 demonstrated superior performance compared to other transition metal dichalcogenides (TMDs), exhibiting an impressive detection threshold of 2.86 nM and a linear measurement range spanning 1 to 1000 nM. Another study by Peng et al. involved the creation of layered MoS_2/graphene nanocomposites through a solvothermal process [45]. These nanomaterial-integrated structures served as substrates for electrochemical sensors targeting the detection of hydroquinone. The collaborative influence of these nanomaterials facilitated excellent electrocatalytic efficiency during the electrochemical reaction of hydroquinone. Optimally configured, the MoS2/graphene-modified electrode exhibited heightened sensitivity for hydroquinone detection within the 1.0–9.0 nM range, boasting a detection limit as low as 0.3 nM. A compilation of various electrochemical sensors employed for pesticide sensing based on two-dimensional materials is presented in Table 1 below.

3.1.2 Nitrate and Nitrite Sensing

Electrochemical sensors have been extensively employed in the assessment of food safety and quality, facilitating the identification of pathogenic microorganisms, heavy metal ions, and pesticide residues [48]. The utilization of inorganic fertilizers in agricultural practices has resulted in a conspicuous augmentation of nitrite levels in the surrounding milieu. Within the realm of the food industry, nitrite finds pervasive usage as a means to impede bacterial proliferation and as a coloring agent in meat-based products. Consequently, nitrite represents a prevalent constituent within these particular domains. The prevalence of nitrites in food materials can bring severely harmful impacts; one of these is the conversion of hemoglobin to methemoglobin. To restrict this activity, the detection of nitrites is necessary. Zhang et al.

Table 1 Electrochemical sensors based on 2DMs towards the detection of pesticides

Pesticides	Electrode material	LOD (μM)	Linear range (μM)	Ref
Glyphosate	GC/rGO/CuNPs	19.0×10^{-2}	0.1–1.1	37
Chlorpyrifos	GO@Fe_3O_4	33.0×10^{-3}	$0.1–10^5$	39
Carbofuran-phenol	AuNPs/GO-SPE	22.0×10^{-3}	1–250	40
Methyl parathion	ERGO-CS	79.0×10^{-3}	0.076–0.988	41, 10
Chlorpyrifos	MXene/AuPt/AChE-CS/GCE	1.5×10^{-3}	0.01–1000	42, 11
Chlorpyrifos	$CuCo_2S_4$/CuS/Ti_3C_2	1.0×10^{-6}	2.8×10^{-6}–2.852	43, 12
Fenitrothion	WS_2	2.8×10^{-3}	$1.0 \times 10^{-6}–1 \times 10^{-3}$	44, 13
Norfloxacin	BP-PEDOT:PSS	2.5×10^{-4}	1×10^{-4}–10	46, 15
Norfloxacin	MIP/BPNS-AuNP/GCE	1.2×10^{-5}	1×10^{-4}–10	47, 16

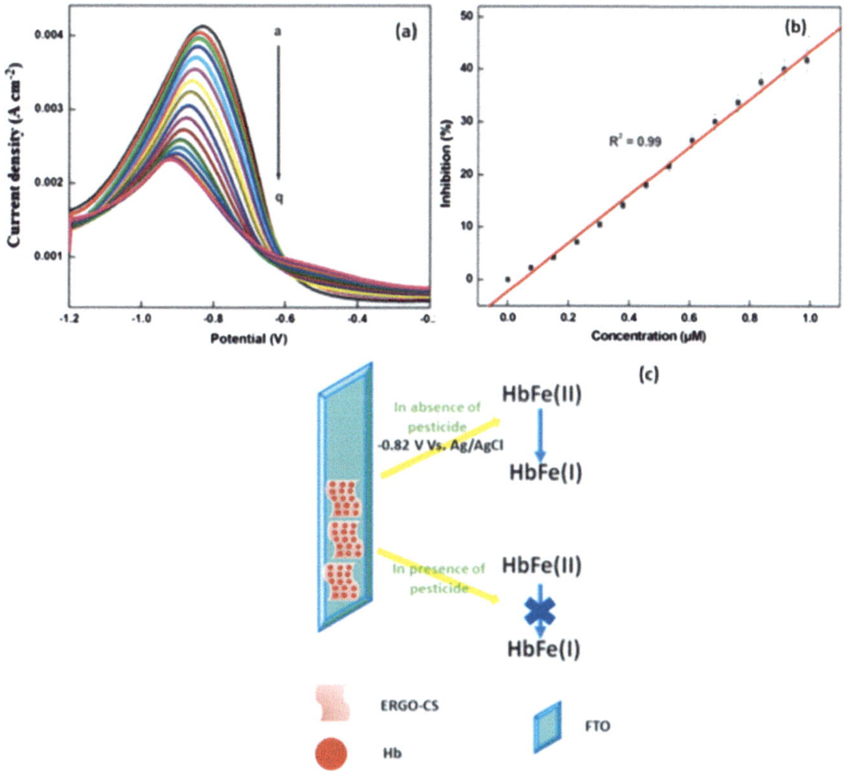

Fig. 5 **a** Square wave voltammogram response of ERGO-CS/Hb/FTO in the presence of MP a to q (0–1.216 μM), **b** calibration plot (linear range of 0.076 μM–0.988 μM) and **c** mechanism in the absence and presence of MP. Reprinted with permission from [41]. Copyright (2020) Elsevier BV

have conducted electrochemical sensing of nitrite ions by using MoS_2 nanosphere coated with AuNPs [49]. Electrochemical studies revealed that the integration of gold nanoparticles (AuNPs) with MoS_2 nanocomposites demonstrated exceptional efficacy in facilitating the oxidation of nitrite. Wang et al. designed an electrochemical sensing electrode consisting of TiO_2–$Ti_3C_2T_X$/CTAB/chitosan composite aimed at detecting nitrite ions [50]. Upon meticulous optimization, the fabricated electrode composed of titanium dioxide (TiO_2) and titanium carbide ($Ti_3C_2T_X$) incorporated with cetyltrimethylammonium bromide (CTAB) and chitosan (CS) exhibited two distinct linear ranges spanning from 0.003 to 0.25 mM and 0.25 to 1.25 mM, with an impressively low detection limit of 0.85 μM. The practical utility of this electrochemical sensor was further substantiated by its successful application in the quantification of nitrite levels in real samples, yielding satisfactory recovery rates.

Zhe et al. designed a nitrite sensing MoO_3/Co_3O_4 nanostructure-based electrochemical senor [51]. For fabricating the sensor hybrid electrochemical approach was followed. The integrated electrode comprising molybdenum trioxide (MoO_3) and cobalt oxide (Co_3O_4) on a carbon cloth substrate demonstrates remarkable electrocatalytic characteristics in the process of nitrite oxidation. This enhanced performance can be attributed to the synergistic effect arising from the heterostructure formed by MoO_3 and Co_3O_4. The electrode exhibits exceptional attributes, including a rapid response time of 2 s, an impressive sensitivity, and an ultralow limit of detection of 0.075 μM. Moreover, the electrode displays promising capabilities for the detection of nitrite in complex food matrices, exemplified by successful measurements in water and sausage samples.

The MoO_3/Co_3O_4 nanohybrid displays a hierarchical heterogeneous morphology on the carbon cloth surface in contrast to Co_3O_4 nanoarrays and MoO_3. The MoO_3/Co_3O_4 heterostructure consists of spherical nanoparticle-dominated nanoclusters. These nanoclusters tightly interconnect to create woven fiber grids, and it is also clear that these fiber grids include dispersed three-dimensional blooming flower-like structures (Fig. 6). Each component of this unique architecture adds exceptional qualities, resulting in an outstanding synergistic effect that improves the overall performance of the nanohybrid (Table 2).

3.2 Sensors for Environmental Applications

During the infancy of the industrialization and urbanization era, there was excessive exploitation of natural resources including fossil fuels, minerals, and water reserves. The waste by-products was dumped either in water bodies or in land reserves. The unregulated waste dumping led to various health extremities which are evident from the past. Minamata River caused many to suffer from neurological disorders as a result of industrial waste disposal [61]. Many reforms were introduced globally to curb the harmful effects of waste dumping in ground and water bodies. Guidelines and standards for permissible pollution levels were set by governments and regulatory organizations and are being renewed in the present also. But it is a matter of highest

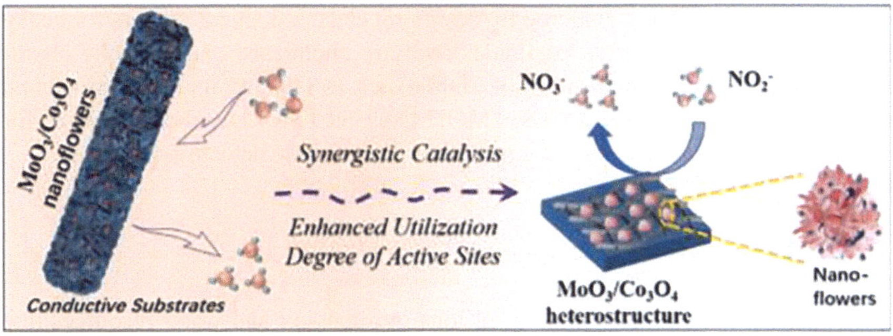

Fig. 6 Scheme illustration of for fabrication of MoO_3/Co_3O_4/CC and its involvement in nitrite sensing. Reprinted with permission from [51]. Copyright (2022) Elsevier B.V

Table 2 Electrochemical sensors based on 2DMs towards the detection of nitrates and nitrites

Analyte	Electrode material	LOD (μM)	Linear range (μM)	References
NO_2^-	Au NPs/MoS_2/GCE	1.67	5.0–27,800	[49]
NO_2^-	TiO_2–$Ti_3C_2T_X$/CTAB/chitosan	0.85	3–1250	[50]
NO_2^-	MoO_3/Co_3O_4/CC	0.075	0.3125–4514	[51]
NO_2^- NO_3^-	Cu/MWCNT/RGO	0.030 0.020	0.1–75 0.1–75	[52]
NO_2^-	Na-CN-300	1.9	10–2000	[53]
NO_2^-	AuNPs/$Ti_3C_2T_x$-PDDA/GCE	0.059	0.1–13,490	[54]
NO_2^-	AuNPs/MXene/ERGO/GCE	0.051	0.5–780	[55]
NO_2^-	Nb_2C@MWCNTs-STAB	0.022	0.1–2000	[56]
NO_2^-	(AuNPs)/$Ti_3C_2T_x$	0.00001	0.00001–5000	[57]
NO_2^-	AuNPs/CS/Ti_3C_2	0.069	0.5–3355	[58]
NO_2^-	Pd-Cu-Mo_2C-modified GCE	0.00035	0.005–0.165	[59]
NO_2^-	NiFe-LDH/Mo_2C	0.0017	0.005–260	[60]

possible concern that to date there has been continuous discharge of unwanted by-products into waterlines and into air in the form of volatile organic compounds and harmful gases such as nitrogen and sulfur oxides which are not only depleting the air quality but harming the life on earth from both biological and ecological perspective [62]. Scientific community has played a major role in suppressing and eliminating excessive exposure to harmful substances by building state-of-the-art technologies that help to control the extensive degradation of environmental resources. In that lieu, researchers have discovered the inexplicable use of 2DMs for environmental applications by manufacturing electrochemical sensors. The potential of old as well as new 2D materials is being explored by designing electrochemical sensors for heavy metal ions, inorganic anions, pesticides, warfare chemicals, and phenolic chemical compounds. Functionalization of 2DMs such as GR and its heterostructures, nanocomposites of MXenes, TMDs, TMO, MNs, and TMCs have been explored for designing electrochemical sensors applicable for real-time detection [63].

3.2.1 Sensing of Heavy Metal Ions

The escalating addition of heavy metal ions into the soil and water resources due to automobiles, industrial discharge, and mining is posing a great risk for both the environment and human well-being. These metals tend to accumulate in the food chain and undergo biomagnification leading to the introduction of these ions into the bloodstreams of living beings. Heavy metal ions can impose really harmful and detrimental effects on human health. Heavy metals, such as lead (Pb^{2+}), arsenic (As^{3+}), cadmium (Cd^{2+}), and mercury (Hg^{2+}) lead to toxicity, neurological disorder, carcinogenicity, organ failure, reproductive inhibitions in addition to other environmental impacts [64]. As a result, it is crucial to create precise and targeted approaches that can evaluate the levels of heavy metal pollution across numerous chemical processes, including biological systems and entire ecosystems. The most common heavy metal ions that are the primary contributors to environmental heavy metal pollution are Hg^{2+}, Cd^{2+}, Pb^{2+}, Cu^{2+}, and As^{3+}. The highly versatile nature of electrochemical sensors motivated many researchers for building electrode materials to monitor heavy metal ion pollution. As they offer portability, fast detection, simultaneous determination [2], and highly sensitive performance, electrochemical sensors have achieved the position of most successful sensing procedures for heavy metal ion determination. The exceptional properties of electrochemical sensors cannot be complemented in a better way than using them in combination with 2D materials. Keeping this in view, the researchers have developed nanostructures-based electrochemical sensors to obtain the highest degree of sensitivity. Mahmoudian et al. have fabricated GO-based voltammetric sensor for the detection of Pb^{2+} metal ions [65]. The in-situ growth of Fe_3O_4 over the reduced graphene oxide was opted to offer higher electrode sensitivity towards Pb^{2+} ion and minimum interference from other heavy metal ions. The sensors presented a response of 0.082 nM as the least possible limit of detection. In another report, Tan et al. presented the use of DNA strands-based aptamer sensors for the selective determination of Hg^{2+} ions [66]. However, the use of linear sweep voltammetry contributed towards the high sensitivity of the sensing electrode.

In a report, Erçarıkcı et al. have fabricated a RGO-based flexible electrochemical sensing membrane by vacuum filtration of the GO/ZnONPs-EDTA suspension followed by hydrothermal treatment. This obtained paper-like material was applied for extraction as well as the detection of Cd^{2+}, Pb^{2+}, Cu^{2+}, and Hg^{2+} metal ion simultaneously [67]. Sun et al. have designed an electrochemical sensor for Cu^{2+} and Pb^{2+} heavy metal ions by exploiting the interactions between MnO_2 and RGO [68]. Under optimized conditions GO/MnO_2/GCE electrode exhibited an excellent electrochemical response towards the analytes offering a linear range of 0.05–1 μM

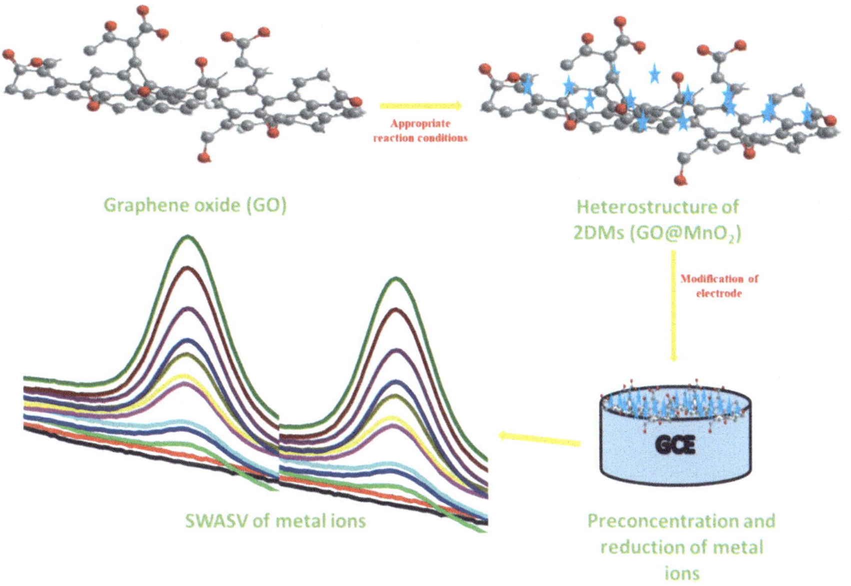

Fig. 7 Schematically represented fabrication of GO and MnO_2-based heterostructures for electrochemical sensing of Cu^{2+} and Pb^{2+}

and a detection limit of 1.67×10^{-3} μM and 3.33×10^{-3} μM for Cu^{2+} and Pb^{2+}, respectively, using square wave anodic stripping voltammetry (SWASV) (Fig. 7). The general mechanism for SWASV of various metal ions can be predicted from Fig. 5, which involves two steps, i.e., preconcentration and reduction of metal ions followed by stripping and oxidation of metal ions.

A polysulfide/graphene (RGO-S)-based electrochemical sensor has been designed for the successful real-time detection of Cd^{2+} and Pb^{2+} [69]. The stripping peak current of RGO-S/PGE varies linearly with heavy metal ion concentration under ideal circumstances, with a range of 17.79 nM–2.66 μM for Cd^{2+} and 4.82 nM–1.44 μM for Pb^{2+}. The estimated limits of detection for Cd^{2+} and Pb^{2+} are 5.96 $\times 10^{-4}$ μM and 82.04 μM, respectively. It is a promising choice for monitoring Cd^{2+} and Pb^{2+} concentrations in aqueous environmental samples because the developed electrochemical heavy-metal-detecting electrode has high sensitivity and good repeatability. A comparison of the limit of detection and linear range of the various types of 2D materials has been done in Table 3.

Zuo et al. devised the synthesis of poly(3,4-ethylenedioxide)nanorods/GR-coated glassy carbon electrode for Hg^{2+} ion sensing [76]. Surface morphological analysis revealed the anchoring of nanorods-like structure over GO sheets resulting in an enhanced number of electrochemically active surface sites. The low detection limit of 2.78 nM and excellent linear range of 10.0 nM-3.0 μM were obtained for the sensors along with real-time applicability for Hg^{2+} analysis in tap water. Ding et al. have reported the synthesis AuNPs decorated GO for the determination of Hg^{2+} using

Table 3 Electrochemical sensors based on 2DMs towards detection of heavy metal ions

Analyte	Electrode material	Detection limit	Range	References
Pb^{2+}	Fe_3O_4/3%RGO/GCE	0.000082 μM	0.00005–0.0015 μM	[65]
Hg^{2+}	DNA aptamer	0.0005 μM	0.005–1.020 μM	[66]
Cd^{2+}, Pb^{2+}, Cu^{2+} Hg^{2+}	GO/ZnO-NPs-EDTA	5.6 μM 6.8 μM 2.5 μM 10 μM	18.5–500 μM 22.4–700 μM 8.3–200 μM 3.3–300 μM	[67]
Cu^{2+} Pb^{2+}	GO/MnO_2/GCE	0.00167 μM 0.0033 μM	0.05–1 μM	[68]
Pb^{2+} Cd^{2+}	BiNPs@Ti_3C_2T	0.01 μM 0.012 μM	60–600 μM 0.08–0.6 μM	[70]
Cd^{2+} Pb^{2+}	Ti_3C_2@N–C/GCE	0.1–0.4 μM 0.05–2 μM	0.0025 μM 1.10 μM	[71]
Cd^{2+} Pb^{2+}	NH_2-$Ti_3C_2T_x$/SPCE	10–500 μM 5 - 500 μM	0.41 μM 0.31 μM	[72]
Cu^{2+} Pb^{2+} Cd^{2+} Hg^{2+}	Co-TMC4R-BDC/GCE	0.05–12.0 μM 0.05–13.0 μM 0.1–17.0 μM 0.75–18.0 μM	0.013 μM 0.011 μM 0.026 μM 0.018 μM	[73]
Pb^{2+}	AuNPs/Ti_3C_2-MXenes	0.5–300 μM	0.3 μM	[74]
Pb^{2+}	Bi/Bi_2O_3@C	0.037–2.0 μM	0.0063 μM	[75]

stripping voltammetry [77]. The large surface area of GO and high affinity between AuNPs and Hg^{2+} resulted in a remarkably sensitive sensor for mercury ion detection in tap water with the least time required for pretreatment. The detection limit turned out to be 0.6 nM within a linear range of 1–150 nM. A GO and zinc metal–organic framework-based exceptionally selective electrochemical sensor has been designed. The high surface area and substantially stable dispersion of GO and explicitly great adsorption capacity of zinc MOF for As^{3+} acted as motivation for designing the nanocomposite [78]. The sensor is equipped with a low detection limit of 0.46 nM within a wide linear range of 1.54 nM to 0.19 μM.

3.2.2 Organic Pollutants Detection

Environmental pollutants that have a major negative effect on natural ecosystems and long-lasting ecological degradation include organic pollutants like bisphenol-A and hydrazine [79]. These compounds have a prolonged environmental persistence, which disturbs the sensitive balance of the ecosystem. Human exposure to these contaminants can happen either topically or by eating tainted food. Long-term exposure to these substances has been associated with a variety of health issues, including neurological diseases, reproductive problems, and respiratory problems [80]. (4,4′-(1-phenylethylamine) bisphenol) commonly known as Bisphenol-A is a persistent organic pollutant that is responsible for adversative effects in human beings because of its use in medicinal, agricultural, and synthetic chemical applications. A large number of electrochemical sensors have emerged based on 2DMs for its detection.

In a recent report, Li et al. proposed a $NiCoMnO_4$ mixed oxide and rGO-based electrochemical sensor for bisphenol-A detection in the industrial wastewater [81]. The microstructural analysis of the nanocomposite revealed that the metallic balls of $NiCoMnO_4$ were coated with rGO thus carrying a larger surface area which resulted in the enhanced electrocatalytic performance of the designed sensor. Linear sweep voltammetry was employed for bisphenol-A detection exhibiting a linear working range of 0.005 μM–7 μM and a detection limit of 2 nM. Rajendran et al. built a 2D hybrid graphene/$Ti_3C_2T_x$ (Gr/MXene) based nanocomposite for the electrochemical detection of bisphenol-A [82]. The sensor was able to convey successful detection of bisphenol-A using DPA and amperometric analysis and exhibited low limit of detection of the order of 4.08 nM and 0.35 μM within two linear ranges of 10 to 180 nM and 1 to 10 μM respectively. The reliability of the sensor was proved by the determination of bisphenol-A in plastic samples. Hydrazine (HY) offers a broad range of applications in the fields of agriculture, pharmaceutical industry, pesticides industry, and for catalysis. Despite these broad ranges of applications, HY offers negative effects on the environment and hence needs detection. Wei et al. designed a hydrazine sensor by fabricating 2D WO_3 over an Au-SiO_2/Si substrate [83]. Upon studying the effect of film thickness on the sensing ability, a major finding was obtained that showed the sensor expressed the highest sensitivity with monolayered 2D WO_3 with a film thickness of 0.78 nm. The linear range for hydrazine detection was observed to be 0.2 to 2100 μM with a detection limit of 0.015 μM. In another report, Durai and Badhulika have presented a voltammetic sensor based on nanospherical niobium pentoxide (Nb_2O_5) anchored on reduced graphene oxide for the simultaneous detection of bisphenol-A and hydrazine [84]. The sensor was able to detect both the analyte in a wide linear range of 10 μM to 300 μM and 10 nM to 50 μM for bisphenol-A and hydrazine, respectively. The sensor has shown successful detection of bisphenol-A up to a detection limit of 0.295 μM and was able to determine hydrazine with a detection limit of 2.709 nM. The real-time analysis of tap water and drinking water showed the reliability of the designed sensor. The synergistic effect of rGO, which provides excellent electrical conductivity, and Nb_2O_5, which provides high electrostatic adsorption of the analyte via the Nb^{4+}/Nb^{5+} redox couple favoring electrooxidation process at the surface of the electrode, can be credited for the excellent performance of the rGO-Nb_2O_5 /GCE sensor. There are various other electrochemical sensors based on 2DMs for electrochemical sensing of BPA and hydrazine listed in Table 4.

3.3 Electrochemical Sensors for Health Monitoring

Apart from food- and environment-based sensors, there is a growing demand for effective and efficient methods of health monitoring. The use of electrochemical

Table 4 Electrochemical sensors based on 2DMs towards detection of BPA and hydrazine

Analyte	Electrode material	LOD (μM)	Linear range (μM)	References
Bisphenol-A	$NiCoMnO_4$@rGO	2.0×10^{-3}	5.0×10^{-3}–7	[81]
Bisphenol-A	Graphene/$Ti_3C_2T_x$	4.1×10^{-3} 35.0×10^{-2}	0.01–0.180 1–10	[82]
Bisphenol-A	Pd-MoS_2-MXene-SPCE	1.20×10^{-4}	0.005–0.175	[85]
Bisphenol-A	V_2O_5@$Ti_3C_2T_x$	0.411–31.2 μM	0.087 μM	[86]
Hydrazine	WO_3	15.0×10^{-3}	$0.2–2.1 \times 10^{+3}$	[83]
Bisphenol-A Hydrazine	rGO-Nb_2O_5/GCE	29.5×10^{-3} 2.7×10^{-3}	10–300 0.01–50	[84]
Hydrazine	FL-Cr_2CT_x	66.0×10^{-2}	1.99–10	[87]
Hydrazine	MoS_2	19.2×10^{-2}	0.1–1	[88]
Hydrazine	Nb_2C/MXene	67.0×10^{-0}	0–13,750	[89]
Hydrazine	RGO/DHB/$Co(OH)_2$	16.5×10^{-2}	5–1700	[90]

sensors has provided numerous advantages to address the key challenges in this field. The electrochemical sensors are able to monitor various kinds of body fluids, saliva, sweat, blood samples, etc. Thus, they enable the use of sensors for non-invasive real-time monitoring of health parameters. Keeping this in mind, different types of electrochemical sensors have been put forward for monitoring glucose, lactate, ascorbic acid, cholesterol, H_2O_2, etc [91].

Gopal et al. used MXenes (Ti_3C_2Tx), a type of 2D TMCs, to create a nanocomposite with Cu_2O using a wet precipitation technique. The nanocomposite was then used as an electrode material to detect glucose using the chronoamperometry technique. The electroanalytical studies of MXene-Cu_2O nanocomposite for glucose show a broad linear range with a very low detection level of 2.83 mM, demonstrating the superior electrochemical properties of 2DTMCs-based nanocomposite. As a result, designed nanocomposite is a potential alternative for glucose detection in clinical diagnosis [92] (Fig. 8)

In a study conducted by Li et al., a nanoporous gold (NPG) needles were employed for directing vertical growth of 2D NiCo bimetal-organic framework nanosheets (NiCo-MOFNs) for highly sensitive non-enzymatic glucose sensing. The synergetic effect among N, Ni, and Co contributed to an enhanced catalytic activity, displaying a low detection limit of 0.29 μM in the linear range of 1 μM to 8 mM. The influence of the Ni to Co ratio on the catalytic activity towards glucose oxidation was also analyzed, concluding a 1:1 ratio of Ni/Co in 2D NiCo-MOFNs dismayed preeminent electrochemical performance with Ni and Co dominating the active sites for glucose determination. The enhanced sensitivity of the NiCo-MOFNs-based material for glucose detection over other previously developed NiCo-based sensors was assigned to the presence of more accessible active sites, thinner 2D NiCo-MOFNs devoid of surfactant, enhanced electrochemical catalytic oxidation of glucose, strong interactions among electrons of N, Ni and Co in 2D NiCo-MOFNs with metal atoms, vertically allied nanosheets endowing the material with excellent surface area and

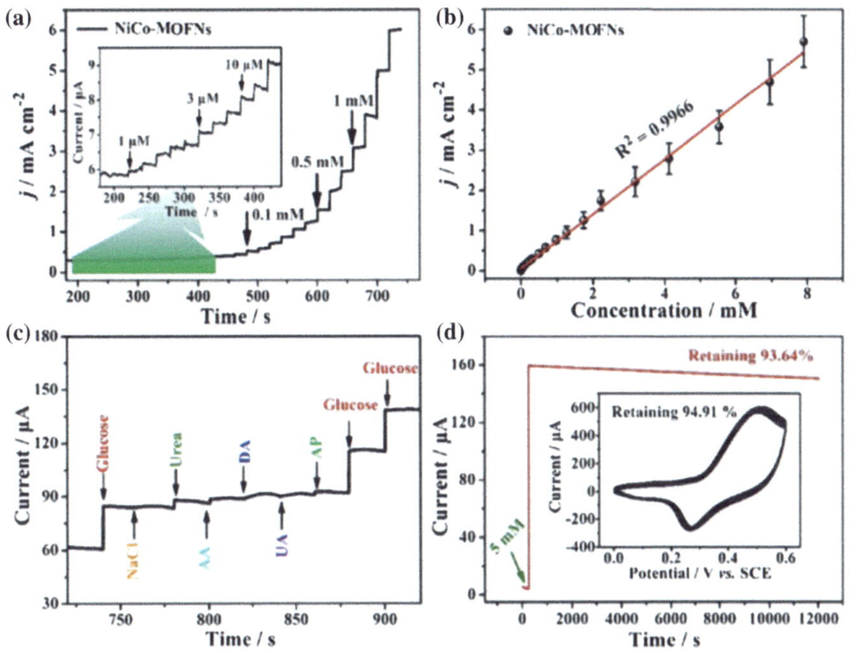

Fig. 8 **a** Amperometric titration of NiCo-MOFNs on successive addition of glucose, **b** Calibration curve showing linear response for amperometric titration of NiCo-MOFNs, **c** Amperometric responses after addition of interferents, **d** stability of electrode. Reprinted with permission from [93]. Copyright (2018) Elsevier B.V.

enhanced mass and charge conveyance for redox reactions. In addition, the practical applicability of the developed sensor for glucose detection in blood serum samples indicates its potential as an efficient healthcare management technology [93].

By taking advantage of the excellent electrochemical activity of MXenes, Li et al. developed a redox sensor using a hydrothermal method by growing NiCo-LDHs nanosheets over Ti_3C_2-based MXene. Due to the synergic interaction between highly porous NiCo-LDH and the high conductivity of MXene, the designed nanocomposite provided very high sensitivity when used as a glucose detector. Using the superior redox assets of the designed nanocomposite, the authors achieved a very low detection limit and a wide linear range towards the detection of glucose [94]. A composite of nickel-based metal organic framework and carbon nanotubes (Ni-MOF/CNTs) was contrived by Zhang et al. via in-situ synthesis of Ni-MOF-containing CNTs solution for efficient glucose detection. The developed Ni-MOF nanostructures offer porosity, increases the charge transfer rate, facilitate the transport of glucose towards the active sites through pores and channels thereby accelerating the diffusion velocity of the glucose molecules towards the electrode. The prepared sensor displayed a detection limit of 0.82 μM in the linear range of 1 μM to 1.6 mM. Moreover, the remarkable efficiency of Ni-MOF/CNTs in the detection of glucose in serum samples indicated its potential application in glucose detection with excellent reproducibility and stability [95].

In another case, MXene-based nanocomposite was used in healthcare monitoring. Li et al. used Cu-Cu_2O NPs on Ti_3C_2Tx-MXene for the detection of hydrogen peroxide [96]. Ali et al. designed a redox-responsive sensor for the detection of H_2O_2 using graphene nanocomposite by green method (electrochemical deposition) using cobalt oxide nanoparticles (CoO_xNPs) on the surface of electrochemically reduced grapheme oxide deposited on glassy carbon electrode (GCE). The electrochemical detection of H_2O_2 conducted at a pH of 7 in a wide linear range showed a low limit of detection of 0.2 M. Both graphene and CoO_xNPs displayed electrocatalytic activity towards H_2O_2 in addition to showing synergistic effects. Also, the presence of ERGO provided a well-organized electrical web for holding highly distributed CoO_xNPS. The sensor also showed practical applicability in determination of H_2O_2 in hydrogen peroxide samples [97].

Acute myocardial infarction is the reason behind the higher rate of clinical death. This can be avoided if myocardial infarction is detected and treated at an early stage. Cardiac troponin I (cTnI) is particularly very sensitive to this type of injury due to its unique structure. In this regard, Li et al. developed a novel redox sensor consisting of a ternary composite electrode made of Ag nanoparticles/MoS_2/reduced graphene oxide (AgNPs/MoS_2/rGO). The proposed EC sensor has very high sensitivity and a low detection limit over a wide linear range of cTnl from 0.3 pg/ml to 0.2 ng/ml due to the synergistic effect among the constituents of the ternary composite [98]. Annadurai et al. contrived a nickel oxide based electrochemical sensor for the detection of 4-acetaminophen in paracetamol (PA) tablets and blood serum samples. 4-Acetaminophen plays an important role in the pharmaceutical field where it has been widely used to treat fever, pain, migraine, headache, etc. However, its excessive use causes kidney and liver damage and can even cause death or coma. Hence to monitor the amount of PA in medical formulations for quality control as well as public health management, development of sensitive sensors for the detection of PA pharmaceutical preparations is of great importance and rigorous work has been done for the same. The prepared NiO-based sensor demonstrated enhanced current response for PA detection owing to its larger electroactive surface area, facilitating effective electron transfer. The sensor displayed a low detection limit of 0.23 μM in a linear range of 7.5–3000 μM while demonstrating excellent efficiency in PA detection in pharmaceutical formulations and clinical diagnosis [99].

The presence of excess cholesterol in the body is associated with many human health complications, such as diabetes, hypertension, coronary heart disease, cardiovascular and cerebrovascular diseases, claiming numerous lives. Thus, the development of efficient cholesterol detection instruments becomes necessary. For this purpose, an electrochemical sensor for highly sensitive cholesterol detection in biological samples was developed by Yan et al. using cubic-shaped Cu_2O as the nanomaterial, Thionine (TH) as an electron mediator, and chitosan (CS) as a crosslinking agent for the immobilization of cholesterol oxidase. Nanomaterials Cu_2O (Nano-Cu_2O) has emerged as a new p-type oxide semiconductor with active electron–hole pairs, large specific surface area, quantum size effect, and volume effect while the tendency of TH to increase the current response density along with chitosan make them highly effective towards cholesterol detection via a reversible diffusion controlled electrochemical process, providing ultrahigh sensitivity with a low detection limit of 0.0018 μM in the linear concentration range of 10–1000 μM. Additionally, the biosensor was applicable in cholesterol detection in human serum samples and food cholesterol quality control [100].

Lactic acid is produced in the skeletal muscles, kidney, brain, and red blood cells, and plays a role in the generation of glycogen. As high concentrations of lactic acid can lead to severe acidosis linked to 80% mortality, the detection and determination of lactic acid for differential diagnosis and management of medical conditions like ischemia, intestinal infection, cardiac arrest become an issue of great importance. Therefore, in the search for effective electrochemical sensors for lactic acid detection, a novel enzyme-free gold (Au)/nickel oxide (NiO@Au) with flower-shaped nanocomposite was prepared. The material possessed highly porous structured NiO flowers providing an excessive surface area while mimicking the enzymatic tendency of lactose dehydrogenase and lactose oxidase for lactose oxidation into pyruvic acid in an alkaline solution. The sensor demonstrated a low detection limit of 11.6 μM in the wide linear range of 100.0 μM–0.5 M. The improved sensing capability of the sensor is accredited to the synergistic effects of Au and NiO, and the excellent electrocatalytic properties of Au amplifying the inherent properties of NiO, displaying potential in biomedical applications [101].

Li et al. developed an electrochemical sensor for the detection of ascorbic acid (AA) by grafting metal–organic frameworks (ZIF-65) onto carboxylated carbon nanotubes. The oxidizing and reducing natures of ZIF-65 and CNT resulted in a synergy that doubled the conductivity of the composite electrode. Furthermore, with detection limits as low as 1.03 nM, the designed ZIF-65@CNT modified glassy carbon electrode sensor has excellent sensing performance towards AA [102]. Similarly, Liu et al. developed a redox sensor for CEA detection using Ag-MOFs containing AuNPs to take advantage of the excellent conductivity and stability offered by gold nanoparticles (AuNPs) and silver nanoparticles (Ag-NPs). When tested to detect CEA using differential pulse voltammetry (DPV), the proposed EC sensor demonstrated an improved detection limit (8.0 fg/ml) having a wide linear range (0.06–120 ng/ml) [103].

3.4 *Electrochemical Sensors for Biomolecular Detection*

Biomolecules, which include carbohydrates, proteins, fats, nucleic acids, and enzymes, are the building blocks of the body and are found in all living systems. They play an important role in the daily activities of all living organisms. As a result, maintaining optimal levels of these biomolecules is necessary to have a longer lifespan and to avoid various biological disorders such as diabetes and heart attacks. This gives rise to the need for the development of electrochemical sensors for their periodical detection in living beings [104].

In this regard, electrochemical sensors are used as the most powerful tools for detecting biomolecules over other traditional methods due to their superior sensitivity, exceptional selectivity, and portability. The exceptional behavior of redox-responsive sensors is due to the use of tailorable electrodes. The electrode material is critical in the detection of various analytes because it can gain or release electrons under controlled potential.

"Levodopa" (L-dopa) was used for the treatment of Parkinson's disease, but it had to be monitored closely to avoid side effects due to overuse. Xiao et al. created a wearable redox sensor by growing a zeolite imidazole-based metal–organic framework/graphene oxide. The ZIF@GO sensor had offered a good detection limit (0.45 μM) and a broad linear range (1–95 μM). Furthermore, the high sensitivity, selectivity, and continuous drug monitoring of the designed sensor made it a prominent device for use in real-time applications [105].

The electrochemical activity, biocompatibility, and stability of noble metals like Pt, Au, and Ag nanoparticles are exceptionally high. With this view, Li et al. developed an electrochemical sensor that capitalized on the synergistic effect of cobalt-MOF and copper–gold nanowires (CuAuNWs). The addition of Au-NPs improved the catalytic activity and stability of the final composite. The designed Co-MOF/CuAuNWs produce an excellent limit of detection of 33×10^{-6} μM when used as a sensor to detect nuclear matrix protein-22 (NMP-22) [106]. Ankhava et al. created a rare electrochemical ultralevel sensor for the detection of leukaemia cells based on graphene nanotemplates. In this study, GO nanotemplates were used to design the electrode for extra guanine withdrawal fabricated in leukaemia cells because of their high conductivity and sheet-like morphology with very sharp edges. The developed reduce graphene oxide nanowalls (rGONW) based electrochemical sensor demonstrated a very low limit of detection (10^{-11} M) for leukaemia detection using the DPV technique, which is 5 times lower than the detection limit of bare GCE. This demonstrates the remarkable sensitivity of rGONW towards leukaemia cells [107].

Saeed et al. created a novel electrochemical sensor for the early-stage detection of breast cancer by combining the conducting properties of noble metal nanoparticles and GO. In this regard, the authors created two different DNA-modified gold nanoparticles (ERBB2c and CD24c) and GO loaded on glassy carbon electrodes. The developed electrodes were utilized for the detection of the breast cancer biomarker ERBB2 and the control marker CD24 with extremely low limits of detection (0.16 nM and 0.23 nM, respectively) [108].

Flavonoids are polyphenolic compounds that help in the reduction of cardiovascular diseases and cancer amongst which baicalin (Bn) is an active molecule in biology and offers versatile physiological activities such as anti-inflammatory-anti-oxidant, anti-cancer, anti-microbial, etc. Owing to its medical and pharmacological importance and the adverse effects of its excessive use have led researchers to search for means to monitor its levels in biological systems. For this purpose, Zhao et al. developed a nanosensor for baicalin using g-C_3N_4-SWCNTs/reduced graphene oxide/cyclodextrin-based metal–organic framework. The designed electrochemical sensor dismayed a low detection limit of 40.46 nM in a wide detection range of

1.0 nM–0.5 μM. The sensor C_3N_4-SWCNTs/rGO/CD-MOF/GCE displayed excellent sensitivity towards Bn detection because of its enhanced electrocatalytic activity. The sensor also had good practical applicability owing to its efficient performance in Bn detection in medical samples [109]. Luteolin which is an example of flavonoid found in abundance in various vegetable or other real samples and shows evident anti-inflamatory, anti-bacterial, and anti- oxidant effects in addition to displaying anti-proliferative effects against cancer cells. Thus, investigating the redox processes and electrocatalytic activity of lutecolin is of great importance and for this purpose sensors for lutecolin detection are being developed one of which involves the sensor developed by Pang et al. using graphene and hydroxyapatite nanocomposite modified electrode displaying a detection limit of 10 nM in the linear range of 20 nM–10 μM [110]. DNA analysis has played a crucial role in the diagnosis of many diseases and genetic mutations and thus highly sensitive and selective DNA detection methods are being developed to monitor DNA sequential abnormalities. Wang et al. designed a highly sensitive redox-responsive DNA biosensor using gold nanoparticle-decorated sheets of reduced graphene oxide. The thiolated capture DNA was covalently linked to the Au NPs and in the presence of target DNA, its 3' terminus interacts with the capture probe and the 5' terminus hybridizes with the signal probe labeled with methylene blue (MB). With an increase in the concentration of target DNA, hybridization between signal probe and target DNA occurs. The prepared sensor displayed a low detection limit of 0.35 μM in the wide detection range of 0.1 μM–1fM. The signal intensities of MB were observed to increase with an increase in the target DNA concentration and the sensitivity of the sensor was attributed to enhancing the surface area provided by Au-NPs [111] (Table 5).

4 Conclusion and Future Perspectives

Utilization of 2DMs in electrochemical sensing has shown tremendous potential for the detection of a range of analytes. The extraordinary properties for instance large surface areas, admirable electrical conductivity, flexibility due to ultra-thin nature, tunability in surface functionalization contributed towards enhanced sensitivity, response time, and selectivity of electrochemical sensors. The synthesis of 2DMs can be accomplished through top-down and bottom-up approaches, each having its own advantages and disadvantages. These 2DMs can be used to detect various food, health, and environmental samples. In all, this chapter provides an outlook on electrochemical sensors based on 2DMs for various sensing applications.

Future outlook: The utilization of 2DMs for electrochemical sensing applications holds unlimited potential for futuristic advancements. The research should focus on the potential of 2DMs for electrochemical applications and explore the following aspects:

- 2DMs and their heterostructures: The authors should focus on developing novel composites of 2DMs and their heterostructures to attain unique properties and

Table 5 Electrochemical sensors based on 2DMs towards detection of health monitoring molecules and biomarkers

Analyte	Electrode material	LOD (μM)	References
Electrochemical sensors for health monitoring			
Glucose	Ti_3C_2Tx-MXene/Cu_2O	2830	[92]
Glucose	NiCo-MOFNs	0.29	[93]
Glucose	Ti_3C_2-MXene/ NiCo-LDH	–	[94]
Glucose	Ni-MOF/CNT	0.82	[95]
H_2O_2	Cu-Cu_2O NPs on Ti_3C_2Tx-MXene	–	[96]
H_2O_2	CoOxNPs/rGO	0.20	[97]
cTn-I	Ag-NPs/MoS2/rGO	0.2×10^{-3}	[98]
4-Acetaminophen	NiO	10^{-2}	[99]
Cholesterol	Nano-Cu_2O	0.0018	[100]
Lactic acid	NiO@AuO	11.60	[101]
Ascorbic acid	ZIF-65@CNT/GCE	1.03×10^{-3}	[102]
CEA	Ag-MOF/Au-NPs	8.0×10^{-6}	[103]
Electrochemical sensors for biomarkers			
Levodopa	ZIF@GO	0.45	[105]
NMP-22	Co-MOF/Cu-NWs	33×10^{-6}	[106]
Leukemia	rGONWs	0.1×10^{-3}	[107]
Breast cancer	rGO/Au-NPs	0.16×10^{-3} (ERB2) 0.23×10^{-3} (CD24)	[108]
Baicalin	C_3N_4-SWCNTs	40.46	[109]
Luteolin	Graph/hydroxyapatite	–	[110]
DNA	AuNPs/rGO	0.35	[111]
Cancer biomarkers	Zn-MOF-Zr-MOF	840×10^{-6}	[112]
Morphine	GO–Co_3O_4	0.54×10^{-3}	[113]
Tramadol	GR–AgNPs-MIP	2.04×10^{-3}	[114]
Cancer biomarkers	Tb-MOF-on-Fe-MOF conjugate	58.0	[115]
Codeine and acetaminophen	GR–$CoFe_2O_4$	25×10^{-3} 11×10^{-3}	[116]
Cocaine	GR–PdNPs-MIP	50.0	[117]
Pholcodine	GR–$(PHL)^{2+}$-(NB)	40.0	[118]
Ketamine and Norketamine	GR–MOF-MIM	0.04×10^{-3} μM	[119]

harness improved electrochemical sensing performance. The development of heterostructures can enhance electrocatalytic performance, good selectivity, and excellent charge transfer kinetics

- Engineering the interface of materials: Optimizing the interfacial properties of 2DMs and their heterostructure will help to control the performance of electrochemical sensors.
- Maneuvering the surface of 2DMs: The surface of 2DMs can be maneuvered by attaching various functional groups which can be utilized to attach other heterostructures to enhance their electrocatalytic activity, stability, and sensitivity towards particular analyte. Dopants, nanomaterials, or composites can be made for good sensing attributes.
- Integrate with other emerging fields: The integration of electrochemical sensing attributes with other fields like electronics, photonics will result in the development of sensing devices with improved performance and miniaturization.

In addition to this, scientists should focus on bridging the gap between academic research and commercialization and thus extending their use for practical applications. By addressing such issues, the field of electrochemical sensing can be further advanced for the development of cutting-edge sensors.

References

1. Feng, L.: Colorimetric sensor array for determination and identification of toxic industrial chemicals. Anal. Chem. **82**, 9433–9440 (2010)
2. Ap, V.S.: Colorimetric sensors for rapid detection of various analytes. Mater. Sci. Eng. C **78**, 1231–1245 (2017)
3. Liu, L., Zhou, Y., Liu, S., Xu, M.: The applications of metal− organic frameworks in electrochemical sensors. ChemElectroChem **5**, 6–19 (2018)
4. Beitollahi, H., Movahedifar, F., Tajik, S., Jahani, S.: A review on the effects of introducing CNTs in the modification process of electrochemical sensors. Electroanalysis **31**, 1195–1203 (2019)
5. Liu, C.-S., Li, J., Pang, H.: Metal-organic framework-based materials as an emerging platform for advanced electrochemical sensing. Coord. Chem. Rev. **410**, 213222 (2020)
6. Lu, S.-M., Peng, Y.-Y., Ying, Y.-L., Long, Y.-T.: Electrochemical sensing at a confined space. Anal. Chem. **92**, 5621–5644 (2020)
7. Joshi, N.: Yolk-shelled $ZnCo_2O_4$ microspheres: Surface properties and gas sensing application. Sens. Actuators B Chem. **257**, 906–915 (2018)
8. Su, S.: Facile synthesis of a MoS_2–Prussian blue nanocube nanohybrid-based electrochemical sensing platform for hydrogen peroxide and carcinoembryonic antigen detection. ACS Appl. Mater. Interfaces **9**, 12773–12781 (2017)
9. Yazyev, O.V., Chen, Y.P.: Polycrystalline graphene and other two-dimensional materials. Nat. Nanotechnol. **9**, 755–767 (2014)
10. Bae, S.H.: 2019 Integration of bulk materials with two-dimensional materials for physical coupling and applications. Nat. Mater. **18**, 550–560 (2019)
11. Aftab, S.: Bulk photovoltaic effect in 2D materials for solar-power harvesting. Adv. Opt. Mater. **10**, 2201288 (2022)
12. Singh, S.: MechElastic: a python library for analysis of mechanical and elastic properties of bulk and 2D materials. Comput. Phys. Commun. **267**, 108068 (2021)

13. Witomska, S., Leydecker, T., Ciesielski, A., Samorì, P.: Production and patterning of liquid phase–exfoliated 2D sheets for applications in optoelectronics. Adv. Funct. Mater. **29**, 1901126 (2019)
14. Yang, Y.: Electrochemical exfoliation of graphene-like two-dimensional nanomaterials. Nanoscale **11**, 16–33 (2019)
15. Akinwande, D.: A review on mechanics and mechanical properties of 2D materials—Graphene and beyond. Extreme Mech. Lett. **13**, 42–77 (2017)
16. Munteanu, R.-E., Moreno, P.S., Bramini, M., Gáspár, S.: 2D materials in electrochemical sensors for in vitro or in vivo use. Anal. Bioanal. Chem. **413**, 701–725 (2021)
17. García-Miranda, F.A., Rowley-Neale, S.J., Banks, C.E.: Recent advances in 2D hexagonal boron nitride (2D-hBN) applied as the basis of electrochemical sensing platforms. Anal. Bioanal. Chem. **413**, 663–672 (2021)
18. Tajik, S.: Transition metal dichalcogenides: synthesis and use in the development of electrochemical sensors and biosensors. Biosens. Bioelectron. **216**, 114674 (2022)
19. Agnihotri, A.S., Varghese, A.M.N.: Transition metal oxides in electrochemical and bio sensing: a state-of-art review. Appl. Surf. Sci. Adv. **4**, 100072 (2021)
20. Shafiee, S.A.: Recent advances on metal nitride materials as emerging electrochemical sensors: a mini review. Electrochem. Commun. **120**, 106828 (2020)
21. Li, Q.: Recent advances in black phosphorus-based electrochemical sensors: a review. Anal. Chim. Acta **1170**, 338480 (2021)
22. Murugan, R., Hwa, K.Y., Santhan, A.: 2D-Graphitic carbon nitride nanosheet/metal nanocomposites for electrochemical sensors of hydroquinone in real sample analysis. ACS Appl. Nano Mater. **6**, 8550–8563 (2023)
23. Zhang, W.: Progress and opportunities for metal–organic framework composites in electrochemical sensors. RSC Adv. **13**, 10800–10817 (2023)
24. Khan, A.F., Brownson, D.A.C., Foster, C.W., Smith, G.C., Banks, C.E.: Surfactant exfoliated 2D hexagonal boron nitride (2D-hBN) explored as a potential electrochemical sensor for dopamine: surfactants significantly influence sensor capabilities. Analyst **142**, 1756–1764 (2017)
25. Huo, C., Yan, Z., Song, X., Zeng, H.: 2D materials via liquid exfoliation: a review on fabrication and applications. Sci. Bull. **60**, 1994–2008 (2015)
26. Zeng, Z.: An effective method for the fabrication of few-layer-thick inorganic nanosheets. Angew. Chem. Int. Ed. **51**, 9052–9056 (2012)
27. Wu, C.: Two-dimensional vanadyl phosphate ultrathin nanosheets for high energy density and flexible pseudocapacitors. Nat. Commun. **4**, 2431 (2013)
28. Zeng, Z.: Single-layer semiconducting nanosheets: high-yield preparation and device fabrication. Angew. Chem. **123**, 11289–11293 (2011)
29. The Rise of Two-Dimensional Materials: Acc. Chem. Res. **48**, 1–2 (2015)
30. Jiang, H., Zheng, L., Liu, Z., Wang, X.: Two-dimensional materials: from mechanical properties to flexible mechanical sensors. InfoMat **2**, 1077–1094 (2020)
31. Jiang, X., Nisar, J., Pathak, B., Zhao, J., Ahuja, R.: Graphene oxide as a chemically tunable 2-D material for visible-light photocatalyst applications. J. Catal. **299**, 204–209 (2013)
32. Munteanu, R.E., Moreno, P.S., Bramini, M., Gaspar, S.: 2D materials in electrochemical sensors for in vitro or in vivo use. Anal. Bioanal. Chem. **413**, 701–725 (2021)
33. Shang, T.L.D., Gao, S., Wang, B., Kong, H., Yang, G., Shu, W., Xu, P., Wei, G.: Two-dimensional material-based electrochemical sensors/biosensors for food safety and biomolecular detection. Biosensors **12**, 314(1–32) (2022)
34. Baranwal, J., Brajesh, B., Gatto, G., Broncova, G., Kumar, A.: Electrochemical sensors and their applications: a review. Chemosensors **10**, 363(1–22) (2022)
35. Sabarwal, A., Kumar, K., Singh, R.P.: Hazardous effects of chemical pesticides on human health–cancer and other associated disorders. Environ. Toxicol. Pharmacol. **63**, 103–114 (2018)
36. Yan, L., Yan, X., Li, H., Zhang, X., Wang, M., Fu, S., Zhang, G., Qian, C., Yang, H., Han, J.: Reduced graphene oxide nanosheets and gold nanoparticles covalently linked to

ferrocene-terminated dendrimer to construct electrochemical sensor with dual signal amplification strategy for ultra-sensitive detection of pesticide in vegetable. Microchem. J. **157**, 105016 (2020)
37. Setznagl, S., Cesarino, I.: Copper nanoparticles and reduced graphene oxide modified a glassy carbon electrode for the determination of glyphosate in water samples. Int. J. Environ. Anal. Chem. **102**, 293–305 (2022)
38. Facure, M.H., Mercante, L.A., Mattoso, L.H., Correa, D.S.: Detection of trace levels of organophosphate pesticides using an electronic tongue based on graphene hybrid nanocomposites. Talanta **167**, 59–66 (2017)
39. Jiao, Y.C., Hou, W.J., Fu, J.Y., Guo, Y.M., Xia, S., Wang, X.Y., Jing, Z.: A nanostructured electrochemical aptasensor for highly sensitive detection of chlorpyrifos. Sens. Actuators B Chem. **243**, 1164–1170 (2017)
40. Jirasirichote, A., Punrat, E., Suea-Ngam, A., Chailapakul, O., Chuanuwatanakul, S.: Voltammetric detection of carbofuran determination using screen-printed carbon electrodes modified with gold nanoparticles and graphene oxide. Talanta **175**, 331–337 (2017)
41. Kaur, R., Rana, S., Lalit, K., Singh, P., Kaur, K.: Electrochemical detection of methyl parathion via a novel biosensor tailored on highly biocompatible electrochemically reduced graphene oxide-chitosan-hemoglobin coatings. Biosens. Bioelectron. **167**, 112486 (2020)
42. Ding, R., Jiang, W., Ma, Y., Yang, Q., Han, X., Hou, X.: A highly sensitive MXene/AuPt/AChE-based electrochemical platform for the detection of chlorpyrifos. Microchem. J. **187**, 108425 (2023)
43. Mi, Y., Zhao, Y., Chen, J., Li, X., Yang, Y., Gao, F.: Ternary heterostructures of 1D/2D/2D $CuCo_2S_4/CuS/Ti_3C_2$ MXene: boosted amperometric sensing for chlorpyrifos. J. Hazard. Mater. **438**, 129419 (2022)
44. Nasir, M.Z.M., Mayorga-Martinez, C.C., Sofer, Z., Pumera, M.: Two-dimensional 1T-phase transition metal dichalcogenides as nanocarriers to enhance and stabilize enzyme activity for electrochemical pesticide detection. ACS Nano **11**, 5774–5784 (2017)
45. Peng, Y., Tang, Z.R., Dong, Y.P., Che, G., Xin, Z.F.: Electrochemical detection of hydroquinone based on MoS_2/reduced graphene oxide nanocomposites. J. Electroanal. Chem. **816**, 38–44 (2018)
46. Li, G., Wu, J., Qi, X., Wan, X., Liu, Y., Chen, Y., Xu, L.: Molecularly imprinted polypyrrole film-coated poly(3,4-ethylenedioxythiophene):polystyrene sulfonate-functionalized black phosphorene for the selective and robust detection of norfloxacin. Mater Today Chem. **26**, 101043 (2022)
47. Li, G., Qi, X., Wu, J., Xu, L., Wan, X., Liu, Y., Chen, Y., Li, Q.: Ultrasensitive, label-free voltammetric determination of norfloxacin based on molecularly imprinted polymers and Au nanoparticle-functionalized black phosphorus nanosheet nanocomposite. J. Hazard. Mater. **436**, 129107 (2022)
48. Su, S., Chen, S., Fan, C.: Recent advances in two-dimensional nanomaterials-based electrochemical sensors for environmental analysis. Green Energy Environ. **3**, 97–106 (2018)
49. Zhang, S., Tang, Y.P., Chen, Y.Y., Zheng, J.B.: Synthesis of gold nanoparticles coated on flower-like MoS_2 microsphere and their application for electrochemical nitrite sensing. J. Electroanal. Chem. **839**, 195–201 (2019)
50. Wang, X., Li, M., Yang, S., Shan, J.: A novel electrochemical sensor based on $TiO_2–Ti_3C_2T_X$/CTAB/chitosan composite for the detection of nitrite. Electrochim. Acta **359**, 136938 (2020)
51. Zhe, T., Li, M., Li, F., Li, R., Bai, F., Bu, T., Jia, P., Wang, L.: Integrating electrochemical sensor based on MoO_3/Co_3O_4 heterostructure for highly sensitive sensing of nitrite in sausages and water. Food Chem. **367**, 130666 (2022)
52. Bagheri, H., Hajian, A., Rezaei, M., Shirzadmehr, A.: Composite of Cu metal nanoparticles-multiwall carbon nanotubes-reduced graphene oxide as a novel and high performance platform of the electrochemical sensor for simultaneous determination of nitrite and nitrate. J. Hazard. Mater. **324**, 762–772 (2017)
53. Jigyasa, P., Rajput, J.K.: Alkali metal (Na/K) doped graphitic carbon nitride (g-C 3 N 4) for highly selective and sensitive electrochemical sensing of nitrite in water and food samples. J. Electroanal. Chem. **878**, 114605 (2020)

54. Wang, Y., Zeng, Z., Qiao, J., Dong, S., Liang, Q., Shao, S.: Ultrasensitive determination of nitrite based on electrochemical platform of AuNPs deposited on PDDA-modified MXene nanosheets. Talanta **221**, 121605 (2021)
55. Wang, T., Wang, C., Xu, X., Li, Z., Li, D.: One-step electrodeposition synthesized AuNPs/MXene/ERGO for selectivity nitrite sensing. Nanomaterials **11**, 1892 (2021)
56. Chen, Y., Waterhouse, L.N., Qiao, X., Sun, Y., Xu, Z.: Sensitive analytical detection of nitrite using an electrochemical sensor with STAB-functionalized Nb_2C@MWCNTs for signal amplification. Food Chem. **372**, 131356 (2022)
57. Cai, Z., Bo, T., Zhou, R., Xia, D., He, H., Gao, N., Chang, G., He, Y.: Ultrathin $Ti_3C_2T_x$ MXene-based electrochemical transistor for highly sensitive determination of nitrite. J. Electroanal. Chem. **928**, 117012 (2023)
58. Wang, T., Xu, X., Wang, C., Li, Z., Li, D.: A novel highly sensitive electrochemical nitrite sensor based on a AuNPs/CS/Ti_3C_2 nanocomposite. Nanomaterials **12**, 397 (2022)
59. Vilian, A.T.E., Umapathi, R., Hwang, S.-K., Huh, Y.S.: Pd–Cu nanospheres supported on Mo_2C for the electrochemical sensing of nitrites. J. Hazard. Mater. **408**, 124914 (2021)
60. Kokulnathan, T., Wang, T.-J., Ahmed, F., Kumar, S.: Deep eutectic solvents-assisted synthesis of NiFe-LDH/Mo 2 C nanocomposite for electrochemical determination of nitrite. J. Mol. Liq. **369**, 120785 (2023)
61. Harda, M.: Minamata Disease: methylmercury poisoning in Japan caused by environmental pollution. Crit. Rev. Toxicol. **25**(1), 1–24 (1995)
62. Manisalidis, I., Stavropoulou, E., Stavropoulos, A., Bezirtzoglou, E.: Environmental and health impacts of air pollution: a review. Front. Public Health **8**, 14 (2020)
63. Khan, M.A., Ramzan, F., Ali, M., Zubair, M., Mehmood, M.Q., Massoud, Y.: Emerging two-dimensional materials-based electrochemical sensors for human health and environment applications. Nanomaterials **13**, 780 (2023)
64. Tyagi, D., Wang, H., Huang, W., Hu, L., Tang, Y., Guo, Z., Ouyang, Z., Zhang, H.: Recent advances in two-dimensional-material-based sensing technology toward health and environmental monitoring applications. Nanoscale **12**, 3535–3559 (2020)
65. Mahmoudian, M.R., Alias, Y., Basirun, W.J., Woi, P.M., Sookhakian, M., Jamali-Sheini, F.: Synthesis and characterization of Fe_3O_4 rose like and spherical/reduced graphene oxide nanosheet composites for lead (II) sensor. Electrochim. Acta **169**, 126–133 (2015)
66. Tan, F., Cong, L.C., Saucedo, N.M., Gao, J.S., Li, X.N., Mulchandani, A.: An electrochemically reduced graphene oxide chemiresistive sensor for sensitive detection of Hg^{2+} ion in water samples. J. Hazard. Mater. **320**, 226–233 (2016)
67. Erçarıkcı, E., Alanyalıoğlu, M.: Dual-functional graphene-based flexible material for membrane filtration and electrochemical sensing of heavy metal ions. IEEE Sens. J. **21**, 2468–2475 (2021)
68. Sun, H.Y., Wang, C.X., Xu, Y.J., Dai, D.M., Deng, X.Y., Gao, H.T.: A novel electrochemical sensor based on a glassy carbon electrode modified with GO/MnO_2 for simultaneous determination of trace Cu(II) and Pb(II) in environmental water. ChemistrySelect **4**, 11862–11871 (2019)
69. Jiang, R.Y., Liu, N.T., Su, Y.H., Gao, S.S., Mamat, X., Wagberg, T., Li, Y.T., Hu, X., Hu, G.Z.: Polysulfide/graphene nanocomposite film for simultaneous electrochemical determination of cadmium and lead ions. NANO **13**, 1850090 (2018)
70. He, Y., Ma, L., Zhou, L., Liu, G., Jiang, Y., Gao, J.: Preparation and application of bismuth/MXene nano-composite as electrochemical sensor for heavy metal ions detection. Nanomaterials **10**, 866 (2020)
71. Zang, X., An, D., Bi, Z., Shan, W., Zhu, B., Zhou, L., Yu, L., Zhang, H., Xia, S., Qiu, M.: Ti_3C_2-MXene@N-doped carbon heterostructure-based electrochemical sensor for simultaneous detection of heavy metals. J. Electroanal. Chem. **911**, 116239 (2022)
72. Chen, Y., Zhao, P., Hu, Z., Liang, Y., Han, H., Yang, M., Luo, X., Hou, C., Huo, D.: Amino-functionalized multilayer Ti_3C_2Tx enabled electrochemical sensor for simultaneous determination of Cd^{2+} and Pb^{2+} in food samples. Food Chem. **402**, 134269 (2023)

73. Wang, F.-F., Liu, C., Yang, J., Xu, H.-L., Pei, W.-Y., Ma, J.-F.: A sulfur-containing capsule-based metal-organic electrochemical sensor for super-sensitive capture and detection of multiple heavy-metal ions. J. Chem. Eng. **438**, 135639 (2022)
74. Zhang, Z., Karimi-Maleh, H.: Label-free electrochemical aptasensor based on gold nanoparticles/titanium carbide MXene for lead detection with its reduction peak as index signal. Adv. Compos. Hybrid Mater. **6**, 68 (2023)
75. Wang, C., Niu, Q., Liu, D., Dong, X., You, T.: Electrochemical sensor based on Bi/Bi_2O_3 doped porous carbon composite derived from Bi-MOFs for Pb^{2+} sensitive detection. Talanta **258**, 124281 (2023)
76. Zuo, Y., Xua, J., Zhua, X., Duana, X., Lub, L., Gaoa, Y., Xinga, H., Yanga, T., Yea, G., Yub, Y.: Poly(3,4-ethylenedioxythiophene) nanorods/graphene oxide nanocomposite as a new electrode material for the selective electrochemical detection of mercury (II). Syth Met. **220**, 14–19 (2016)
77. Ding, L., Liu, Y., Zhai, J., Bond, A.M., Zhang, J.: Direct electrodeposition of graphene-gold nanocomposite films for ultrasensitive voltammetric determination of mercury(II). Electroanalysis **26**, 112–128 (2014)
78. Baghayeri, M., Ghanei-Motlagh, M., Tayebee, R., Fayazi, M., Narenji, F.: Application of graphene/zinc-based metal-organic framework nanocomposite for electrochemical sensing of As(III) in water resources. Anal. Chim. Acta **1099**, 60–67 (2020)
79. Benjamin, S.R., Ribeiro Júnior, E.J.M.: Graphene-Based electrochemical sensors for detection of environmental pollutants. Curr. Opin. Environ. Sci. Health. **29**, 100381 (2022)
80. Bhunia, S.K., Jana, N.R.: Reduced graphene oxide-silver nanoparticle composite as visible light photocatalyst for degradation of colorless endocrine disruptors. ACS Appl. Mater. Interfaces **6**, 20085–20092 (2014)
81. Li, H., Guo, S., Wang, L.: Voltammetric sensor detection of bisphenol AP in industrial wastewater. Int. J. Environ. Anal. Chem. **101**, 1220–1230 (2021)
82. Rajendran, J., Kannan, T.S., Dhanasekaran, L.S., Murugan, P., Atchudan, R., ALOthman ZA, Ouladsmane M, Sundramoorthy AK.: Preparation of 2D Graphene/MXene nanocomposite for the electrochemical determination of hazardous bisphenol A in plastic products. Chemosphere **287**(2), 132106 (2022)
83. Wei, Z., Hai, Z., Akbari, M.K., Hu, P.J., Hyde, L., Depuydt, P.S., Verpoort, P.F., Zhuiykov, P.S.: Ultrasensitive, sustainable, and selective electrochemical hydrazine detection by ALD-developed two-dimensional WO_3. ChemElectroChem **5**, 266–272 (2017)
84. Durai, L., Badhulika, S.: One-Pot synthesis of rGO supported Nb_2O_5 nanospheres for ultra-selective sensing of bisphenol A and hydrazine in water samples. IEEE Sens. J. **21**(4), 4152–4159 (2021)
85. Vilian, A.T.E., Tiwari, J.N., Alhammadi, M., Bhaskaran, G., Hwang, S.-S., Kim, S., Kumar, K., Kumar, A.S., Hug, Y.S., Han, Y.-K.: Palladium nanoparticles anchored MoS_2-MXene composite modified electrode for rapid sensing of toxic bisphenol A in aqueous media. J. Chem. Eng. **469**, 144017 (2023)
86. Thukkaram, M.P., Chakravorty, A., Mini, A.A., Ramesh, K., Grace, A.N., Raghavan, V.: Titanium carbide MXene and V_2O_5 composite-based electrochemical sensor for detection of bisphenol A. Microchem. J. **193**, 109004 (2023)
87. Soundiraraju, B., Anthony, A.M., Pandurangan, P., George, B.K.: Electrochemical behaviour of chromium carbide MXene modified electrodes: hydrazine sensing. Mater. Today Commun. **32**, 103982 (2022)
88. Rana, D.S., Thakur, N., Thakur, S., Singh, D.: Electrochemical determination of hydrazine by using MoS2 nanostructure modified gold electrode. Nanofabrication. **7**, 79–86 (2022)
89. Gul, S., Serna, M.I., Zahra, S.A., Arif, N., Iqbal, M., Akinwande, D., Rizwan, S.: Un-doped and Er-adsorbed layered Nb_2C MXene for efficient hydrazine sensing application. Surf Interfaces. **24**, 101074 (2021)
90. Mohiuddin, A.K., Ahmed, M.S., Roy, N., Jeon, S.: Electrochemical determination of hydrazine in surface water on $Co(OH)_2$ nanoparticles immobilized on functionalized graphene interface. Appl. Surf. Sci. **540**(2), 148346 (2021)

91. Yan, A., Yan, F.: Flexible electrochemical biosensors for health monitoring. ACS Appl. Electron. Mater. **1**, 53–67 (2021)
92. Gopal, T.S., Jeong, S.K., Alrebdi, T.A., Pandiaraj, S., Alodhayb, A., Muthuramamoorthy, M., Grace, A.N.: MXene based composite electrdes for efficient electrochemical sensing of glucose by non-enzymatic method. Mater. Today Chem. **24**, 100891 (2022)
93. Li, W., Lv, S., Wang, Y., Zhang, L., Cui, X.: Nanoporous gold induced vertically standing 2D NiCo bimetal-organic framework nanosheets for non-enzymatic glucose biosensing. Sens. Actuators B:Chem. **281**, 652–658 (2018)
94. Li, M., Fang, L., Zhou, H., Wu, F., Lu, Y., Luo, H., Zhang, YHu.B.: Three-dimensional porous MXene/NiCo-LDH composite for high performance non-enzymatic glucose sensor. Appl. Surf. Sci. **495**, 143554 (2019)
95. Zhang, X., Xu, Y., Ye, B.: An efficient electrochemical glucose sensor based on porous nickel based metal organic framework/carbon nanotubes composite (Ni-MOF/CNTs). J. Alloys Compd. **767**, 651–656 (2018)
96. Khan, M.A, Ramzan, F., Ali, M., Zabair, M., Mehmood, M.Q., Massoud, Y.: Emerging two-dimensional material-based electrochemical sensors for human health and environment applications. J. Nanomater. **780** (2023)
97. Li, S., Du, J., Zhang, J., Zhang, M., Chen, J.: A glassy carbon electrode modified with a film composed of cobalt oxide nanoparticles and graphene for electrochemical sensing of H_2O_2 ddb. Microchim. Acta **181**, 631–638 (2014)
98. Li, J., Zhang, S., Zhang, L., Zhang, Y., Zhang, H., Zhang, C., Xuan, X., Wang, M., Zhang, J., Yuan, Y.: A novel graphene based nanomaterial modified electrochemical sensor for the detection of cardiac troponin I. Front. Chem. **9**, 680593 (2021)
99. Annadurai, K., Sudha, V., Murugadoss, G., Thangamuthu, R.: Electrochemical sensor based on hydrothermally prepared nickel oxide for the determination of 4-acetaminophen in paracetamol tablets and human blood serum samples. J. Alloys Compd. **852**, 186911 (2021)
100. Yan, Q., Wu, R., Chen, H., Wang, H., Nan, W.: Highly sensitive cholesterol biosensor based on electron mediator thionine and cubic-shaped Cu_2O nanomaterials. Microchem. J. **185**, 108201 (2022)
101. Maduraiveeran, G., Chen, A.: Design of an enzyme-mimicking NiO@Au nanocomposite for the sensitive electrochemical detection of lactic acid in human serum and urine. Electrochim. Acta **368**, 137612 (2020)
102. Li, Y., Ye, W., Cui, Y., Li, B., Yang, Y., Quan, G.: A metal-organic framework carbon nanotubes based electrochemical sensor for highly sensitive and selective determination of ascorbic acid. Int. J. Mol. Sci. **209**, 127986 (2020)
103. Liu, Y., Shang, Y., Zhu, Q., Zhang, X., Zheng, J.: Solamargine inhibits proliferation and promotes apoptosis of CM-319 human chordoma cells through suppression of notch pathway. Microchim. Acta **186**, 509 (2019)
104. Tsai, T.-H., Thiagarajan, S., Chen, S.-M.: Detection of melamine in milk powder and human urine. J. Agric. Food Chem. **58**, 4537 (2010)
105. Xiao, J., Fan, C., Xu, T., Su, L., Zhang, X.: An electrochemical wearable sensor for levodopa quantification in sweat based on a metal-organic framework/graphene oxide composite with integrated enzymes. Sens. Actuators B:Chem. **359**, 131586 (2022)
106. Li, S., Yue, S., Yu, C., Chen, Y., Yuan, D., Yu, Q.: A label-free immunosensor for the detection of nuclear matrix protein-22 based on a chrysanthemum-like Co-MOFs/CuAu NWs nanocomposite. Analyst **144**, 649–655 (2019)
107. Akhavan, O., Ghaderi, E., Hashemi, E., Rahighi, R.: Ultrasensitive detection of leukemia by graphene. Nanoscale **47**, 777–780 (2014)
108. Saeed, A.A., Sánchez, J.L.A., O'sullivan, C.K., Abbas, M.N.: DNA biosensors based on gold nanoparticles-modified graphene oxide for the detection of breast cancer biomarkers for early diagnosis. Biochem. J. **118**, 91–99 (2017)
109. Zhao, P., Huang, L., Wang, H., Wang, C., Chen, J., Yang, P., Ni, M., Chen, C., Li, C., Xie, Y., Fei, J.: An ultrasensitive high-performance baicalin sensor based on C_3N_4-SWCNTs/reduced graphene oxide/cyclodextrin metal-organic framework nanocomposite. Sens. Actuators B:Chem. **350**, 30853 (2021)

110. Pang, P., Liu, Y., Zhang, Y., Gao, Y., Hui, Q.: Electrochemical determination of luteolin in peanut hulls using graphene and hydroxyapatite nanocomposite modified electrode. Sens. Actuators B Chem. **194**, 397–403 (2017)
111. Wang, J., Shi, A., Fang, X., Han, X., Zhang, Y.: Ultrasensitive electrochemical supersandwich DNA biosensor using a glassy carbon electrode modified with gold particle-decorated sheets of graphene *oxide*. Microchim. Acta **181**, 935–940 (2014)
112. Zhou, N., Su, F., Guo, C., He, L., Jia, Z., Wang, M., Jia, Q., Zhang, Z., Lu, S.: Two-dimensional oriented growth of Zn-MOF-on-Zr-MOF architecture: a highly sensitive and selective platform for detecting cancer markers. Biosens. Bioelectron. **123**, 51–58 (2019)
113. Atta, N.F., Galal, A., El-Ads, E.H., Hassan, S.H.: Cobalt oxide nanoparticles/graphene/ionic liquid crystal modified carbon paste electrochemical sensor for ultra-sensitive determination of a narcotic drug. Adv. Pharm. Bull. **9**, 110 (2019)
114. Bagheri, H., Shirzadmehr, A., Rezaei, M., Khoshsafar, H.: Determination of tramadol in pharmaceutical products and biological samples using a new nanocomposite carbon paste sensor based on decorated nano graphene/tramadol-imprinted polymer nanoparticles/ionic liquid. Ionics **24**, 833–843 (2018)
115. Wang, M., Hu, M., Li, Z., He, L., Song, Y., Jia, Q., Zhang, Z., Du, M.: Construction of Tb-MOF-on-Fe-MOF conjugate as a novel platform for ultrasensitive detection of carbohydrate antigen 125 and living cancer cells. Biosens. Bioelectron. **142**, 111536 (2019)
116. Afkhami, A., Khoshsafar, H., Bagheri, H., Madrakian, T.: Facile simultaneous electrochemical determination of codeine and acetaminophen in pharmaceutical samples and biological fluids by graphene-$CoFe_2O_4$ nancomposite modified carbon paste electrode. Sens. Actuators B Chem. **203**, 909–918 (2014)
117. Florea, A., Cowen, T., Piletsky, S., De Wael, K.: Electrochemical sensing of cocaine in real samples based on electrodeposited biomimetic affinity ligands. Analyst **44**, 4639–4646 (2019)
118. Abd-Rabboh, H.S., Amr, A.E.-G.E., Elsayed, E.A., Sayed, A.Y., Kamel, A.H.: Paper-based potentiometric sensing devices modified with chemically reduced graphene oxide (CRGO) for trace level determination of pholcodine (opiate derivative drug). RSC Adv. **11**, 12227–12234 (2021)
119. Fu, K., Zhang, R., He, J., Bai, H., Zhang, G.: Sensitive detection of ketamine with an electrochemical sensor based on UV-induced polymerized molecularly imprinted membranes at graphene and MOFs modified electrode. Biosens. Bioelectron. **143**, 111636 (2019)

2D Materials for Gas Sensing Application

Nainsi Sharma, Raeesah Islam, Harish Mudila, and Anil Kumar

Abstract The need for reliable radiation detection technology in environmental monitoring, industrial safety, and health-related research has led to advancements in gas sensing technology. Two-dimensional materials, such as graphene oxide and MoS_2, have been explored for gas detection due to their minimal atomic thickness and high surface-to-volume ratio. The renewal of MoS_2 is attributed to its historical application as a lubricant. Sulfides such as WS_2 have found utility in gas-sensing technologies and nano-electronic devices. Single atomic species materials like phosphorene have been employed in a variety of applications including industrial design control, health inspection, environmental monitoring, agriculture, and safety, particularly in the development of gas sensors. These sensors have shown impressive capabilities in detecting various gases, including toxic pollutants, volatile organic compounds, flammable gases, and biological gases. This chapter provides a comprehensive overview of gas sensors utilizing 2D materials, with a focus on recent advancements in graphene oxide, exfoliated MoS_2, WS_2, and Phosphorene for gas detection applications.

Keywords Graphene · Molybdenum disulphide · Tungsten disulphide · Gas sensors

1 Introduction

The use of gas sensors in both indoor and outdoor environments, as well as in human exhalation, has become increasingly important for applications such as environmental monitoring, medical diagnosis, and industrial safety. Various materials like conducting polymers, carbon nanotubes, and metal oxides have been utilized for gas detection, with metal oxides demonstrating superior characteristics in terms of response and stability. Despite their advantages, metal oxide-based gas sensors often lack selectivity and sensitivity at low gas concentrations. Similarly, conducting

N. Sharma · R. Islam · H. Mudila · A. Kumar (✉)
Department of Chemistry, Lovely Professional University, Phagwara, Punjab, India
e-mail: rsanil.nit@gmail.com

A. Katoch et al. (eds.), *2D Materials*, Engineering Materials,
https://doi.org/10.1007/978-981-97-6258-3_6

polymer-based sensors are effective at room temperature but are sensitive to changes in humidity and prone to degradation with prolonged exposure to air. As a result, there is a growing emphasis on researching new materials that can detect gases at room temperature under optimal environmental conditions while exhibiting high selectivity. The emergence of two-dimensional (2D) materials has garnered significant attention in gas sensor research since the discovery of graphene. These materials, including transition metal dichalcogenides [1], black phosphorus, MXenes, and layered metal oxides [2], offer unique physical and chemical characteristics that make them promising candidates for gas sensing applications [3, 4]. Some of the recent work of gas sensing materials has been shown in Table 1. Efforts are being made to enhance the sensitivity, recovery rate, stability, and detection limit of 2D material-based gas sensors to meet practical requirements. Various types of gas sensors, such as resistive-type, optical, electrochemical, and field effect transistor (FET)-type sensors, have been developed, each with its advantages and limitations. Among these, chemiresistive-type sensors, which utilize aggregated nanosheets to form a conductive channel, are widely used for their simplicity and sensitivity. FET-type gas sensors, on the other hand, rely on gas adsorption to modulate their electrical characteristics for targeted gas detection.

The current chapter focuses on the recent developments in 2D materials-based gas sensors, with specific emphasis on graphene oxide, MoS_2, WS_2, and phosphorene. These materials play crucial roles as sensitive elements for gas detection, and this chapter provides a comprehensive summary of the latest research findings in this field.

Table 1 Different gases sensed by 2D Materials

Sr No	Material	Method	Gas sensed	Year	Reference
1	rGO/SnO_2	Spraying process	NO_2	2022	[5]
2	Polypyrrole	Electrochemical polymerization, Chemical oxidative polymerization, plasma polymerization	NO_2, NH_3, CO, H_2, CH_4	2022	[6]
3	In_2O_3	Hydrothermal method	CO	2021	[7]
4	ZnO	Chemiresistive sensors	O_2, CO, CO_2	2021	[8]
5	ZnO	Hydrothermal method	H_2S	2020	[9]
6	WS_2 and MoS_2	Thermal sulfurization	NH_3	2019	[10]

2 Two-Dimensional Materials for Gas Sensing

2D Materials are only one layer thick and sometimes have a thickness of nanometres. The vast number of reactive sites in 2D materials is based on their large surface area. Therefore, they are ideal for sensing, catalytic, and energy storage applications. Also, 2D materials contain large surface areas, high electronic conducting ability, and low noise. Thus, they become key players in prospective gas detectors. The small change in carrier concentration caused by gas may bring a huge difference in 2D material conductivity with low electric noise but high electric conductivity [4]. The charge transfer, characteristic of adsorption of the gas molecules upon interaction in the 2D material, as well as features of these materials makes them one of the most suitable candidates for 2D material sensors design. Scientists have discovered the qualities of 2D materials to comprehend their properties. As a result, there is a need to look at the benefits of exploring and applying two-dimensional and three-dimensional materials in research areas such as gas sensing materials as well as graphene.

2.1 *Graphene*

Graphene is considered an extremely promising sensing material for its superior electronic characteristics and mechanical attributes comprising sp^2 bonded carbon's only monolayer in two-dimensional space [11]. This material has a notable feature in that each carbon atom has surface area orientation, and therefore it is sensitive to adsorption molecules in two dimensions. Both mechanical and even chemical synthesis of it is possible. First of all, mechanical exfoliation was commonly used until recently when people started using chemical vapor deposition and the epitaxial growth has been utilized in the manufacturing process [9]. Nevertheless, the required high temperatures or ultra-high vacuum conditions might hinder its practicality as well as the zero or quasi-zero bandgap of graphene [4]. However, this could be a challenge if used as a sensitive layer on devices. However, many writers suggest incorporating or decorating it with graphene oxide (GO) that was synthesized by Hummer's method [12]. GO is viewed as a non-conductive material and hence, an inappropriate gas sensor. To solve this problem, low-cost reduced graphene oxide and its derivatives as shown in Fig. 1 have been generated that have conductivity and utilize various reduction methods such as chemical reduction and thermal reduction oxides of graphite. Even though the interlayer spacing increases compared with that in graphite, the layered structure is maintained. It then results in single-layer GO flakes as graphite oxide is dispersed in aqueous basic solution and exfoliated at the bulk surface [13]. Thus, these are highly soluble in water. There exist different routes for the synthesis of graphene. Another method used is chemical vapor deposition (CVD). Graphene is produced when the hydrocarbon vapes are broken down at the contact point between a metal plate and a stream of gas (in CVD). This means that

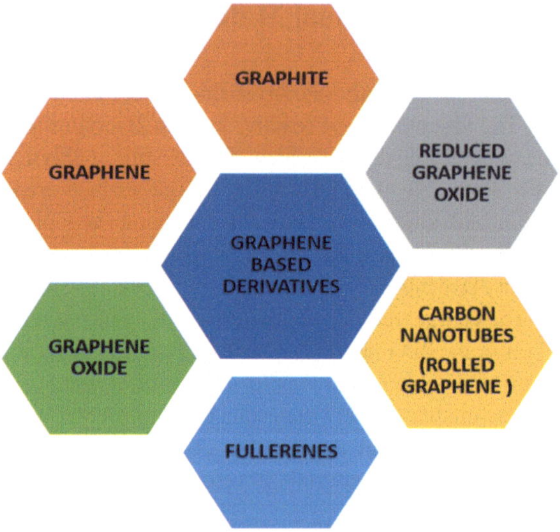

Fig. 1 Graphene-based derivatives

one can produce an adequate amount of high-quality graphene across a large area; however, this method requires precise management of process parameters. Another way of making graphene involves thermolysis of silicon carbide [11, 12]. The sublimation of silicon molecules from the surface of silicon carbide and the disappearance of carbon give rise to the graphene lattice.

This straightforward and reasonable process often produces less graphene than that obtained when using mechanical processes such as strip peeling of graphite grains into single-layer graphene. A third technique used in developing graphene is the chemical and thermal reduction of graphene oxide. Hence, the flexibility and the capability to produce graphene with various functional groups give rise to the creation of selective sensors for its construction. In this process, the oxygen-containing functional groups in graphene oxide are destroyed by chemical or thermal treatment to produce reduced graphene oxide that behaves similarly to pure graphene [11]. Finally, graphene's outstanding properties render it ideal in these areas of gas-sensing applications [14].

2.2 *Transitional Metal Dichalcogenides*

Because of the unique dielectric features, the two-dimensional material TMDs (such as MoS_2, WS_2, and Black Phosphorus) are among the many materials that have been studied extensively within the gas sensing circle. In Fig. 2, the application of TMDs in various domains such as electronics, optoelectronics, sensing, and energy storage [15]. Typically used for gas sensing applications, TMDs mostly utilize WS_2-based gas sensors which are very sensitive in their physisorption mechanism in response to

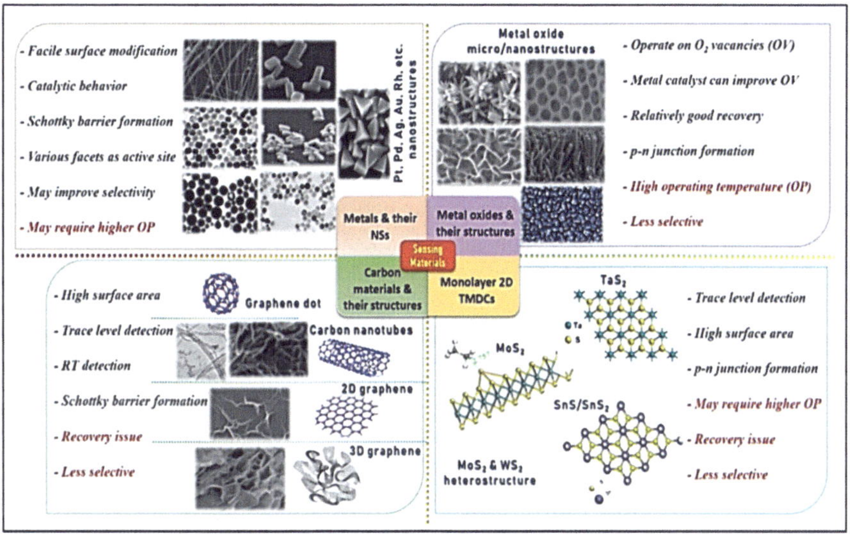

Fig. 2 Micro/nanostructured sensing materials, their advantages and limitations. Reprinted with permission from U. Yaqoob et al. [4]. Copyright 2021 MDPI

NH_3 sensing [16]. They have a large surface area; this leads to several binding sites or external moieties making them highly sensitive. There are many ways to adjust the bandgap of the TDM and make it match the composite one. Tunable characteristics of TMDs make them even more sensitive and specific regarding their Schottky defect formation [16]. Thus, TMDs are ideal materials with desirable properties required for the development of future gas sensor technology. These sensors exhibit high room temperature sensitivity, are wide-band gas detectors, and have very good selectivity for those gases.

2.2.1 Tungsten Disulphide (WS_2)

WS_2 is being increasingly recognized as a promising nanomaterial for use in chemiresistive gas sensors due to its responsiveness to hydrogen gas [16]. Its capability to detect ammonia molecules through physical adsorption is particularly remarkable, since it creates a high density of active sites on the WS_2 surface for the adsorption of external particles, benefiting from its high surface area to volume ratio [17]. Researchers have developed both single and multilayer gas sensors based on WS_2 and have tested their reactivity to various gases including NO_2, CO, NH_3, and CH_4. Their research suggests that the multilayer WS_2 sensor outperforms the single-layer WS_2 sensor in detecting odor at low temperatures [18]. The hydrogen sensing mechanism of nanostructured WS_2 layers has been studied using both near-ambient pressure X-ray photoelectron spectroscopy (NAP-XPS) and density functional theory (DFT). These investigations have shown that hydrogen is physically adsorbed on the WS_2

active surface at room temperature while chemical adsorption occurs at temperatures above 150 °C [19].

2.2.2 Molybdenum Disulphide (MoS_2)

Recent research continues to improve the performance of MoS_2 by integrating different quantum dots (QDs) such as MoO_3, ZnO, and SnO. These efforts aim at providing multiple active sites in the interior of MoS_2 including sulfur sites, defects, and edge regions will be used to improve the gas sensing capability. For example, studies have shown that QD-enhanced MoS_2 sensing in gas sensors has a fast response time and recovery after ambient heating [20]. The improved efficiency is attributed to the increased surface area and the improved efficiency of the quantum dots providing the interface between the sensor and the fuel molecules. Furthermore, the integration of the quantum dots reveals additional MoS_2 interactions that improve its sensitivity to gas molecules. The combination of MoS_2 and quantum dots gives good results in the detection of various gases, leading to surprising results in obtaining low limits [21]. For example, the limit of detection of nitrogen dioxide (NO_2) is calculated as small as 94 parts per billion (ppb) using sensitive MoS_2 air sensors.

2.2.3 Black Phosphorus

The two-dimensional compound Black Phosphorus (BP), also known as Phosphorene, has received considerable interest in radiation sensing [18, 19]. The unique biochemical properties of phosphorene make it a good candidate for sensing applications [19] and the reason for the interest in this compound is characterized by high phosphorene surfaces and concentrations one of the most important [22, 23]. Its sensitivity as a gas sensor is increased due to the greater interaction with gas molecules enabled by this property. Moreover, each phosphorene atom has only one electron pair [24]. This property makes it easier for gas molecules to adhere to the phosphorene surface, increasing the sensing properties of the material [24, 25]. Studies have shown that phosphorene has the potential to be a gas sensor in comparison to Other 2D materials such as graphene or MoS_2 for CO_x, and NO_x [25]. Small molecules such as NH_3 have been found to have high absorbance This suggests that phosphorene-based sensors may have a low detection limit and high selectivity [24]. In conclusion, phosphorene is a potential next gas-sensing material due to its outstanding properties [24]. There are several methods to produce black phosphorus through complex chemical steps during recrystallization from bismuth (Bi) streams. The products of this reaction are deposited on the cooler surface of the system after a high-temperature chemical reaction in the gas phase. White phosphorus (WP) is typically converted to BP_4 by the high-pressure process, which involves placing the WP in a high-pressure chamber.

3 Fabrication Techniques

3.1 Chemical Vapor Deposition (CVD)

A key approach for the controlled synthesis of extraordinary 2D materials, such as graphene and transition metal dichalcogenides is chemical vapor deposition (CVD). This approach gives amazing management over the 2D fabric's thickness, first-rate, and scalability. Gaseous precursors comprising the fundamental components of 2D substances are first added right into an excessive-temperature reactor to start the method [26]. Precursor choice is important and varies depending on the substance being sought. The substrate regularly consisting of a silicon wafer, silicon dioxide, or metallic foil is produced [22, 26]. To stimulate the boom manner, this usually entails putting a small layer of catalytic material (inclusive of copper for graphene) onto the substrate. The reactor temperature is cautiously maintained to permit the formation of 2D layers at the substrate. The elements of increased time, gasoline drift rate, and stress are vital. To create a 2D substance, the precursor gasoline's atoms acquire right into a lattice structure in the course of growth. After growth, earlier than being included, the synthesized 2D cloth is usually transported from the boom substrate to the goal fuel sensing substrate [22, 26]. This switch segment is essential due to the truth that substrate traits can also have a huge effect on how properly materials are painted. Both Wet transfer (dissolving the increased substrate) and dry transfer (the usage of a polymer help layer) are one-of-a-kind transfer strategies that are hired.

3.2 Liquid-Phase Exfoliation

A simple method for fabricating two-dimensional materials in dispersed form, liquid-phase sliding is suitable for solution-based fabrication techniques and inkjet printing Firstly, it is fabricated with bulk 2D materials such as graphite or TMD crystals physical or chemical dispersion. In this process, the bulk material is divided into several layers [23]. Ultrasound is often used for this purpose. Surfactants or polymers may be added to the solution to prevent agglomeration of crystallized particles [23]. This modification results in better dispersion of the 2D material in the fluid. When printed or coated, the resulting dispersion produces an ink or paste that can be applied to the gas sensor surface. The ink or paste is applied to the gas sensor surface by several methods such as ink-printing, spin coating, drop molding, or screen printing. In this step, the shape and size of the two-dimensional deposits must be accurately controlled.

3.3 *Transfer Methods*

Moving 2D components from their growing substrate to the gas-sensing substrate is a crucial stage in the production process. Transfer wet is a typical transmission technique [4, 27]. The growth substrate is dissolved or etched in the wet transfer procedure, releasing the 2D material. It is subsequently transported to the sensor substrate, Once the substance is floating or submerged in a liquid medium like water or isopropyl alcohol. The substance can be dried on the intended surface after transfer. The 2D material is raised off the growing substrate using a polymer support layer in the dry transfer process [27]. The 2D material can tear off the growing substrate because the support layer, commonly comprised of poly (methyl methacrylate) (PMMA) or other polymers, is attached to it.

4 Sensing Mechanisms of 2D Materials

The particular structural, chemical, and electrical properties of two-dimensional (2D) accessories make them perfect for viewing operations [28, 29]. The three forms of chemical adsorption, charge transfer, and face reactions can be astronomically distributed as the seeing mechanisms of 2D appurtenances.

- Chemical Adsorption—The conformation of a chemical link between the detector material and the target patch or ion is what causes chemical adsorption, also known as chemisorption [29]. The characteristics of the 2D material and the target patch/ion play a significant role in this process.
- Transfer of Charge—Another essential seeing medium in 2D accessories is charge transfer. There may be a charge transfer when an ion or target patch comes into contact with the 2D material [28].
- Surface reactions—Surface reactions are modifications done to the physical or chemical components of the face of the 2D material as a result of interaction with target ions or motes [29]. The material's electrical or optical properties, which may be measured to reveal the presence of certain atoms or ions, may be impacted by these alterations [29].

Perceptivity, selectivity, and response/recovery time are a few of the metrics that may be used to gauge how well a detector constructed of 2D components is working. Perceptivity is the capacity of the detector to detect minute variations in the volume of the target patch or ion. High perceptivity is preferable since it enables the finding of target substances with less attention. A detector's capacity to respond only to particular ions or motes while ignoring others is known as selectivity [30]. In complicated environments where numerous distinct chemicals may be present, high selectivity is crucial. Response time is the amount of time a detector takes to react to a change in the target substance's attention. Recovery time is the period needed for the detector to recover to its initial condition following the removal of

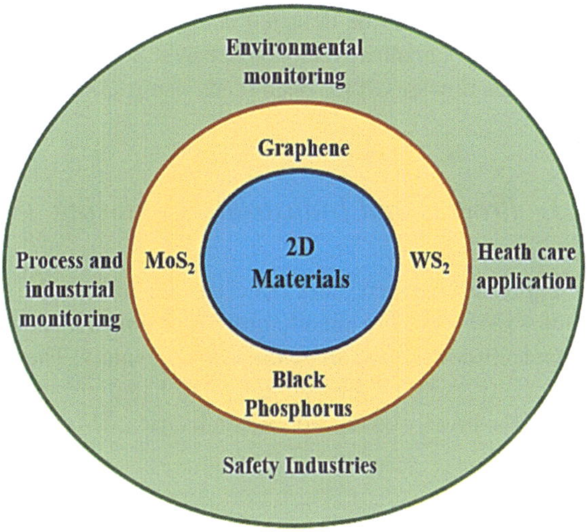

Fig. 3 Application of 2D Materials in numerous areas

the target material [30]. The speed with which a detector can provide reliable data depends on each of these timeframes. Due to their distinctive physical and chemical characteristics, two-dimensional (2D) materials have received a lot of interest and are appealing platforms for several applications as shown in Fig. 3.

5 Application of Gas Sensing in Numerous Areas

5.1 Environmental Monitoring

Air quality surveillance highly effective sensors that are compatible with current manufacturing technologies may be made using 2D materials. These sensors can efficiently monitor health and environmental situations by identifying dangerous and poisonous substances, gases, microbes, and radiation [31]. Future gas sensing materials include 2D materials like graphene, transition metal chalcogenides, boron nitride, transition metal carbide/nitride, metal–organic frameworks, and metal oxide nanosheets. They are very helpful in medical equipment like respiratory detection devices [32].

5.2 Healthcare Applications

A promising non-invasive diagnostic technique is breath analysis. Exhaled volatile organic compounds (VOCs) and inorganic gases, which are significant biomarkers for many species, may be detected using 2D material-based sensors. various diseases [33]. Electrochemical nano biosensors, which are used to identify and diagnose infectious illnesses, have been made using 2D materials [34].

5.3 Process and Industrial Monitoring

In applications where bulk materials are unsuitable, 2D materials are an excellent choice [35]. They are already utilized in sensors, energy storage and production, and optoelectronic devices in manufacturing, safety. The dielectric layer may deteriorate because of the high temperatures and harsh growth chemicals needed to build 2D materials directly on amorphous dielectrics. This is a problem with production safety [35].

5.4 Safety Industries

The safety industries have shown great interest in gas sensing utilizing 2D materials because of their high sensitivity, selectivity, and quick reaction time. Graphene, black phosphorus, and transition metal dichalcogenides (TMDs) are examples of 2D materials that have been researched for the detection of different gases, including explosive, poisonous, and volatile organic substances [36]. These materials provide a high surface area for gas adsorption, which improves the performance of the sensor. Gas sensors based on 2D materials are used in the safety sectors for early detection of dangerous gases, pipeline leak detection, and air quality monitoring. The advancement of gas sensors based on two-dimensional materials has enormous potential to enhance industry safety protocols [37].

6 Conclusion and Future Aspects

2D materials have garnered significant interest for their potential applications in gas detection. Among the various 2D materials, graphene oxide, MoS_2, WS_2, and phosphorene show promising characteristics for gas sensing due to their high surface-to-volume ratio and semiconductive nature. However, challenges such as the lack of selectivity and stability under different environmental conditions need to be addressed before these sensors can be widely adopted [31]. Despite these challenges, 2D materials-based gas sensors hold great potential to revolutionize gas detection in various fields including environmental and human safety, emission control, industrial applications, and medical diagnostics [38]. Nevertheless, the lack of selectivity remains a key challenge as gases with similar chemical properties can be difficult to differentiate. Therefore, further research and development efforts are necessary to meet the sensitivity, selectivity, and stability (3S) requirements for gas sensors. Many researchers have been working towards developing 2D materials-based gas sensors with the hope of eventually replacing metal oxides [39] for gas detection, leveraging the promising properties of 2D materials to create cost-effective and reliable sensors [40].

References

1. Sood, Y., Lawaniya, S.D., Mudila, H., Awasthi, K., Kumar, A.: Polypyrrole-tungsten disulphide 2D nanocomposites for ammonia sensing. Sens. Actuators B Chem. **394**, 134298 (2023). https://doi.org/10.1016/j.snb.2023.134298
2. Bhati, V.S., Kumar, M., Banerjee, R.: Gas sensing performance of 2D nanomaterials/metal oxide nanocomposites: a review. J. Mater. Chem. C **9**(28), 8776–8808 (2021). https://doi.org/10.1039/D1TC01857D
3. Zhang, J., Liu, L., Yang, Y., Huang, Q., Li, D., Zeng, D.: A review on two-dimensional materials for chemiresistive and FET-type gas sensors. Phys. Chem. Chem. Phys. **23**(29), 15420–15439 (2021). https://doi.org/10.1039/d1cp01890f
4. Yaqoob, U., Younis, M. I.: Chemical gas sensors: Recent developments, challenges, and the potential of machine learning—a review. Sensors **21**(8), MDPI AG (2021). https://doi.org/10.3390/s21082877
5. Zhang et al.: A flexible and wearable NO_2 gas detection and early warning device based on a spraying process and an interdigital electrode at room temperature. Microsyst. Nanoeng. **8**(1), (2022), https://doi.org/10.1038/s41378-022-00369-z
6. Sood, Y., Pawar, V.S., Mudila, H., Kumar, A.: Influence of polypyrrole morphology on gas sensing. J. Phys. Conf. Ser. Inst. Phys. (2022). https://doi.org/10.1088/1742-6596/2267/1/012050
7. Van Tong, P., Hoang Minh, L., Van Duy, N., Manh Hung, C.: Porous In_2O_3 nanorods fabricated by hydrothermal method for an effective CO gas sensor. Mater. Res. Bull. **137**, 111179 (2021). https://doi.org/10.1016/j.materresbull.2020.111179
8. Radhakrishnan, J.K., Kumara, M., Geetika: Effect of temperature modulation, on the gas sensing characteristics of ZnO nanostructures, for gases O_2, CO and CO_2. Sensors Int. **2** (2021). https://doi.org/10.1016/j.sintl.2020.100059
9. Yu, Z. et al.: Fabrication of lettuce-like ZnO gas sensor with enhanced H2S gas sensitivity. Crystals **10**(3), (2020). https://doi.org/10.3390/cryst10030145
10. Jrvinen, T. et al.: WS_2 and MoS_2 thin film gas sensors with high response to NH_3 in air at low temperature. Nanotechnology **30**(40), (2019). https://doi.org/10.1088/1361-6528/ab2d48
11. Joshi, N., Braunger, M.L., Shimizu, F.M., Riul, A., Oliveira, O.N.: Two-dimensional transition metal dichalcogenides for gas sensing applications. pp. 131–155 (2020). https://doi.org/10.1007/978-3-030-38101-1_4
12. Jimenez-Cervantes, E., López-Barroso, J., Martínez-Hernández, A.L., Velasco-Santos, C.: Graphene-based materials functionalization with natural polymeric biomolecules. Recent Adv. Graph. Res. INTECH (2016). https://doi.org/10.5772/64001
13. S. P. b, B. R. b, V. K. L. a c, D. K. a d, K. L. B. c, D. N. e, D.-V. N. V. f, A. S. d, S. D. b, S. B. a, P. P. S. g Hema Singh a.: Ultrasonically assisted solvohydrothermal synthesis of nanocrystalline Zn-Ni ferrite advanced material for EMI shielding
14. Kumar, S., Pavelyev, V., Mishra, P., Tripathi, N., Sharma, P., Calle, F.: A review on 2D transition metal di-chalcogenides and metal oxide nanostructures based NO_2 gas sensors. Mater. Sci. Semicond. Process. **107**, 104865 (2020). https://doi.org/10.1016/J.MSSP.2019.104865
15. Wang, H., et al.: Gas sensing materials roadmap. J. Phys. Conden. Matter **33**(30) (2021). IOP Publishing Ltd. https://doi.org/10.1088/1361-648X/abf477

16. Minezaki, T. et al.: Hydrogen sensing mechanism of WS_2 gas sensors analyzed with DFT and NAP-XPS. Sensors **23**(10) (2023). https://doi.org/10.3390/s23104623
17. Bhattacharyya, P., Acharyya, D.: Impact of device configurations on sensing performance of WS_2-based gas sensors: a review. IEEE Sens. J. **21**(20), 22414–22425 (2021). https://doi.org/10.1109/JSEN.2021.3104615
18. Agrawal, A.V., Kumar, N., Kumar, M.: Strategy and future prospects to develop room-temperature-recoverable NO_2 gas sensor based on two-dimensional molybdenum disulfide. Nano-Micro Lett. **13**(1). Springer Science and Business Media B.V. https://doi.org/10.1007/s40820-020-00558-3
19. Wu, J., Tan, Y., Lyu, J.: Improving gas sensing performance of hydroxyapatite by incorporating black phosphorene. IEEE Sens. J. **22**(11), 10370–10380 (2022). https://doi.org/10.1109/JSEN.2022.3170930
20. Li, W., Shahbazi, M., Xing, K., Tesfamichael, T., Motta, N., Qi, D.C.: Highly sensitive NO_2 gas sensors based on MoS_2@MoO_3 magnetic heterostructure. Nanomaterials **12**(8) (2022). https://doi.org/10.3390/nano12081303
21. Liu, J. et al.: MoS_2 nanosheets sensitized with quantum dots for room-temperature gas sensors. Nanomicro. Lett. **12**(1), (2020). https://doi.org/10.1007/s40820-020-0394-6
22. Nassef, B.G., Nassef, G.A., Daha, M.A.: Graphene and its industrial applications. A review. Int. J. Mater. Eng. **10**(1), 1–12 (2020). https://doi.org/10.5923/j.ijme.20201001.01
23. Johra, F.T., Jung, W.G.: Synthesis of black phosphorus via a facile vapor transfer method. Electron. Mater. Lett. **15**(5), 639–644 (2019). https://doi.org/10.1007/s13391-019-00162-7
24. Akbari, E., Nilashi, M., Buntat, Z.: Analytical investigation of superior gas sensor based on phosphorene. Microsyst. Technol. **25**(3), 897–903 (2019). https://doi.org/10.1007/s00542-018-4054-4
25. Gajić, N., Kamberović, Ž., Anđić, Z., Trpčevská, J., Plešingerova, B., Korać, M.: Synthesis of tribological WS_2 powder from WO_3 prepared by ultrasonic spray pyrolysis (USP). Metals (Basel) **9**(3), (2019). https://doi.org/10.3390/met9030277
26. Urade, A.R., Lahiri, I., Suresh, K.S.: Graphene properties, synthesis and applications: a review. JOM **75**(3), 614–630 (2023). https://doi.org/10.1007/s11837-022-05505-8
27. Sharma, A., Rout, C.S.: Advances in understanding the gas sensing mechanisms by in situ and operando spectroscopy. J. Mater. Chem. A **9**(34), 18175–18207 (2021). https://doi.org/10.1039/D1TA05054K
28. Yang, T., et al.: iScience Mechanical sensors based on two-dimensional materials: Sensing mechanisms, structural designs and wearable applications (2022). https://doi.org/10.1016/j.isci
29. Lee, C.W., Suh, J.M., Jang, H.W.: Chemical sensors based on two-dimensional (2D) materials for selective detection of ions and molecules in liquid. Front. Chem. **7**, (2019). https://doi.org/10.3389/fchem.2019.00708
30. Mathew, M., Shinde, P.V., Samal, R., Rout, C.S.: A review on mechanisms and recent developments in p-n heterojunctions of 2D materials for gas sensing applications. J. Mater. Sci. **56**(16), 9575–9604 (2021). https://doi.org/10.1007/s10853-021-05884-4
31. Tyagi, D. et al.: Recent advances in two-dimensional-material-based sensing technology toward health and environmental monitoring applications. Nanoscale **12**(6), 3535–3559 (2020). https://doi.org/10.1039/c9nr10178k
32. Nikolic, M. V., Milovanovic, V., Vasiljevic, Z. Z., Stamenkovic, Z.: Semiconductor gas sensors: materials, technology, design, and application. Sensors (Switzerland) **20**(22), 1–31 (2020). https://doi.org/10.3390/s20226694
33. Kaloumenou, M., Skotadis, E., Lagopati, N., Efstathopoulos, E., Tsoukalas, D.: Breath analysis: a promising tool for disease diagnosis—the role of sensors. Sensors **22**(3), (2022). https://doi.org/10.3390/s22031238
34. Singh, N., Dkhar, D.S., Chandra, P., Azad, U.P.: Nanobiosensors design using 2D materials: implementation in infectious and fatal disease diagnosis. Biosensors **13**(2), (2023). https://doi.org/10.3390/bios13020166

35. Rohaizad, N., Mayorga-Martinez, C.C., Fojtů, M., Latiff, N.M., Pumera, M.: Two-dimensional materials in biomedical, biosensing and sensing applications. Chem. Soc. Rev. **50**(1), 619–657 (2021). https://doi.org/10.1039/D0CS00150C
36. Athawale, A., et al.: Synthesis of Ag_2O coated TiO_2 nanoparticles by sonochemically activated methods for enhanced photocatalytic activities. Top. Catal. **63**(11–14), 1056–1065 (2020). https://doi.org/10.1007/s11244-020-01374-0
37. Márquez-Sánchez, S., Huerta-Muñoz, J., Herrera-Santos, J., Arrieta, A. G., De la Prieta, F.: Gas sensing in industry. A case study: train hangar. Ad Hoc Networks **148**, (2023). https://doi.org/10.1016/j.adhoc.2023.103205
38. Naikoo, G. A. et al.: 2D materials, synthesis, characterization and toxicity: a critical review. Chem.-Biol. Interact. **365**, (2022). https://doi.org/10.1016/j.cbi.2022.110081
39. A. K. b, A. T. c, D.-V. N. V. d, A. S. a, D. K. a Aiswarya Moharana a.: Chapter 6—Role of nanostructured metal oxides in photocatalysis: an overview.
40. Zhou, L. et al.: Nonlinear optical characterization of 2d materials. Nanomaterials **10**(11), 1–38 (2020). https://doi.org/10.3390/nano10112263

Organic Linkers and Their Modified Nanoparticles for Colorimetric Monitoring Devices for Inorganic Water Contaminants

Nancy Sidana

Abstract A large number of inorganic contaminants have been discharged into the surrounding waters as a result of expanding industry, mining industries, and urbanization so they have a fatal influence on the well-being of people and developed significant environmental harm. Due to their strong interactions with the earth's framework and increased danger to the surroundings, heavy metals make up the majority of inorganic contaminants. Even Nevertheless, when ingested below permitted levels, several of these metal particles are crucial for biological activities. Earlier, many traditional spectroscopic like ICP-MS, AAS were used to quantify these contaminants. However, these methods of analysis demand intricate preparation of samples often are laborious, expensive, as well as tedious. Their application for in-the-field testing for heavy-metal ions is constrained by these drawbacks. Because of its excellent specificity, sensitivity, and price effectiveness, unaided colorimetric sensing is now widely used for monitoring pollution to allow quick in-the-field identification of contaminants. This chapter will examine the newly developed organic linkages and their modified nanoparticles utilized to create inorganic pollution sensors and will provide an overview of the novel approaches for the onsite evaluation of these metallic ions utilizing reagent kits and test strips.

1 Introduction

Water that is clean and safe is crucial for both the environment and human health [1]. But inorganic water pollutants including toxic metals, metalloids, and insecticides may seriously harm both human health and water quality. Numerous health issues, including stroke, failure of the kidneys, problems with the immune system, weakened muscles, limb paralysis, cardiovascular concerns, as well as nervous system disorders, can be brought on by these harmful heavy metals. For many years, scientists

N. Sidana (✉)
Department of Applied Sciences, Chandigarh Engineering College-CGC, Landran, Mohali, Punjab 140307, India
e-mail: nancysidana18@gmail.com

A. Katoch et al. (eds.), *2D Materials*, Engineering Materials,
https://doi.org/10.1007/978-981-97-6258-3_7

have worked to create precise, trustworthy analytical techniques for identifying heavy metal ions and significantly reducing their acute toxicity [2–5]. These pollutants can enter the drinking water supply from a number of different sources, including industrial operations, geological processes, and agricultural runoff. For the purpose of maintaining safe and clean water, identification along with tracking of these pollutants is essential. Inductively coupled plasma mass spectrometry and atomic absorption spectroscopy are two common traditional detection techniques for inorganic water pollutants, however, they can be time-consuming, costly, and expensive to set up [6]. These procedures also call for qualified staff, which disqualifies them for on-site monitoring. Additionally, they frequently produce toxic waste, which adds to environmental damage.

Colorimetric monitoring tools have become a potential substitute for identifying and keeping track of inorganic water pollutants in recent years. To detect particular toxins in water, these devices employ a color change response. In comparison to conventional detection techniques, colorimetric monitoring tools provide a number of benefits, including portability, ease of use, and excellent sensitivity and selectivity [7]. Additionally, because they produce less hazardous waste, they are ecologically beneficial. Two types of materials have been employed in colorimetric monitoring systems for inorganic water contaminants: organic linkers and altered nanoparticles. The signal needed for detection can be produced by organic linkers, which are molecules of organic matter that can bond to inorganic compounds. In comparison to conventional detection techniques, the application of organic linkages in colorimetric monitoring equipment has a number of benefits. Organic linkers may be made to be very sensitive and selective to a particular kind of inorganic pollutant. They are also excellent for on-site monitoring since they may deliver a visually clear signal [8]. Numerous thiocarbazide groups are well-known sensor molecules that can improve coordination between transition and heavy metals. These receptor moieties serve as binding sites for cation detection and are composed of hydroxy, thio, and imine groups. The binding site spacer and signaling unit methodologies, which are utilized to recognize various heavy metal transition metal ions, formed the basis for the ligands' design [9].

Due to their tiny size, nanoparticles are materials with distinctive physical and chemical characteristics. Nanoparticles that have been chemically functionalized with certain molecules to improve their characteristics or permit particular interactions are referred to as modified nanoparticles [10]. When utilized in colorimetric detectors, altered nanoparticles provide a number of advantages, including exceptionally high sensitivity and specificity, ease of production and customization, and compatibility with a range of detecting substrates [11]. Colorimetric nanosensors' greater degree of modification, larger volume-to-surface ratio, size-dependent characteristics, and strong reactivity are the sources of their enhanced selectivity and sensitivity. Furthermore, the synthesis of nanocomposite materials, covalent functionalization, and organic ligand anchoring have a bigger influence on the increased sensitivity and selectivity in the identification of heavy metal ions. Additionally, the addition of nanomaterials has significantly raised the colorimetric nanosensors' repeatability and decreased their detection limit for identifying heavy metal ions [12–15].

A mixture containing organic compounds and altered nanoparticles can boost the sensitivity and selectivity of colorimetric monitoring systems even further. To improve the way, they interact with certain inorganic contaminants, nanoparticles are able to be modified utilizing organic linkers. Furthermore, tailored nanoparticles can be used to enhance the information that organic bonds provide, increasing responsiveness [16]. Colorimetric testing techniques that use organic bonds and modified nanoparticles have shown great potential for identifying and tracing contaminants in mineral water. These gadgets provide a variety of advantages, including superior sensitivity, selectivity, and ease of use. However, further research is needed to address the current limitations and challenges associated with the production of such devices. Colorimetric monitoring tools may be a useful tool for assuring clean and secure water with further improvement [17].

2 Inorganic Pollutants

Chemical substances that come from natural or man-made sources and do not include carbon-hydrogen bonds are known as inorganic pollutants. These pollutants can get into the environment in a number of ways, involving commercial and farming activities, erosion and weathering brought on by nature, and the usage of certain consumer items. Long-lasting inorganic pollutants can pose serious threats to the health of both people and our surroundings [18].

Inorganic pollutants include substances like heavy metals including arsenic, lead, mercury, arsenic, and cadmium. When swallowed or breathed, these metals can build up in the air we breathe and result in a variety of health issues, including cancer, organ damage, and brain impairment. Additionally, they have the potential for accumulation in the nutritional chain and damage animals and other non-target creatures [19].

Metalloids, including arsenic and selenium, are an additional category of inorganic contaminants because they possess both metallic and non-metallic characteristics. Both natural processes like the aging of minerals and human actions like mining and smelting can release these chemicals into the environment. For instance, arsenic, a frequent groundwater pollutant, can result in a variety of health issues, including skin sores, cancer, and mental damage [20].

Nitrates and phosphates are two additional inorganic pollutants that are frequently linked to agricultural practices like the usage of fertilizers. These substances can induce eutrophication, which lowers the oxygen content of the water and can result in the mortality of fish and other aquatic life when they enter water bodies. Chemicals-derived pesticide, products are employed for agricultural purposes and can harm non-target creatures when they reach the environment, are examples of inorganic chemicals [21].

Due to their low quantities and the complicated chemical and physical characteristics of the natural setting in which they are located, inorganic contaminants can be challenging to identify and detect. Aspects of inductively coupled plasma mass spectrometry and atomic absorption spectroscopy are conventional techniques for identifying and tracking inorganic contaminants. These procedures frequently need a lot of time and money as well as specialized tools and skilled labor [22].

Instruments for colorimetric surveillance are currently a viable option for identifying and keeping track of inorganic contaminants. To detect particular toxins in water, these devices employ a color change response. In comparison to conventional detection techniques, colorimetric tracking instruments provide a number of benefits, including mobility, simplicity of use, and high accuracy and selectivity. Additionally, because they produce less dangerous materials, they are ecologically beneficial [23].

The incorporation of organic linkages and altered nanoparticles has aided in the creation of colorimetric monitoring systems for inorganic pollutants. A signal for detection can be produced by organic linkers, which are molecules of organic matter that can bond to inorganic compounds. Nanoparticles that have been chemically functionalized with certain molecules to improve their characteristics or permit particular interactions are referred to as modified nanoparticles. In colorimetric systems for tracking, a mixture of organic linkages and modified nanoparticles can further improve their responsiveness and specificity [24].

To sum up, inorganic contaminants can seriously jeopardize the health of humans and the environment. For the purpose of maintaining safe and clean water, the identification and continual tracking of these contaminants are essential. The introduction of organic linkages and altered nanoparticles can further improve the sensitivity and selectivity of colorimetric monitoring devices, which present a viable alternative to conventional detection techniques for inorganic contaminants. The creation of more effective monitoring tools might result from additional studies and research in this field, protecting watersheds and the health of the population. Some of the commercially important pollutants are discussed as follows.

2.1 Mercury

- Mercury has an atomic number of 80 naturally found metal components with its molecular symbol Hg. Because of its distinctive qualities, including its low point of melting, dense structure, and conductivity to electricity, it has been utilized for countless years for a variety of uses. Mercury usage has been limited in numerous countries due to the possibility of harm to human beings and their surroundings [19].
- The manufacturing of chlorinated and sodium hydroxide, thermometers as well dentistry fillings, electronic switches, and circuits are just a few of the commercial applications for mercury. In small-scale businesses gold mining, it has also been employed to extract gold from different minerals. The combustion of energy sources, especially coal, constitutes one of the main causes of mercury contamination. Small quantities of mercury explode into the sky after the burning of coal, where they can travel great distances without landing on water or land. Mercury

may go through an action called methylation in the environment that turns it into methylmercury, an extremely hazardous compound. Methylmercury may build up in marine organisms, which can subsequently be eaten by people and have a negative impact on their health [25].

- Numerous health issues, including brain impairment, damage to organs, and developmental difficulties, can be brought on by prolonged exposure to excessive mercury levels. Because of how sensitive the nervous system is to mercury, exposure can cause shaking hands, loss of memory, and other neurological problems. Additionally, damaging to fetal development, exposure to mercury during pregnancy can result in birth abnormalities such as developmental delays and cognitive disabilities. There is no degree of mercury exposure that is safe, and even small levels can have negative effects. The World Health Organization (WHO) advises adults to consume no more than 1.6 mg of mercury per kilogram of body weight every week. Young children and pregnant women are particularly susceptible to the negative effects of mercury, thus great care should be taken to prevent exposure [26].
- Many nations have banned mercury usage and put in place rules to manage its leakage into the environment in an effort to reduce exposure. To stop the negative effects of mercury on people and the environment, a worldwide agreement known as the Minamata Protocol on Mercury was signed in 2013 [27]. The agreement establishes guidelines for reducing the use of mercury in industrial processes, controlling mercury emissions, and restricting the disposal of trash that includes mercury. Mercury is a versatile metal with unique properties that have been used in a number of applications. Due to the potential harm that mercury might do to both people and the environment, its use has been restricted in many countries. To protect the environment and human health, mercury exposure must be maintained to the very least, and attempts to limit its use and emissions must continue [28].

2.2 *Arsenic*

- Arsenic is an element that occurs naturally with the atomic number 33 and its chemical name As. It is found in the crust of the earth together with numerous other minerals, such as arsenopyrite, realgar, and orpiment [29]. Exposure to arsenic can be dangerous since it affects human health both negatively and favorably. Arsenic has been used for a number of purposes throughout history, including the production of alloys, pesticides, and wood preservatives. Arsenic was occasionally used as medicine to treat a number of diseases, including syphilis and skin disorders. Although arsenic was formerly employed in medicine, it has since been discovered to be exceedingly dangerous [30].
- The main sources of arsenic exposure for people are contaminated food and water. Arsenic may infiltrate water sources naturally as well as enter into the atmosphere through operations in agriculture and industry. Arsenic may accumulate in some crops, such as rice, grown on contaminated soil or irrigated with poisonous water.

Heart attacks, cancers, and skin rashes are among the many of the health problems that can arise from long-term exposure to high arsenic levels. Long-term exposure to water containing arsenic has been linked to an increased likelihood of the bladder, skin, and cancers of the lungs. Arsenic is a renowned carcinogen, it's exposure can lead to heart disease, diabetes, and other chronic medical issues [31].

- The World Health Organization (WHO) has established a 10 μg per liter intake recommendation for arsenic in water [32]. However, in some parts of the world, particularly in impoverished countries, the amount of arsenic in drinking water can exceed this limit. By ensuring that these communities have a source of clean water for consumption and encouraging agricultural practices that minimize arsenic exposure, efforts are being undertaken to limit arsenic exposure. Arsenic is an element that exists naturally that can have both positive and negative impacts on human health. Because of its usage in agriculture and a variety of industrial operations, it has the potential to contaminate water and food supplies, exposing people to high doses of arsenic. Arsenic exposure over an extended period of time can result in a number of health issues, including tumors and heart failure. Arsenic exposure must be decreased to safeguard the public's health, which may be accomplished by taking steps such as ensuring that everyone has access to clean water and encouraging farming methods that limit arsenic uptake by plants [33].

2.3 Copper

- Throughout the past, different uses for copper's malleability, ductility, and high conductivity have been made. It is frequently obtained from copper ore by mining and smelting procedures and is typically found in minerals like chalcopyrite, bornite, and malachite. Electrical wire, plumbing pipelines, and electronic equipment are just a few examples of the industrial applications for copper. Coins, jewelry, and other ornamental objects as well as building materials like roofs and gutters also make use of it [34]. Despite the fact that copper has numerous beneficial qualities, excessive copper exposure can be dangerous to one's health. Ingestion of tainted food or water, absorption of dust containing copper, or coming into contact with copper-containing items can all result in copper poisoning [35].
- In addition to nausea, vomiting, diarrhea, and abdominal discomfort, copper intoxication can also cause liver damage. Anemia, renal damage, and neurological diseases are just a few of the more significant health issues that can result from prolonged exposure to excessive copper levels. Regulating bodies have created standards for safe exposure to copper levels in various contexts in order to safeguard public health [36]. For instance, the Environmental Protection Agency (EPA) has established a maximum contamination threshold for copper in drinking water of 1.3 mg per liter [37]. Additionally, workplace tolerances for copper were

developed to safeguard employees who could be exposed to excessive copper levels at work.

- In the final analysis, copper is a versatile metal with a variety of beneficial qualities. High quantities of copper exposure, however, can be detrimental to human health and cause a variety of signs and conditions. Guidelines for acceptable copper levels of exposure have been developed in many contexts to safeguard public health.

2.4 Lead

- During past times, lead, an extremely hazardous metal, has been utilized for a variety of things. It is frequently recovered from lead ore via mining and smelting techniques, and is typically found in minerals like galena and cerussite. Paint, petrol, pipes, batteries, and other things have all included lead at some point. Additionally, it has been utilized in firearms, fisheries sinkers, and other pieces of hunting and fishing gear [38].
- Lead exposure can have negative impacts on health, especially in youngsters and pregnant women. Lead exposure can harm the brain and neurological system, which can lead to behavioral and learning issues, lowered IQs, and developmental delays. Adult exposure to lead can result in high blood pressure, renal damage, and issues with reproduction. Regulatory organizations have created standards for acceptable amounts of lead exposure in various contexts in order to safeguard the public's health [39, 40]. For instance, lead in drinking water has a maximum contamination level of 15 parts per billion as specified by the Environmental Protection Agency (EPA). Regulations are also in place to restrict the consumption of lead in products for consumers [38].
- Additionally, efforts are being undertaken to clean up lead-contaminated regions, such as the soil surrounding former industrial sites and the paint containing lead in older homes. It is feasible to avoid or lessen the negative health effects of lead poisoning by limiting lead exposure.

2.5 Silver

- Over the years, silver, a soft, lustrous, and highly conductive metal, has had a variety of uses. It is frequently produced from silver ore using mining and refining techniques and is typically present in minerals like argentite and chlorargyrite. Numerous products, including jewelry, coinage, silverware, and electrical parts, have been made with silver. Additionally, it serves as an antibiotic in medical settings and on photographic film [41].
- Although silver has many beneficial qualities, excessive silver exposure can be hazardous to human health. Ingestion of tainted food or drink, inhalation of dust

containing silver, or coming into contact with silver-containing items can all result in silver poisoning. Silver poisoning can cause brain issues, renal damage, and gastrointestinal discomfort. Continuous exposure to high silver concentrations can potentially cause more severe health issues, such as argyria, a disorder where the skin becomes blue-gray.

- Regulatory bodies have developed standards for acceptable amounts of silver exposure in various contexts in order to safeguard the public's health. Additionally, job-specific limits of exposure to silver are being developed to safeguard employees who could be exposed to excessive concentrations of silver at work [42].

3 Organic Linkers for Colorimetric Monitoring Devices

The signal needed for detection can be produced by organic linkers, which are molecules of organic matter that can bond to inorganic compounds. In comparison to conventional detection techniques, the application of organic linkages in colorimetric devices for tracking has a number of benefits. Organic linkers may be made to be very sensitive and selective to a particular kind of inorganic pollutant. They are also excellent for on-site monitoring since they may deliver a visually clear signal. The various organic linkers utilized in colorimetric monitoring systems and their benefits are covered in this section [43]. A schematic representation of the binding mode of organic linkers with metal ions is shown in Fig. 1.

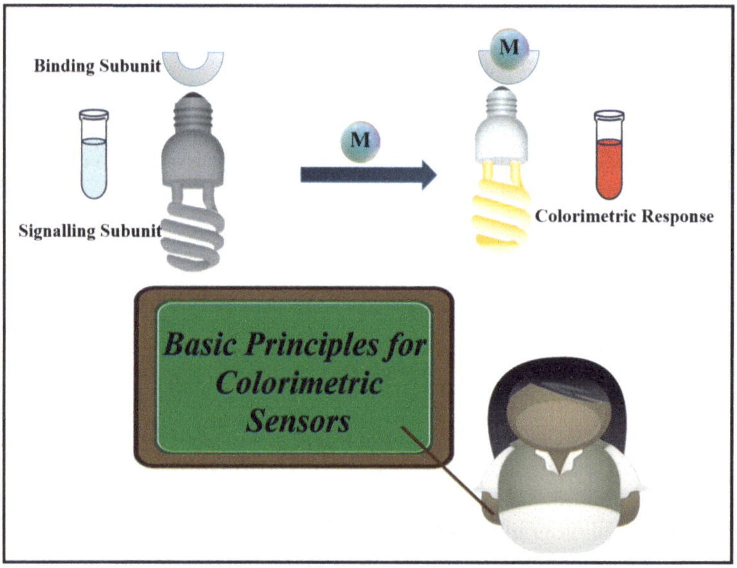

Fig. 1 Schematic representation of the Binding mode of organic linkers with metal ions

3.1 Schiff Base Organic Ligands

Hugo Schiff's invention of Schiff bases led to their widespread usage in both industrial and pharmacological applications [44]. Schiff bases' many biological actions, such as their antifungal, anti-inflammatory, anti-cancer, and antioxidant capabilities, are facilitated by the azomethine connection [45]. These substances are useful in therapeutic applications because they function as good chelating ligands and may create stable complexes with different metal ions.

Schiff bases also have exceptional qualities as chemosensors for identifying metallic ions, and organisms lacking in electrons. To identify certain metal ions like Cu^{2+}, Co^{2+}, Fe^{2+}, and Zn^{2+}, researchers have created selective probes employing Schiff bases. Chemosensors, as opposed to traditional detection techniques, provide a straightforward and effective tool to measure the presence of metal and ions within surroundings and biological materials.

Colorimetric or fluorescence sensors are two different types of chemosensors. Colorimetric detectors rely on color alterations brought on by outside stimuli to provide sensitive and focused identification of different species. These sensors provide instrument-free detection that is rapid, repeatable, affordable, and visible to the human eye. When fluorescent chemosensors interact with analytes, they vary in emission strength, wavelength, and lifespan decay, which makes them suited for biological environment sensing applications.

Internal charge transfer, fluorescence quenching, and fluorescence amplification are a few examples of the mechanisms that are responsible for the absorption and fluorescence changes in chemosensors. The chemosensor and the analyte collide or chelate to cause these processes, which include intramolecular charge transfer, fluorescence resonance energy transfer, and photo-induced electron transfer.

The creation of colorimetric and fluorescent chemical sensors can be greatly aided by the use of Schiff compounds and their altered derivatives in particular which enable the sensitive and precise identification of metallic ions, and other analytes in ambient and biological materials.

(E)-1-((p-tolylimino)methyl)naphthalen-2-ol is a fluorescent and colorimetric chemosensor known as SB. Several spectroscopic methods, including 1H NMR, FT-IR, and fluorescence spectroscopy, were used to examine the structure of SB. A variety of metal ions, including Mn^{2+}, Cd^{2+}, Al^{3+}, Cu^{2+}, Na^{+}, K^{+}, Ca^{2+}, Ag^{+}, Zn^{2+}, Co^{2+}, Ni^{2+}, Pb^{2+}, Cr^{3+}, Hg^{2+}, and Mg^{2+} were studied for the chemosensor's detecting capabilities. In MeOH, SB displayed an extraordinary colorimetric reaction (yellow to yellowish brown) as well as a fluorescence "turn-on" when exposed to Cu^{2+} in a solution of MeOH/Water (10/90, v/v).

The method of sensing for Cu^{2+} was examined using FT-IR, ^{1}H NMR titration, DFT studies, and Job's plot analysis. The detection limit of Cu^{2+} was found to be very low, at 0.0025 g mL^{-1} (0.0025 ppm). Furthermore, when added to a test strip, SB demonstrated excellent sensitivity and selectivity for Cu^{2+} detection both in solution and on solid surfaces [46]. In another study, 4-Methylbenzenesulfonohydrazide

(R1), a new sulfonohydrazide Schiff base colorimetric probe, was created and characterized for the purpose of selectively detecting Cu^{2+} ions in a semi-aqueous media. Exceptional specificity for Cu^{2+} above the other metal ions was demonstrated by R1, which also displayed a significant color shift in sunlight from colorless to crimson yellow.

Employing UV–vis titration and Job's plot analysis, the mixture produced between R1 and Cu^{2+} was shown to have a stoichiometric proportion of 2:1. A paper strip that had already been pre-stained with R1 solutions was created and showed a high sensitivity to Cu^{2+} with a small identification limit. Theoretical and experimental experiments were used to further validate the sensing mechanism, demonstrating a strong correlation between the predicted HOMO–LUMO level of energy and the measured colorimetric response. The probe is suited for on-site investigation of Cu^{2+} in samples of water due to its mobility and user-friendliness [47]. In a different research. S. Khan et al. developed SB1, a thiazole-based Schiff base chemosensor enabling the identification of Cu^{2+} ions. As shown by notable spectrum alterations and color transformation, the chemosensor displayed strong sensitivity and reactivity towards Cu^{2+} ions. Other cations barely interfered with it at all. Various methodologies were used to explore the sensing process, and computer analyses validated it. It was discovered that the 2:1 interaction mode involved SB1 and Cu^{2+} ions [48]. A luminous colorimetric chemosensor called H2L was created by Sahu et al. for the quick recognition of Cu^{2+} ions in solution form. H2L was produced in an easy and ecologically responsible manner. When the instrument was bound to Cu^{2+} ions, its color changed from colorless to a bright yellow, and its fluorescence was enhanced. According to colorimetric and fluorometric measurements, the limit of quantification for Cu^{2+} ions is 7.1×10^8 M and 3.8×10^8 M, respectively [49].

In a pioneering study, PM a brand-new fluorescent probe made up of two diaminomaleonitrile moieties attached to a scaffold made of naphtholphthalein. Through colorimetric fluorescence, PM can sensitively and specifically identify Hg^{2+} in an aqueous THF solution. 20 nM was found to be the Hg^{2+} detection limit for PM. As demonstrated by several experimental approaches including UV–Vis, HRMS, ^{1}H NMR titrations, and DFT calculations, the sensing mechanism of PM includes the breakdown of the diaminomaleonitrile-based Schiff-base skeleton brought on by Hg^{2+}. Additionally, a portable platform powered by a smartphone was created for the quick and affordable detection of Hg^{2+} leveraging PM's superior fluorometric capabilities. The technology produced good recoveries when used to identify Hg^{2+} in ambient water samples on-site [50]. A chromoionophoric sensor was impregnated onto a porosity polymeric monolithic framework to produce another sensor. When the sensor complexed with Hg^{2+}, its color changed from white to yellow. Various methods were used to characterize the sensor's surface and structural characteristics. At pH 8.0, the improved sensor performed admirably, with a quick reaction time and a large linear detection range. It had a high degree of Hg^{2+} detection selectivity and was generable with EDTA. The sensor has several cycles of reusability and accurate data. Its applicability and viability for local evaluation of samples from the environment were demonstrated by the fact that it performed well using a water and smoke sample. The created sensor has promise for other uses in distant heavy metal detection and is both affordable and environmentally safe [51]. Gangatharan et al. investigated the capacity of an intriguing cyanide-colored probe (L) to precisely detect Hg^{2+} ions using a colorimetric technique. Hg^{2+} ions induced noticeable color changes in the sample from yellow to orange, although other metal ions had little influence on the probe's colorimetric response. During the binding procedure, a charge transfer combination with a low detection limit of 0.62 M was developed. Through ^{1}H NMR titration, Job's plot, and FTIR investigations, it was determined that the binding stoichiometry among the probe and Hg^{2+} ions was 1:1. With the use of ethylenediaminetetraacetic acid (EDTA), the chromogenic probe's reversibility was proven. Additionally, the reversible probe shows promise as a molecular logic gate. Quantum mechanical study supported the binding mechanism. The probe could identify minute levels of mercury ions in water samples, and a strip for testing was created to make it simple to identify Hg^{2+} ions without the use of extra equipment [52]. A diethylamino-substituted pyrazole imine fluorescent probe was created in another investigation for the detection of Hg^{2+} ions. After interacting with Hg^{2+} ions, the probe had a colorimetric and fluorescence response, with a tenfold increase in fluorescence. With a low limit of detection and a high binding association constant, it demonstrated exceptional specificity for Hg^{2+} ions in a CH_3CN/H_2O solution. Several investigations verified the establishment of a 3+Hg^{2+} compound with a 1:1 interaction mode [53].

To detect several ions using one sensor, other multi-ion sensors have also been developed in addition to these ones. Nancy and colleagues successfully synthesized a novel colorimetric sensor (sensor L) based on thiophenol amines in a single step using condensation. In particular, sensor L performed remarkably better in recognizing Cu^{2+} and Hg^{2+} ions than other heavy-metal ions. The sensor clearly displayed a color shift in a mixed aqueous medium, going from colorless to orange for Hg^{2+} ions along with colorless to yellow for Cu^{2+} ions as shown in Fig. 2. The binding stoichiometry for the sensors with Cu^{2+} and Hg^{2+} ions was found to be 2:1 using Job's Plot. It was discovered that the detection limits for Cu^{2+} and Hg^{2+} ions were 50 $\times$ 10^{-9} M and 18 $\times$ 10^{-8} M, respectively. The sensor responded quickly, and EDTA titrations demonstrated its reversibility. The sensor responded quickly, and EDTA titrations demonstrated its reversibility [11]. Another effort produced two acyclic Schiff base receptors, CS-1 and CS-2, which were used as selective fluorescence turn-on sensors for Zn^{2+} ions and as colorimetric chemical sensors for Zn^{2+}, Cu^{2+}, and Co^{2+} ions. When these receptors are diluted 1:1 with acetonitrile and HEPES buffer (pH 7.4), they alter color from colorless to yellow, even in the presence of other screening metal ions [54]. A Schiff base made of imidazoles called IMP has been developed and is known for its ability to detect Co^{2+} and Zn^{2+} fluorometrically. The probe operates on the basis of photo-induced electron transfer, or PET. When Zn^{2+} ions are present, the normally pale yellow solution of IMP turns bright yellow, with increased absorption at 410 nm and a noticeable fluorescence emission at 612 nm. Furthermore, IMP responds to Co^{2+} ions in a highly selective manner in terms of colorimetry, providing an obvious pink color without any further ion interference right away [55]. For the selective and straightforward colorimetric identification of two metal ions, Cu(II) and Zn(II), as well as the anion F, in the presence of multiple competing cations and anions, P. das et al. developed the Schiff base chemosensor known as HL based on a piperazine platform. When Zn(II) is added to HL, its mild fluorescence activity increases and takes on a rich green hue when exposed to UV light (= 365 nm). However, when Cu(II) is added, the HL fluorescence becomes colorless. The identification of these ions in drinking water also made use of the chemosensor's capacity to detect F^- ions [56].

In an additional study, colorimetric and spectrum alterations enabled the novel NB probe, based on naphthaldehyde, to demonstrate dual sensing behavior for Hg^{2+} and Cr^{3+} ions. It was discovered that the Hg^{2+} and Cr^{3+} ions have detection limits of 2.7×10^8 M and 1.1×10^8 M, respectively. The existence of hydroxyl and imine groups is thought to be responsible for the probe's capacity to identify these ions in a 1:1 ratio. In addition, the practical effectiveness of a paper strip test for the detection of Hg^{2+} and Cr^{3+} ions was demonstrated [57].

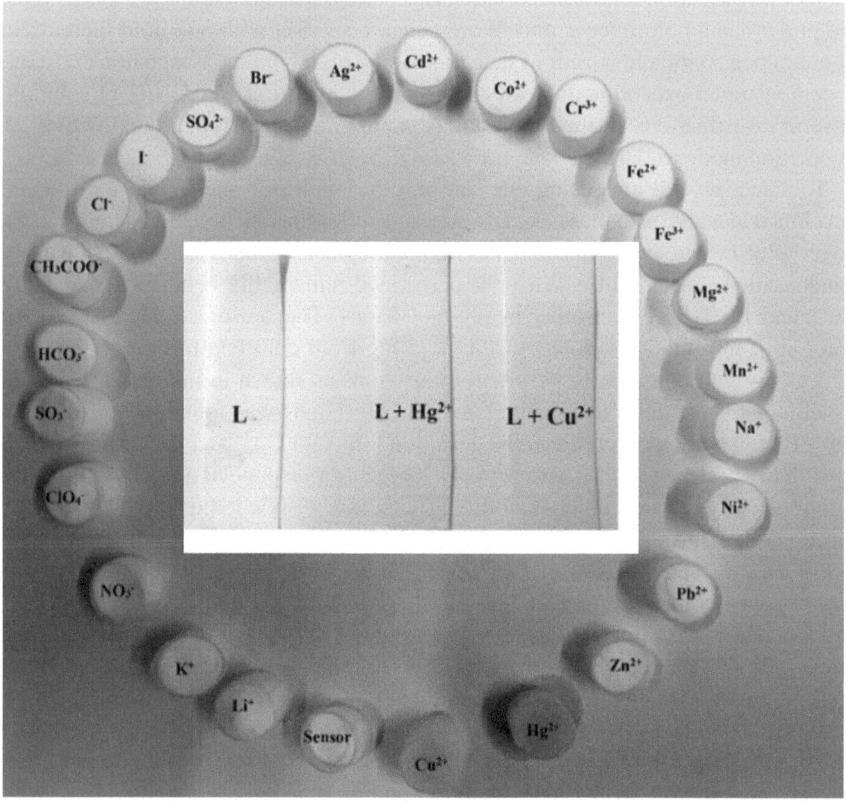

Fig. 2 Digital response of sensor L on the addition of different metal ions using test strips indicating orange color for mercury ions, and yellow color for copper ions (center image), and silica gel-based solid substrate indicating selective color for copper and mercury ions over other ions (arranged in a circular pattern)

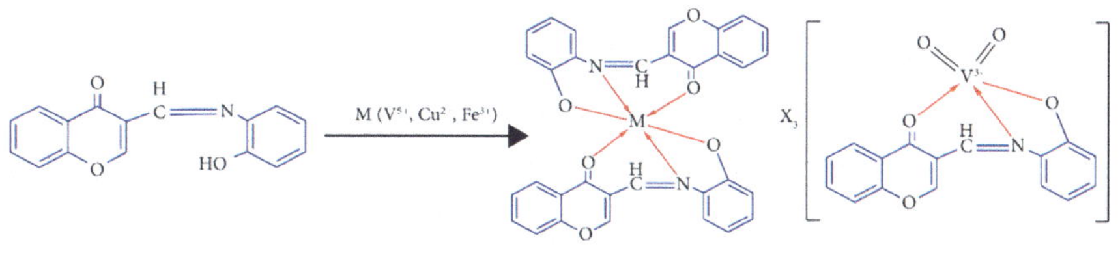

Fig. 3 Proposed structures for the SL-Cu^{2+}, SL-Fe^{3+}, and SL-V^{5+} complexes

A novel ligand called SL was created and then tested using colorimetric analysis as a chemosensor for ions of metal in aqueous solutions. The outcomes showed selectivity and sensitivity for Cu^{2+}, Fe^{3+}, and V^{5+} ions. Using a ligand-to-metal binding stoichiometry of 2:1 for the SL-Cu^{2+} and SL-Fe^{3+} complexes and 1:1 for the SL-V^{5+} complex as shown in **Fig. 3**, stable complexes were generated between SL and these metal ions. Cu^{2+} (7.03 M, 1.37×10^4 M^{-1}), Fe^{3+} (5.16 M, 2.01×10^4 M^{-1}), and V^{5+} (5.94 M, 1.82×10^4 M^{-1}) were the LOD and binding constants found [58]. A different research demonstrated the use of a colorimetric and fluorometric sensor (RA) for the efficient and concurrent detection of multi-metal ions. Rhodamine hydrazide and 9-anthraldehyde underwent a Schiff base reaction to create the sensor. It demonstrated exceptional sensitivity for Cu^{2+} with a detection limit of 4.81×10^8 M and great sensibility for Hg^{2+} with an ultrasensitive measurement limit of 9.978×10^9 M. Through a variety of experimental approaches and theoretical computations, the system for sensing and interactions among RA and metal ions were validated [59].

3.2 *Modified Nanoparticles-Based Colorimetric Ligands*

Due to their tiny size, nanoparticles are materials with distinctive physical and chemical characteristics. Nanoparticles that have been modified with certain molecules to improve their characteristics or permit particular interactions are referred to as modified nanoparticles. Modified nanoparticles have significant benefits when used in colorimetric monitoring systems, including excellent sensitivity and selectivity, simplicity in synthesis and modification, and compatibility with various detection platforms. The various kinds of modified nanoparticles utilized in colorimetric monitoring systems are covered in this section, along with their benefits.

Due to their special physical and chemical characteristics, gold nanoparticles (GNPs) are perfect for use in biosensors. They feature a wide surface area for improved sensitivity, can be synthesized utilizing a variety of techniques, and are extremely stable. GNPs have unique optical and electrochemical characteristics, with size and shape having an impact on the color and absorption maxima. GNPs may readily be coupled with ligand molecules in order to detect certain analytes by altering their surface. The creation of sensors has widely utilized GNPs, notably for the detection of metallic ions in samples of water. A noticeable color change and shift in the absorption peak are the results of alterations in the aggregation of particles or disassembly, which are the basis for colorimetric GNP sensors. These sensors are capable of detecting analytes by monitoring fluctuations in the dielectric constant of the medium around them as well as particle aggregation or disaggregation. GNP sensors' colorimetric response is based on changes in the analyte's aggregation state and dielectric constant that result in variations in photon emission [60]. Schematic representation of the binding of gold nanoparticles-based sensor with metal ion is shown in Fig. 4.

For the simultaneous recognition of chromium ions (Cr^{3+} and Cr^{6+}), a colorimetric technique utilizing sodium hyaluronate modified gold nanoparticles (SH-AuNPs) has been devised. It is readily apparent to the unaided eye that the SH-AuNPs rapidly combine and transform from wine-red to blue-gray when interacting with chromium ions. It was discovered that the suggested value for Cr^{3+}, Cr^{6+}, and mixed chromium

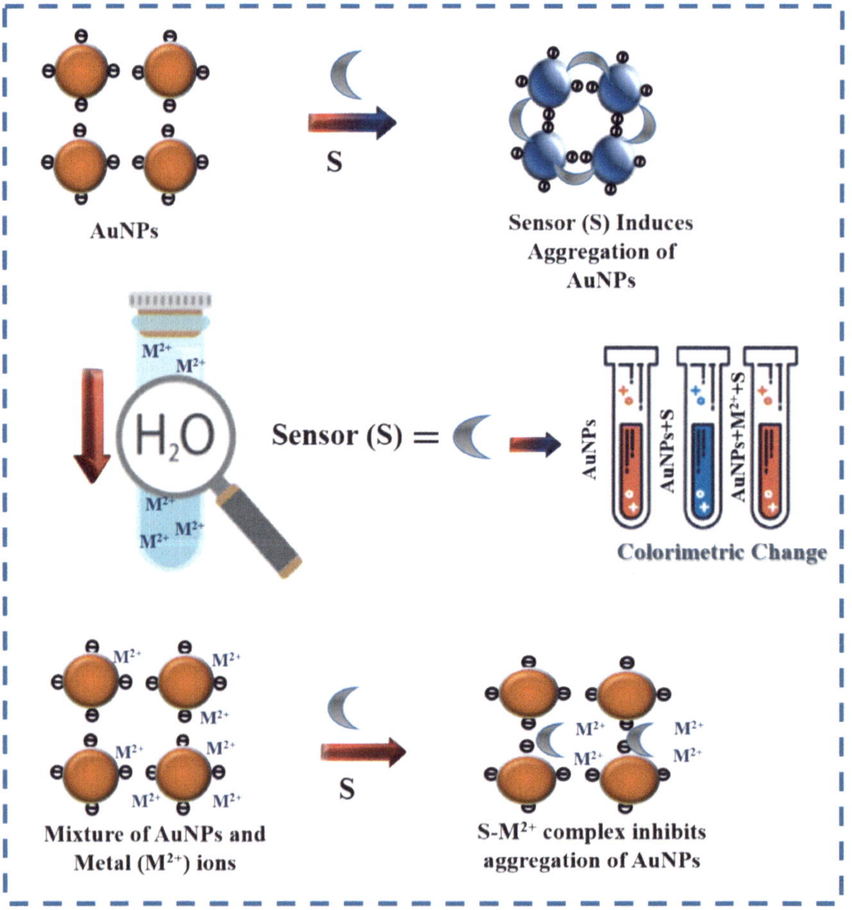

Fig. 4 Schematic representation of binding of gold nanoparticles based sensor with metal ion

ions by the World Health Organization (WHO) was below the limit of detection. By effectively detecting chromium ions in a sample of actual lake water, the usefulness of SH-AuNPs was demonstrated [61].

With the aid of gold nanoparticles (AuNPs), a fresh technique for optically identifying Hg^{2+} has been created. Employing poly(diallyldimethylammonium) chloride (PDDA) as a reducer and stabilizer, positively charged AuNPs were created. Cysteine caused the AuNPs to aggregate, which altered the color of the solution. On the other hand, extremely stable complexes were formed when Hg^{2+} and cysteine were pre-incubated; this reduced the number of free cysteine molecules and stopped AuNPs from aggregating. As the Hg^{2+} content increased, the AuNPs solution's color changed from purple to red, with a limit of detection of 2.5×10^{8} M. As a result, Hg^{2+} in the range of concentrations of 5.0×10^{8} to 1.0×10^{5} M could be seen. This approach is practical, economical, and doesn't call for sophisticated equipment [62].

A hazardous metal that pollutes the environment and endangers human health is cadmium. For the safety of the general population, effective testing procedures are required for cadmium contamination of milk and dairy products. Using l-cysteine-altered gold nanoparticles (AuNPs), a straightforward and inexpensive colorimetric test was created in this work to find cadmium in food. Strong surface plasmon resonance (SPR) absorptions in the visible wavelength region were seen in the synthesized AuNPs. The AuNPs aggregated and changed from red to deep blue once metal ions were added. The target metal ions were bound by ligands on the surface of the AuNPs, causing regulated aggregation and color change. A substantial interaction between cadmium ions and spherical AuNPs with an average size of 31 nm was observed. For the purpose of identifying harmful metal ions in food, this methodology provides a possible substitute for spectroscopy-based methods [63].

In this work, unaltered gold nanoparticles are used as the probe in a simple and accurate colorimetric detection approach for mercury ions (Hg^{2+}). The test is based on the interaction of bare gold nanoparticles with an anti-Hg^{2+} aptamer, which when exposed to Hg^{2+} takes on a certain conformation (T-Hg^{2+}-T). The Hg^{2+}-induced conformational shift of the aptamer may be clearly distinguished by the color change or adsorption ratio of the gold nanoparticles. The test responds linearly to Hg^{2+} concentrations between 1×10^{4} mol L^{-1} and 1×10^{9} mol L^{-1}, with spectrometric detection limits as low as 0.6 nM. It exhibits exceptional Hg^{2+} selectivity over other metal cations. The technique has several benefits that make it a potentially helpful tool for Hg^{2+} detection, including water solubility, simplicity, cheap cost, optical colorimetry, and high sensitivity [64].

In their comprehensive study, the authors describe a very sensitive and specific colorimetric sensor assay for Hg^{2+} ion detection. Gold nanoparticles (AuNPs) and an amine-based sensor (S) are used in the assay as part of an anti-aggregation sensing strategy. Normally, AuNPs aggregate when a sensor (S) is added to them, changing the color from red to blue. But when Hg^{2+} ions are present, sensor (S) prefers to attach to Hg^{2+} ions rather than AuNPs, blocking their aggregation and color change. To obtain excellent sensitivity and selectivity, the method is optimized for a number of parameters including incubation duration, pH, and sensor titration. With a predicted detection limit of 17 nM, below the permitted limit of Hg^{2+} ions in drinking water (30 nM), UV–visible investigations support the assay's efficacy. In addition to the presence of other ions, interference experiments show the method's selectivity for Hg^{2+} ions. Computational density functional theory (DFT) investigations are used to substantiate the experimental findings. The technique decreases the requirement for labor-intensive operations, eliminates complicated synthetic procedures, and is affordable [28].

In another interesting work, glutathione-stabilized silver nanoparticles (GSH-AgNPs) are synthesized in water and used in a colorimetric test to find Ni^{2+} ions. Transmission electron microscopy (TEM), UV–visible spectroscopy, and IR spectroscopy were all used to characterize the GSH-Ag NPs. Because of their surface plasmon absorption, newly created GSH-Ag NPs have a yellow color. The color of the GSH-AgNPs solution shifts to a deep orange in the presence of Ni^{2+} ions, and the surface plasmon absorption band widens and shifts to red. Except for Co^{2+}, which exhibits mild interference, the sensitivity of GSH-AgNPs to other metal ions, such as transition-metal ions, alkaline metals, and alkaline earth metals, is minimal. The reaction of the silver nanoparticles was shown to be affected by the Ni^{2+} ion concentration, and the detection limit for Ni^{2+} utilizing GSH-AgNPs was established to be 7.5×10^5 mol L^{-1}. The likely mechanism behind this reaction is examined [24]. In addition, the study noticed an intriguing phenomenon where 1,2-ethylenediamine induced the aggregated nanoparticles to disperse [65].

This study uses a water-based extract of Vigna mungo beans to synthesize silver nanoparticles (AgNPs) in an ecologically friendly manner. The extract functions as a reducing and stabilizing agent and contains proteins, phenolic acids, and flavonoids. UV–visible spectroscopy, Fourier transform infrared spectroscopy, and transmission electron microscopy were used to characterize the synthesized AgNPs. The AgNPs had a spherical shape and an approximately 430 nm-wide plasmon resonance absorption band. Then, without first surface modification with organic ligands, the AgNPs were employed for colorimetric identification of mercuric (Hg^{2+}) ions. The yellowish-brown colloidal solution containing AgNPs changed color upon the addition of Hg^{2+} ions, becoming colorless with a blue shift and widening of the plasmon band as a result of the breakdown of Ag nuclei. The interference of different metal ions was researched along with the specificity of the AgNPs for Hg^{2+} ions. A detection limit (LOD) of around 0.13 M (130 nM) for Hg^{2+} ions was obtained after testing several Hg^{2+} ion dilutions. The biologically made AgNPs have the potential to be used for inexpensive, one-step Hg^{2+} ion monitoring in wastewater reservoirs eliminating the necessity for pre-functionalization of the nanoparticles with organic compounds or ligands [66].

Keeping green synthesis in consideration, a biologically green synthesized silver nanoparticle (Ag NPs) to develop a colorimetric device for selectively identifying and tracking mercuric ions. Surface plasmon resonance (SPR), UV–vis spectroscopy, SEM, and X-ray diffraction analysis (XRD) techniques were used to characterize the Ag NPs. Due to their strong SPR absorption band, the fresh Ag NPs had a yellowish-brown color. The yellow Ag NPs solution became colorless when Hg^{2+} ions were added, and the SPR band also widened and shifted to blue. The green synthesized Ag NPs' receptivity and selectivity to other metal ions were assessed. Mercury (II) ions were found to have a LOD of 2.2×10^6 mol L^{-1}. The suggested technique was successfully used to determine the presence of mercury (II) ions in many water samples [67].

Utilizing unaltered silver nanoparticles (AgNPs) were used to create selective colorimetric sensors for the quick detection of Hg (II) ions. The diameter of the synthesized AgNPs was measured using transmission electron microscopy (TEM), which gave an average value of 8.3–1.4 nm. As the quantity of Hg (II) grew, the absorbance of the AgNPs reduced and slightly moved to the blue wavelength, suggesting the oxidation of Ag (0) to Ag(I). AgNPs were broken up into tiny pieces and Hg (II) adsorbed on their surface was reduced to Hg (0) as a result of their contact with Hg (II). With a detection level of detection (LOD) of 0.06 ppm, the detection range of this approach was proven to be between 0.5 and 7 ppm. Interestingly, the sensitivity of the analysis for Hg (II) was greatly improved by the addition of Cu (II) to the AgNPs, decreasing the detection limit to 0.008 ppm. When compared to the other metal ions evaluated, the technique showed a higher selectivity for Hg (II). The method worked well and showed high accuracy and consistency with results from the graphite furnace atomic absorption spectrometer (GFAAS). It was successfully used to detect Hg (II) in water samples. Overall, this work offers an AgNP-based technique for the quick, easy, sensitive, and focused detection of Hg (II) in water samples [68].

In this work, silver nanoparticles (AgNP) coated with glutathione (GSH) are used to provide an extremely sensitive colorimetric sensing platform for the targeted measurement of Co^{2+} ions. Metallic nanoparticles with shapes like nanospheres, nanoplates, and nanorods that are utilized in colorimetric detection may be more or less discriminating in what they can detect. GSH-AgNPs with a spherical shape have good sensitivity to a variety of metal ions but lack Co^{2+}-specific recognition. But rod-shaped GSH-AgNPs in AgNPs exhibit a special reaction to Co^{2+} ions, that is ascribed to the synergistic action of CTAB and GSH. Potential applications for this rod-shaped GSH-AgNPs-based selective Co^{2+} ion detection include trace analysis [69].

Dopamine dithiocarbamate functionalized silver nanoparticles (DDTC-Ag NPs) have been used as a colorimetric sensor in a colorimetric assay that has been designed for the quick and accurate detection of Co^{2+}. AgNPs aggregate when Co^{2+} ions are present due to coordinating covalent connections formed with the DDTC functional groups, changing their color from yellow to reddish. This color shift can be immediately seen with the unaided eye or quantified via UV–vis spectrophotometry. With a detection limit of 14 M, the test exhibits uniformity in a concentration range of 1.0 mM to 15 mM. The procedure is straightforward, sensitive, and enables visual inspection-based direct cobalt ion monitoring [70].

4 Future Directions and Challenges

Despite the positive outcomes of colorimetric monitoring tools, there are still a number of issues to be solved. The existing restrictions and difficulties in the creation of colorimetric monitoring systems are discussed in this section, including their range of sensitivity and selectivity towards certain pollutants. The integration of colorimetric monitoring tools with other detection platforms is one of the probable future approaches and developments that are addressed.

5 Conclusion

A comprehensive discussion of several colorimetric sensing techniques for the identification of heavy metal ions has been provided in this book chapter's conclusion. The research that has been emphasized has shown the potential of colorimetric assays as quick, sensitive, and focused ways to analyze heavy metals in a range of samples. The usage of various kinds of nanoparticles, functionalization with certain ligands or biomolecules, and the development of anti-aggregation or anti-coagulation techniques are just a few of the cutting-edge tactics that have been investigated. These techniques make use of the unique optical properties of plasmonic nanoparticles and how they connect with the target analytes to create color changes that are easy to detect and measure.

With detection limits ranging from nanomolar to micromolar levels, the assays that have been published have shown extraordinary sensitivity, making it feasible to analyze heavy metal ions at a minimal level. Overall, the research results discussed in this book chapter provide insights into the development of colorimetric sensing techniques for heavy metal ion identification as well as the design and optimization of sensing platforms.

References

1. Shakoor, M.B., et al.: Human health implications, risk assessment and remediation of As-contaminated water: a critical review. Sci. Total. Environ. **601–602**, 756–769 (2017). https://doi.org/10.1016/j.scitotenv.2017.05.223
2. Ravindran, A., Elavarasi, M., Prathna, T.C., Raichur, A.M., Chandrasekaran, N., Mukherjee, A.: Selective colorimetric detection of nanomolar Cr (VI) in aqueous solutions using unmodified silver nanoparticles. Sens. Actuators B Chem. **166–167**, 365–371 (2012). https://doi.org/10.1016/j.snb.2012.02.073
3. Hyder, A., et al.: Identification of heavy metal ions from aqueous environment through gold, silver and copper nanoparticles: an excellent colorimetric approach. Environ. Res. **205**, 112475 (2022). https://doi.org/10.1016/j.envres.2021.112475
4. Buledi, J.A., Amin, S., Haider, S.I., Bhanger, M.I., Solangi, A.R.: A review on detection of heavy metals from aqueous media using nanomaterial-based sensors. Environ. Sci. Pollut. Res. Int. **28**(42), 58994–59002 (2021). https://doi.org/10.1007/s11356-020-07865-7

5. Chen, Z., Zhang, Z., Qi, J., You, J., Ma, J., Chen, L.: Colorimetric detection of heavy metal ions with various chromogenic materials: Strategies and applications. J. Hazard. Mater. **441**, 129889 (2023). https://doi.org/10.1016/j.jhazmat.2022.129889
6. Lindberg, A.-L., Goessler, W., Grandér, M., Nermell, B., Vahter, M.: Evaluation of the three most commonly used analytical methods for determination of inorganic arsenic and its metabolites in urine. Toxicol. Lett. **168**(3), 310–318 (2007). https://doi.org/10.1016/j.toxlet.2006.10.028
7. Umapathi, R., et al.: Colorimetric based on-site sensing strategies for the rapid detection of pesticides in agricultural foods: New horizons, perspectives, and challenges. Coord. Chem. Rev. **446**, 214061 (2021). https://doi.org/10.1016/j.ccr.2021.214061
8. Sidana, N., Kaur, H., Devi, P.: Chapter 8—organic linkers for colorimetric detection of inorganic water pollutants. In: Devi, P., Singh, P., S. K. B. T.-I. P. in W. Kansal, Eds. Elsevier, pp. 135–152 (2020)
9. Momidi, B.K., Tekuri, V., Trivedi, D.R.: Multi-signaling thiocarbohydrazide based colorimetric sensors for the selective recognition of heavy metal ions in an aqueous medium. Spectrochim. Acta Part A Mol. Biomol. Spectrosc. **180**, 175–182 (2017). https://doi.org/10.1016/j.saa.2017.03.010
10. Veiseh, O., Gunn, J.W., Zhang, M.: Design and fabrication of magnetic nanoparticles for targeted drug delivery and imaging. Adv. Drug Deliv. Rev. **62**(3), 284–304 (2010). https://doi.org/10.1016/j.addr.2009.11.002
11. Kar, A., Rout, S. R., Giri, L., Sahebkar, A., Kesharwani, P., Dandela, R.: 8—Aptamer-functionalized nanoparticles for targeted cancer therapy. In: P. B. T.-A. E. N. for C. T. Kesharwani, Ed., Woodhead Publishing Series in Biomaterials, Woodhead Publishing, pp. 191–218 (2023)
12. Rasheed, T., Shafi, S., Ali, J., Sher, F., Rizwan, K., Khan, S.: Recent advances in chemically and biologically synthesized nanostructures for colorimetric detection of heavy metal. J. King Saud Univ. Sci. **34**(2), 101745 (2022). https://doi.org/10.1016/j.jksus.2021.101745
13. Rasheed, T., Nabeel, F., Li, C., Zhang, Y.: Rhodol assisted alternating copolymer based chromogenic vesicles for the aqueous detection and quantification of hydrazine via switch-on strategy. J. Mol. Liq. **274**, 461–469 (2019). https://doi.org/10.1016/j.molliq.2018.11.014
14. Borah, S.B.D., Bora, T., Baruah, S., Dutta, J.: Heavy metal ion sensing in water using surface plasmon resonance of metallic nanostructures. Groundw. Sustain. Dev. **1**(1), 1–11 (2015). https://doi.org/10.1016/j.gsd.2015.12.004
15. Kaur, B., Srivastava, R., Satpati, B.: Ultratrace detection of toxic heavy metal ions found in water bodies using hydroxyapatite supported nanocrystalline ZSM-5 modified electrodes. New J. Chem. **39**(7), 5137–5149 (2015). https://doi.org/10.1039/C4NJ02369B
16. Du, J., et al.: Colorimetric detection of mercury ions based on plasmonic nanoparticles. Small **9**(9–10), 1467–1481 (2013). https://doi.org/10.1002/smll.201200811
17. Gabriel, E.F.M., Garcia, P.T., Cardoso, T.M.G., Lopes, F.M., Martins, F.T., Coltro, W.K.T.: Highly sensitive colorimetric detection of glucose and uric acid in biological fluids using chitosan-modified paper microfluidic devices. Analyst **141**(15), 4749–4756 (2016). https://doi.org/10.1039/C6AN00430J
18. Wang, B., Lan, J., Bo, C., Gong, B., Ou, J.: Adsorption of heavy metal onto biomass-derived activated carbon: review. RSC Adv. **13**(7), 4275–4302 (2023). https://doi.org/10.1039/d2ra07911a
19. Briffa, J., Sinagra, E., Blundell, R.: Heavy metal pollution in the environment and their toxicological effects on humans. Heliyon **6**(9), e04691 (2020). https://doi.org/10.1016/j.heliyon.2020.e04691
20. Khatun, J., Intekhab, A., Dhak, D.: Effect of uncontrolled fertilization and heavy metal toxicity associated with arsenic(As), lead(Pb) and cadmium (Cd), and possible remediation. Toxicology **477**, 153274 (2022). https://doi.org/10.1016/j.tox.2022.153274
21. Vurro, M., Miguel-Rojas, C., Pérez-de-Luque, A.: Safe nanotechnologies for increasing the effectiveness of environmentally friendly natural agrochemicals. Pest Manag. Sci. **75**(9), 2403–2412 (2019). https://doi.org/10.1002/ps.5348

22. Beauchemin, D.: Environmental analysis by inductively coupled plasma mass spectrometry. Mass Spectrom. Rev. **29**(4), 560–592 (2010). https://doi.org/10.1002/mas.20257
23. Jarujamrus, P., et al.: Use of a smartphone as a colorimetric analyzer in paper-based devices for sensitive and selective determination of mercury in water samples. Anal. Sci. **34**(1), 75–81 (2018). https://doi.org/10.2116/analsci.34.75
24. Azzouz, A., Vikrant, K., Kim, K.-H., Ballesteros, E., Rhadfi, T., Malik, A.K.: Advances in colorimetric and optical sensing for gaseous volatile organic compounds. TrAC Trends Anal. Chem. **118**, 502–516 (2019). https://doi.org/10.1016/j.trac.2019.06.017
25. Nanseu-Njiki, C.P., Tchamango, S.R., Ngom, P.C., Darchen, A., Ngameni, E.: Mercury(II) removal from water by electrocoagulation using aluminium and iron electrodes. J. Hazard. Mater. **168**(2), 1430–1436 (2009). https://doi.org/10.1016/j.jhazmat.2009.03.042
26. Nunes, E., Cavaco, A., Carvalho, C.: Exposure assessment of pregnant portuguese women to methylmercury through the ingestion of fish: cross-sectional survey and biomarker validation. J. Toxicol. Environ. Heal. Part A **77**(1–3), 133–142 (2014). https://doi.org/10.1080/15287394.2014.867200
27. Spiegel, S., Keane, S., Metcalf, S., Veiga, M.: Implications of the Minamata Convention on Mercury for informal gold mining in Sub-Saharan Africa: from global policy debates to grass-roots implementation? Environ. Dev. Sustain. **17**(4), 765–785 (2015). https://doi.org/10.1007/s10668-014-9574-1
28. Sidana, N., Rajput, P., Devi, P., Kaur, H.: A simple and highly selective colorimetric sensing of Hg2+ ions based on anti-aggregation of gold nanoparticles. IEEE Sens. J. **22**(23), 22549–22556 (2022). https://doi.org/10.1109/JSEN.2022.3213686
29. Garelick, H., Jones, H., Dybowska, A., Valsami-Jones, E.: Arsenic pollution sources BT—reviews of environmental contamination volume 197: international perspectives on arsenic pollution and remediation. New York, NY: Springer New York, pp. 17–60 (2008)
30. Saha, J.C., Dikshit, A.K., Bandyopadhyay, M., Saha, K.C.: A review of arsenic poisoning and its effects on human health. Crit. Rev. Environ. Sci. Technol. **29**(3), 281–313 (1999). https://doi.org/10.1080/10643389991259227
31. Tchounwou, P.B., Patlolla, A.K., Centeno, J.A.: Invited reviews: carcinogenic and systemic health effects associated with arsenic exposure—a critical review. Toxicol. Pathol. **31**(6), 575–588 (2003). https://doi.org/10.1080/01926230390242007
32. Abejón, A., Garea, A., Irabien, A.: Arsenic removal from drinking water by reverse osmosis: minimization of costs and energy consumption. Sep. Purif. Technol. **144**, 46–53 (2015). https://doi.org/10.1016/j.seppur.2015.02.017
33. Shrivas, K., Shankar, R., Dewangan, K.: Gold nanoparticles as a localized surface plasmon resonance based chemical sensor for on-site colorimetric detection of arsenic in water samples. Sens. Actuators B Chem. **220**, 1376–1383 (2015). https://doi.org/10.1016/j.snb.2015.07.058
34. Aksuner, N., Henden, E., Yilmaz, I., Cukurovali, A.: A highly sensitive and selective fluorescent sensor for the determination of copper(II) based on a schiff base. Dye Pigment. **83**(2), 211–217 (2009). https://doi.org/10.1016/j.dyepig.2009.04.012
35. Sidana, N., Kaur, M., Kaur, H., Devi, P.: Pyrrole-based schiff-base sensor for copper ions. Chem. Select **5**(46), 14776–14782 (2020). https://doi.org/10.1002/slct.202003701
36. Mergu, N., Gupta, V.K.: A novel colorimetric detection probe for copper(II) ions based on a Schiff base. Sens. Actuators B Chem. **210**, 408–417 (2015). https://doi.org/10.1016/j.snb.2014.12.130
37. Fitzgerald, D.J.: Safety guidelines for copper in water. Am. J. Clin. Nutr. **67**(5), 1098S-1102S (1998). https://doi.org/10.1093/ajcn/67.5.1098S
38. Métivier, R., Leray, I., Valeur, B.: Lead and mercury sensing by calixarene-based fluoroionophores bearing two or four dansyl fluorophores. Chem. A Eur. J. **10**(18), 4480–4490 (2004). https://doi.org/10.1002/chem.200400259
39. Syal, R., Sharma, P., Singh, S., Kumar, S.: Investigation of energy storage properties in lead-free BZT-40BCT relaxor ceramic. Mater. Today Proc. (2023). https://doi.org/10.1016/j.matpr.2023.02.057

40. Syal, R., et al.: Chemical engineering at morphotropic phase boundary: interplay with thermally stabilized ultra-high energy storage and electrocaloric response. Solid State Sci. **140**, 107190 (2023). https://doi.org/10.1016/j.solidstatesciences.2023.107190
41. Medici, S., Peana, M., Nurchi, V.M., Zoroddu, M.A.: Medical uses of silver: history, myths, and scientific evidence. J. Med. Chem. **62**(13), 5923–5943 (2019). https://doi.org/10.1021/acs.jmedchem.8b01439
42. Kagan, D., et al.: Chemical sensing based on catalytic nanomotors: motion-based detection of trace silver. J. Am. Chem. Soc. **131**(34), 12082–12083 (2009). https://doi.org/10.1021/ja905142q
43. Nayan Kumar, H.N. et al.: Recent advances in on-site monitoring of heavy metal ions in the environment. Microchem. J. **182**, 107894 (2022). https://doi.org/10.1016/j.microc.2022.107894
44. Boulechfar, C., et al.: Schiff bases and their metal Complexes: a review on the history, synthesis, and applications. Inorg. Chem. Commun. **150**, 110451 (2023). https://doi.org/10.1016/j.inoche.2023.110451
45. Hameed, A., al-Rashida, M., Uroos, M., Abid Ali, S., Khan, K.M.: Schiff bases in medicinal chemistry: a patent review (2010–2015). Expert Opin. Ther. Pat. **27**(1), 63–79 (2017). https://doi.org/10.1080/13543776.2017.1252752
46. Khan, S., Muhammad, M., Kamran, A.W., Al-Saidi, H.M., Alharthi, S.S., Algethami, J.S.: An ultrasensitive colorimetric and fluorescent 'turn-on' chemosensor based on Schiff base for the detection of Cu2+ in the aqueous medium. Environ. Monit. Assess. **195**(6), 633 (2023). https://doi.org/10.1007/s10661-023-11260-3
47. Tamizhselvi, R., Arumugam Napoleon, A.: A simple sulfonohydrazide Schiff base molecular probe for colorimetrically selective recognition of Cu2+ ion in semi-aqueous medium. Inorg. Chem. Commun. **139**, 109330 (2022). https://doi.org/10.1016/j.inoche.2022.109330
48. Khan, S., Muhammad, M., Al-Saidi, H.M., Hassanian, A.A., Alharbi, W., Alharbi, K.H.: Synthesis, characterization and applications of schiff base chemosensor for determination of Cu2+ ions. J. Saudi Chem. Soc. **26**(4), 101503 (2022). https://doi.org/10.1016/j.jscs.2022.101503
49. Sahu, M., Manna, A.K., Patra, G.K.: A fluorescent colorimetric vanillin di-Schiff base chemosensor for detection of Cu(II) and isolation of trinuclear Cu(II)-dihydrazide. Mater. Adv. **3**(5), 2495–2504 (2022). https://doi.org/10.1039/d1ma01227d
50. Jiang, D., et al.: An α-naphtholphthalein-derived colorimetric fluorescent chemoprobe for the portable and visualized monitoring of Hg2+ by the hydrolysis mechanism. New J. Chem. **46**(24), 11695–11705 (2022). https://doi.org/10.1039/D2NJ01051H
51. Gigi, G., Mohan, A.M.: Probe-impregnated monolithic polymer as a robust solid-state colorimetric chemosensor for selective sensing of Hg2+ in environmental water and cigarette samples. Environ. Res. **220**, 115210 (2023). https://doi.org/10.1016/j.envres.2023.115210
52. Gangatharan Vinoth Kumar, G., Bhaskar, R., Harathi, J., Jayaprakash, N.: Selective colorimetric signaling of mercury (II) ions using a quinoline-based probe with INHIBIT logic gate behavior and test strip. Inorg. Chem. Commun. **148**, 110364 (2023). https://doi.org/10.1016/j.inoche.2022.110364
53. Krishnan, U., Iyer, S.K.: A pyrazolo imine-based colorimetric and turn-on fluorescent sensor probe for determination of Hg2+ ion and its application in test paper strips. Photochem. Photobiol. **98**(4), 843–855 (2022). https://doi.org/10.1111/php.13538
54. Jayaraj, A., Gayathri, M.S., Sivaraman, G., C.A.S.P.: A highly potential acyclic Schiff base fluorescent turn on sensor for Zn2+ ions and colorimetric chemosensor for Zn2+, Cu2+ and Co2+ ions and its applicability in live cell imaging. J. Photochem. Photobiol. B Biol. 226, 112371 (2022). https://doi.org/10.1016/j.jphotobiol.2021.112371
55. Gopalakrishnan, A.K., Angamaly, S.A., Pradeep, S.D., Madhusoodhanan, D.T., Manoharan, D.K., Mohanan, P.V.: A novel imidazole bound schiff base as highly selective 'turn-on' fluorescence sensor for Zn2+ and colorimetric Kit for Co2+. J. Fluoresc. **32**(1), 189–202 (2022). https://doi.org/10.1007/s10895-021-02839-5

56. Das, P., et al.: Easy, selective and colorimetric detection of Zn(II), Cu(II), F-ions by a new piperazine based Schiff base chemosensor along with molecular logic gate formation and live cell images study. J. Photochem. Photobiol. A Chem. **427**, 113817 (2022). https://doi.org/10.1016/j.jphotochem.2022.113817
57. Bhalla, P., Bhagat, P., Malhotra, S.: A simple naphthaldehyde based sensor as optical and colorimetric for the detection of Hg2+/Cr3+ in real samples. J. Mol. Struct. **1282**, 135130 (2023). https://doi.org/10.1016/j.molstruc.2023.135130
58. Alorabi, A.Q., Zabin, S.A., Alam, M.M., Abdelbaset, M.: Schiff base ligand 3-(-(2-Hydroxyphenylimino) Methyl)-4H-Chromen-4-One as colorimetric sensor for detection of Cu^{2+}, Fe^{3+}, and V^{5+} in aqueous solutions. Int. J. Anal. Chem. **2022**, 4899145 (2022). https://doi.org/10.1155/2022/4899145
59. Hu, J.-P., Lin, Q., Yao, H., Zhang, Y.-M., Wei, T.-B.: Fluorescence and colorimetric-based sensing of Hg2+ and Cu2+ using rhodamine derivative. Appl. Organomet. Chem. **37**(2), e6938 (2023). https://doi.org/10.1002/aoc.6938
60. Priyadarshini, E., Pradhan, N.: Gold nanoparticles as efficient sensors in colorimetric detection of toxic metal ions: a review. Sens. Actuators, B Chem. **238**, 888–902 (2017). https://doi.org/10.1016/j.snb.2016.06.081
61. Li, S., Wei, T., Ren, G., Chai, F., Wu, H., Qu, F.: Gold nanoparticles based colorimetric probe for Cr(III) and Cr(VI) detection. Colloids Surf. A Physicochem. Eng. Asp. **535**, 215–224 (2017). https://doi.org/10.1016/j.colsurfa.2017.09.028
62. Ding, N., et al.: A simple colorimetric sensor based on anti-aggregation of gold nanoparticles for Hg2+ detection. Colloids Surfaces A Physicochem. Eng. Asp. **395**, 161–167 (2012). https://doi.org/10.1016/j.colsurfa.2011.12.024
63. Sonia, Seth, R.: L-cysteine functionalized gold nanoparticles as a colorimetric sensor for ultra-sensitive detection of toxic metal ion cadmium. Mater. Today Proc. **24**, 2375–2382 (2020). https://doi.org/10.1016/j.matpr.2020.03.767
64. Li, L., Li, B., Qi, Y., Jin, Y.: Label-free aptamer-based colorimetric detection of mercury ions in aqueous media using unmodified gold nanoparticles as colorimetric probe. Anal. Bioanal. Chem. **393**(8), 2051–2057 (2009). https://doi.org/10.1007/s00216-009-2640-0
65. Li, H., Cui, Z., Han, C.: Glutathione-stabilized silver nanoparticles as colorimetric sensor for Ni2+ ion. Sens. Actuators B Chem. **143**(1), 87–92 (2009). https://doi.org/10.1016/j.snb.2009.09.013
66. Choudhary, M.K., Garg, S., Kaur, A., Kataria, J., Sharma, S.: Green biomimetic silver nanoparticles as invigorated colorimetric probe for Hg2+ ions: a cleaner approach towards recognition of heavy metal ions in aqueous media. Mater. Chem. Phys. **240**, 122164 (2020). https://doi.org/10.1016/j.matchemphys.2019.122164
67. Farhadi, K., Forough, M., Molaei, R., Hajizadeh, S., Rafipour, A.: Highly selective Hg2+ colorimetric sensor using green synthesized and unmodified silver nanoparticles. Sens. Actuators B Chem. **161**(1), 880–885 (2012). https://doi.org/10.1016/j.snb.2011.11.052
68. Jarujamrus, P., Amatatongchai, M., Thima, A., Khongrangdee, T., Mongkontong, C.: Selective colorimetric sensors based on the monitoring of an unmodified silver nanoparticles (AgNPs) reduction for a simple and rapid determination of mercury. Spectrochim. Acta Part A Mol. Biomol. Spectrosc. **142**, 86–93 (2015). https://doi.org/10.1016/j.saa.2015.01.084
69. Sung, H.K., Oh, S.Y., Park, C., Kim, Y.: Colorimetric detection of Co2+ ion using silver nanoparticles with spherical, plate, and rod shapes. Langmuir **29**(28), 8978–8982 (2013). https://doi.org/10.1021/la401408f
70. Mehta, V.N., Mungara, A.K., Kailasa, S.K.: Dopamine dithiocarbamate functionalized silver nanoparticles as colorimetric sensors for the detection of cobalt ion. Anal. Methods **5**(7), 1818–1822 (2013). https://doi.org/10.1039/C3AY26150F

2D Material-Based Textile Sensors for Human Health Monitoring Applications

Pradeep Kumar, Monika Gupta, Varun Jeoti, and Goran M. Stojanović

Abstract Two-dimensional (2D) materials such as graphene, transition metal dichalcogenides (TMDs), transition metal carbides, and nitrides have outstanding electronic, optical, thermal, chemical, and mechanical properties. Their employability in textiles has been extensively investigated for the realization of a new class of efficient wearable sensors exploited for health monitoring. In this chapter, a brief overview of different 2D materials will be given at the beginning with a discussion of their merits and properties. Next, different recently demonstrated textile sensors based on 2D materials will be discussed. This is followed by a detailed review of their state-of-the-art designs, methods, and performance evaluation. Later, the progress and recent development of wearable textile sensors toward monitoring human health will be highlighted, wherein different monitoring analytes and diagnosis methods will be emphasized. Finally, a summary with a brief discussion of the prospects of 2D materials-based textile sensors will be presented.

Keywords 2D materials · Wearable sensors · Textile · Health monitoring · Diagnosis · Analytes · Signals

1 Introduction

In recent years, industrialization has increased along with population growth. This raises the human's exposure to different chemical, biological, and physical threats. Due to these threats, people from all sensible age groups have seriously started paying attention to their health, thereby increasing the demand for healthcare monitoring devices and systems. For daily health evaluation, it is not convenient and

P. Kumar (✉) · V. Jeoti · G. M. Stojanović
Faculty of Technical Sciences, University of Novi Sad, Trg Dositeja Obradovića 6, 21000 Novi Sad, Serbia
e-mail: pradeep.kumar@uns.ac.rs

M. Gupta
Department of Electrical and Electronic Engineering, Universiti Teknologi PETRONAS, 32610 Seri Iskandar, Perak, Malaysia

A. Katoch et al. (eds.), *2D Materials*, Engineering Materials,
https://doi.org/10.1007/978-981-97-6258-3_8

economical for individuals to visit the clinic or hospital every day [1]. Therefore, it is essential to seek easy, affordable, user-friendly, sensitive, and selective sensors [2]. With technological advancement, wearable electronics have recently increased people's interest in developing biosensors and monitors that are capable of quantifying human metabolites and physiology [3, 4]. These biomarkers' information is very crucial to get early symptoms or detection of the disease. The development of high-performance, wearable electronics with high flexibility, multi-functionality, and lightweight is the demand to achieve the precise detection of different biomarkers [5]. The typical wearable sensors are based on silicon and organic materials, which has some drawbacks such as rigid hard substrate to be attached to human skin, poor biological compatibility, and not being very comfortable [6].

For fulfilling the aforementioned requirements numerous skin-attachable, portable wearable devices have been widely explored, which can be employed at some body parts such as the arm wrist, and inside of the mouth to detect the biomarkers like sweat, tears, and saliva [7]. Such wearable skin attachable biosensors are very successful in health monitoring applications owing to their miniaturized form factors, light weight, flexibility, and facile integration with the transmission/receiving units [7, 8]. However, major wearable biosensing devices use plastic thin films or papers which are limited by air or moisture permeability and not very stretchable. Due to these reasons, wearing comfort and durability are still some challenges [7, 9]. To resolve these technical challenges, researchers have recently started to explore textile substances like fibers, yarn, and fabrics because the textiles are light in weight, soft, biocompatible, and offer air/moisture permeability, high surface area, and comfortability [10]. The wearer will find textile-based sensors to be more comfortable and less invasive when they are readily incorporated into clothing or other textiles [11, 12]. For continuous and non-invasive physiological parameter monitoring, textile-based wearable biosensors have the benefit of being conveniently integrated into a variety of medical wearables, such as patient gowns or vests. Biosensors made of silicone and hard plastic, on the contrary, might not be as soft and flexible to the body's movements as textile-based sensors, making them less comfortable. Biosensors made of silicone and hard plastic might not be as adaptable in terms of integration, which could limit their practical applications in healthcare monitoring. Furthermore, different textile technologies such as weaving, sewing, and knitting can facilitate the accurate integration of textile-based sensors with the clothes. These features have increased the viability of textile-based biosensors in human health monitoring and disease diagnosis [13]. Numerous textile sensors have been developed for monitoring human health [14–20].

To realize high-performance sensors, various nanomaterials, and polymers have been widely explored [21–23]. Among them, two-dimensional (2D) materials such as graphene, MXene, and TMDs (transition metal dichalcogenides) have gained popularity owing to their outstanding physical, chemical, mechanical, and optical properties (i.e. large surface area, high transparency, flexibility, stretchability, and biocompatibility) [6, 24]. After the advent of graphene in 2004, different 2D materials have been exploited to successfully realize high-performance and efficient electronic [25], optical [26], optoelectronic [27, 28], energy storage [29], and sensing devices

[30]. The excellent stretchability and functionality with good electrical conductivity of 2D materials make them a highly suitable and ideal choice for sensing material in wearable biosensors [31].

In this chapter, we cover the 2D materials-based textile biosensors for monitoring human health. We highlight the different 2D materials along with their properties and merits in the beginning. Afterward, we discuss the different state-of-the-art textile biosensors based on 2D materials to detect different biomarkers such as sweat-glucose, cortisol, sodium ion, lactate, electrocardiogram (ECG), motion, and exhaled breath vapors as well as toxic gases. We reviewed the role of 2D materials in biomonitoring devices and their sensing performance. A summary is presented at the end with the prospects to be considered for future opportunities.

2 Research Methodology of the Chapter

A systematic literature review has been conducted to evaluate the body of the chapter on the 2D materials-based textile sensor for human health monitoring. An iterative process was used to conduct this systematic literature review, which involved defining relevant research terms, reviewing literature, analyzing data, and finalizing classification results. The following outlines the research methodology used in this chapter:

2.1 Methodology for Searching Literature

The literature for the chapter considers publications about 2D materials, textile-sensors based on the 2D materials for healthcare monitoring from the year 2010 to 2023. It includes features and properties of the 2D materials, device design, and behavior of the textile sensors based on the 2D materials followed by an extensive and systematic search of the peer-reviewed literature. The chapter contains empirical and non-empirical studies. An initial search was carried out through the Web of Science, Google Scholar databases to identify the relevant research papers. The list of papers was obtained from different publishers such as the Elsevier/Science Direct, ACS (American Chemical Society) publications, Springer Nature, Taylor & Francis, IEEE, Wiley, MDPI (multidisciplinary digital publishing institute), Nature, RSC (royal society of chemistry), AIP (American institute of physics), IOP (institute of physics). In order to ensure recent publications have been covered in this chapter, the same procedure was implemented to locate papers published between the years 2022 and 2023 according to set keywords such as textile biosensors, 2D materials, wearable sensors, textile-based monitors, sensor analytes, signals, sensors for health care, 2D materials for health monitoring. To make sure that recent publications have been covered in this chapter, the same process was used to find papers published between 2022 and 2023 according to predetermined keywords like textile

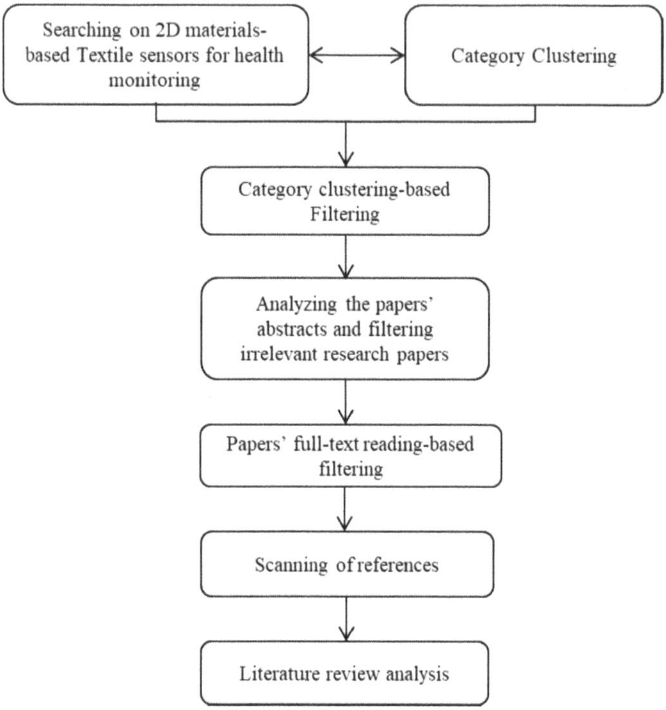

Scheme 1 Flowchart of the methodology for literature review for chapter formation

biosensors, 2D materials, wearable sensors, textile-based monitors, sensor analytes, signals, senor for healthcare, and 2D materials for health monitoring (Scheme 1).

2.2 Implementation of the Methodology

The research papers and literature documents were searched using combinations of the Boolean keywords—'AND', 'OR', and 'NOT' and field selection was set in 'All fields'. The keywords such as 'textile sensors', '2D materials', 'sweat sensors', 'ECG sensors' 'motion sensors', 'breath sensors' 'healthcare monitors' were used during the search of the literature for the chapter.

2.3 Reviewing, Refining, and Filtering Database

The search produced some results, and each article's full text was read and analyzed, including its abstract. After a thorough review, irrelevant papers were removed using

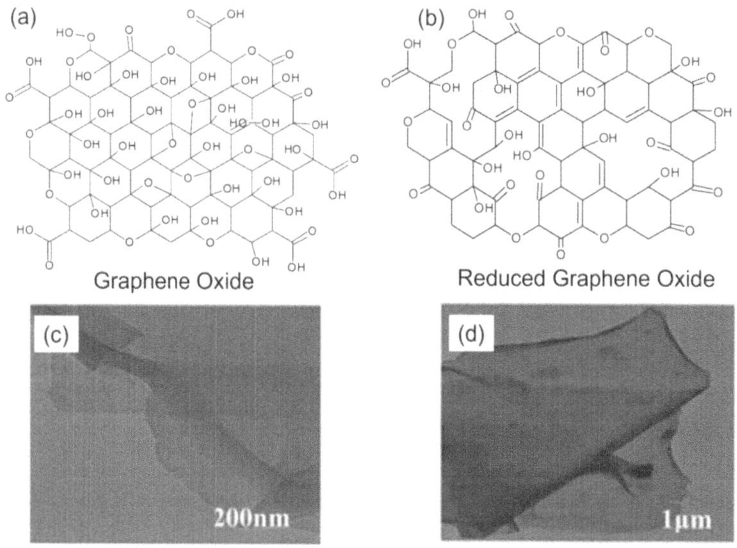

Fig. 1 Representation of **a** GO **b** rGO and TEM image of **c** GO and **d** rGO. Reprinted with permission from Shen et al. [38]. Copyright 2018 RSC

scholarly judgment. This process determined the scope of coverage and captured significant aspects of textile sensors based on 2D materials used for human health monitoring. This chapter includes reviews of a total of 68 papers. A flowchart representing the scheme of the methodology for the chapter is shown in Fig. 1. The results of the literature review enable the clustering of journal and keyword information, as well as data visualization.

3 Two Dimensional (2D) Materials

3.1 Graphene

Graphene is a single-layer thick carbon atom arranged in a honeycomb lattice (Fig. 1a and b). The SEM image of graphene oxide (GO) and reduced graphene oxide (rGO) thin films (Fig. 1c and d) has slight wrinkles with a crumpled surface. Graphene has been widely explored as wearable sensors and implantable devices in textile and health monitoring owing to extraordinary properties such as high surface area (2630 m^2 g^{-1}), ultrahigh carrier mobility (200,000 cm^2V^{-1} s^{-1}) [32], good electrical conductivity, high optical transmittance (97.7%) [33], and superior thermal conductivity (5000 Wm^{-1} K^{-1}) [34–36] with good mechanical properties. Graphene is a competitively low-cost choice because it is efficient, cheap, suitable for mass production, and compatible with recent technology. Therefore, graphene as a wearable sensor has a high potential in contributing to the development of fast and efficient sensors. The high surface area and single atomic thickness of graphene layers make entire carbon atoms more accessible for the analytes. Moreover, being just carbon, graphene can be used in intimate contact with human organs (skin, brain, eyes) without any risk [37]. Also, its excellent mechanical properties and ultrathin thickness allow it to be used in motion objects as a flexible wearable sensor without irritation or contamination. Moreover, the high optical transparency and high electroconductivity of graphene make it an outstanding material for clear visual images for bio-tissue observations.

3.2 *Transition Metal Carbides and Nitrides-MXene*

2D transition metal carbides and nitrides are generally termed as MXene. When the precursor transforms itself into an MXene molecule, there is an observation of color change that is very distinctive. The precursors (MAX phase) are usually grey colored and the products given possess different colors that are dependent on the overall structure and composition of the molecules. The ternary carbides and nitrides with a wide-ranging formula MAX ($M_{n+1}AX_n$) have been augmented to the family of two-dimensional materials. In this general formula, early transition metals such as V, Ti, Sc, Cr, Nb, Zr, Hf, Mo, Ta) are denoted by M. n is an integer 1, 2, or 3. A represents the element of group IIIA or IVA, and X refers to carbon and/or nitrogen (C and/or N). MXenes was first produced in 2011 by Naguib et al. [39]. MXenes are synthesized by selective etching of layer A from the MAX phase precursors (see Fig. 2) [40]. The resultant synthesized material that is MXene has the general formula of $M_{n+1}X_nT_x$, where T represents to the functional groups like –F, –OH, and –O). These surface functional groups can be manipulated using calcination and alkalization treatments depending on the application [29]. Monolayer MXene is metallic in nature due to conductivity comparable to the metallic material. However, O-terminated MXene is considered a semiconductor. Therefore, multilayered MXene shows extraordinary properties such as chemical stability, large active surface area, good conductivity (~4000 S/cm) [41], excellent hydrophilicity, open paths towards various applications such as catalysis, transparent conducting electrodes (TCEs), energy storage and gas sensors [42].

Transition Metal Dichalcogenides—TMDs

Another important category of 2D material is TMDs, particularly represented by MX_2 formula, where M represents transition metal elements (molybdenum-Mo, tungsten-W, etc.) and X is chalcogen (sulphur-S, selenium-Se, or tellurium-Te). The properties of TMDs greatly depend on the transition metal and structure phase. The band gap energy ($E_g = 0.2$–2 eV) of TMDs depends on their thickness [43]. The TMD phases have two types: either trigonal prismatic with a honeycomb motif (2H) or octahedral with a centered-honeycomb motif (1 T) [44]. TMDs could reveal semiconducting (WSe_2, WS_2, MoS_2, and $MoSe_2$ with 2H phase) or metallic (niobium disulfide

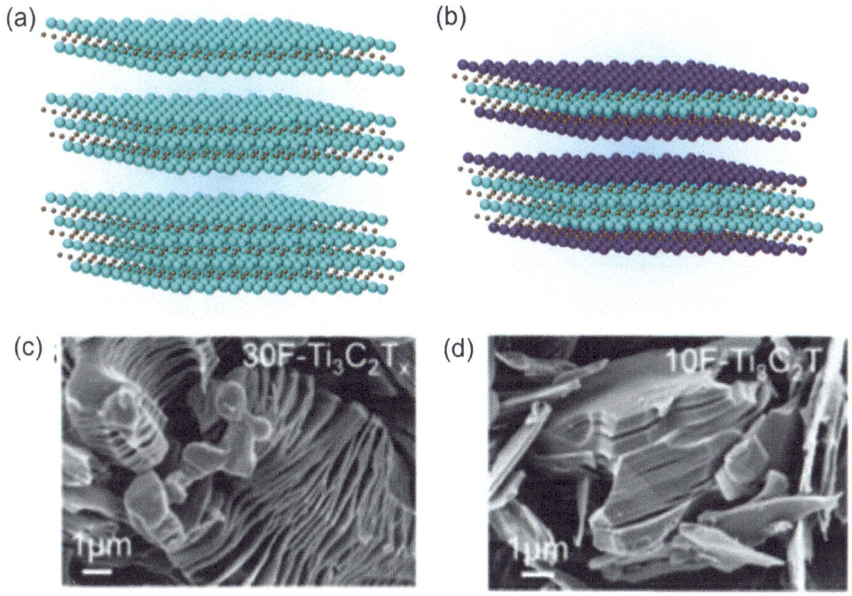

Fig. 2 **a** MXenes structures: M_2X, M_3X_2, and M_4X_3. **b** Two new families of MXenes: ordered double transition metal carbides with $M'_2 M''C_2$ and $M'_2 M''_2C_3$ formula. M, M′, and M″ are early transition metals and X is carbon and/or nitrogen. SEM image of multilayered $Ti_3C_2T_x$ powder synthesized with **c** 30 wt%, **d** 10 wt%. Reprinted with permission from Hajian et al. [40]. Copyright 2022 Frontiers

NbS_2 and TaX_2 with 1 T phases) behaviors. The most commonly explored TMDs are molybdenum sulfide (MoS_2), molybdenum diselenide ($MoSe_2$), Tungsten disulfide (WS_2), Tungsten diselenide (WSe_2), Rhenium disulfide (ReS_2), and Rhenium diselenide ($ReSe_2$) [45–48]. Among them, MoS_2 (Fig. 3a–c, high-resolution transmission electron microscopy HRTEM) is widely explored, and owing to excellent properties such as fast electron transport, a quenching ability, and good conductivity makes it a promising candidate in various areas such as storage devices and sensors [49].

4 2D Material-Based Textile Sensors/Systems for Monitoring of Human Health

The integration of 2D materials into textiles has shown great promise in developing sensors and systems for monitoring human health. These sensors can provide real-time data on various physiological parameters, enabling continuous and non-invasive health monitoring. Here are some key aspects and applications of 2D material-based textile sensors/systems in the context of human health monitoring:

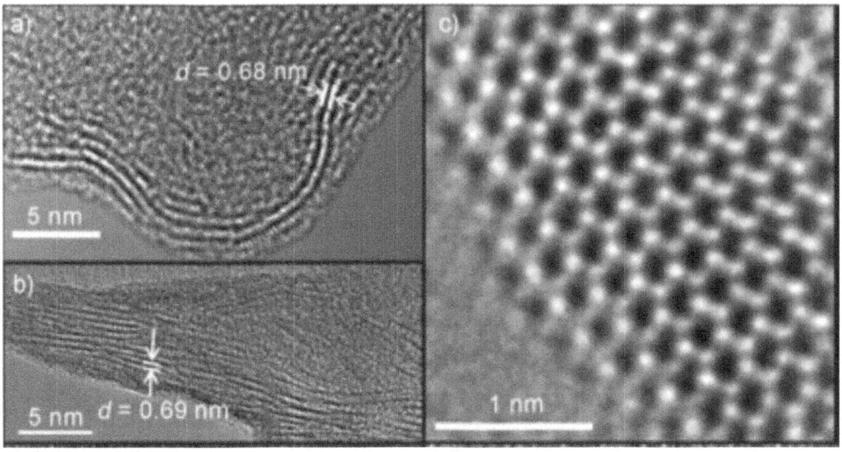

Fig. 3 **a** and **b** TEM images of MoS_2 layers **c** High-resolution TEM image of layered MoS_2. Reprinted with permission from Matte et al. [49]. Copyright 2010 Wiley

4.1 2D Materials-Based Textile Sweat Sensors

The 2D material–based textile sensors have been demonstrated as very effective detectors in identifying sweat biomarkers such as electrolytes, trace elements, small molecules, and metabolites. For example, the wearable sweat sensor socks developed using graphene ink-based stretchable electrodes indicated high potentiometric sensing ability to detect the sodium ions concentration in human sweat [50]. These textile sensors based on ecoflex-modified graphene ink electrodes performed well during the testing under different mechanical deformations. It also exhibited high sensitivity, selectivity, repeatability, and stability (Fig. 4). During the on-body measurement, the sensor realized a high sensing response in terms of electromotive force (EMF ~ 92–175 mV). Moreover, the graphene-based textile sweat sensor indicates excellent biocompatibility. During the test, there was no clinical evidence of any symptoms of any kind of skin irritation.

Alike graphene, another 2D material MXene-based textile smart healthcare system can effectively monitor the individual's thermal comfort, disease conditions, and nervous system activity by evaluating the thermal regulation and skin barrier functions of the human body. Liu et al. [51] demonstrated an MXene composite based on a noninvasive, highly sensitive, hydrophobic sweat sensor as shown in Fig. 5, which can constantly detect the sweat vapor from insensible sweat with rapid response. While validating the textile sensors for on-body testing, they were able to effectively detect the insensible sweat from different body parts such as the arm (6–15 g/m^2), forehead (29–38 g/m^2), and palms (67–72 g/m^2). The insensible sweat rate is a very informative tool for evaluating the reliability of skin barrier function. During the wound healing test, it was observed that the rate of insensible sweat gradually decreased from 106.18 g/m^2/ h to 45.77 g/m^2/h on day 10. On the 60th day, the

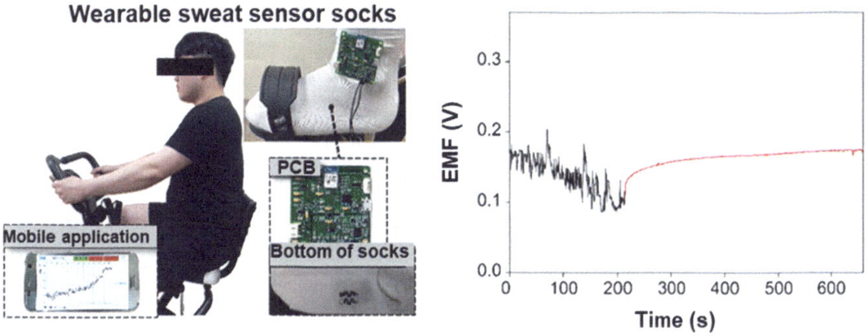

Fig. 4 Wearable sweat sensor sock (Eco – Gr sensors with an electrochemical analyzer) with its EMF response (right) obtained from a subject's sweat during indoor cycling. Reprinted with permission from Kil et al. [50]. Copyright 2022 ACS

insensible sweat rate reached a low value of 22.01 g/m^2/h. Besides monitoring the human sweat biomarkers, the sensor device system was capable of diagnosis of the disease, and detection of emotional and gustatory stimulations. The sensing system was highly suitable for different weather conditions. In the sensor design, the usage of super hydrophobic textiles as the encapsulation layer has the advantage of preventing the sensor's behavior from external hydrated conditions like sensible sweat, washing, etc.

In another novel approach, transition metal oxide (TMO) integrated carbon yarn is used to design a non-invasive electrochemical sensor for cortisol monitoring. Cortisol is a small molecule in the sweat. The cortisol level is extremely important in analyzing stress and related diseases. Madhu et al. [52] developed the titanium dioxide (TiO_2) anchored conductive carbon-coated textile yarn (CCY) as the sensing platform to detect the cortisol level in the sweat as depicted in Fig. 6. They found the TMO-CCY-based immunosensor as a highly sensitive cortisol detector which offered a lower detection limit (6 fg/ml) and wide response range (10 fg/ml–1 μg/ml). The sensor also demonstrated the high selectivity towards cortisol concentration with less cross-reactivity (±3 μA) while examining it in the presence of other interferents such as cholesterol, testosterone, cortisone, BSA, progesterone, and corticosterone. During the real-time examination, the TMO-CCY was found to be highly accurate, indicating its capability for highly sensitive and selective non-invasive POC (point-of-care) cortisol immunosensor.

Innovations in materials and sensor design can improve the quality of wearable bioelectronics textile monitors. 2D materials have been explored to extract behavioral information from sweat metabolites. Glucose and lactate levels in sweat are extremely important biomarkers for monitoring human health. Graphene and MXene-based biosensors have been demonstrated as very efficient and effective sensing platforms smartly integrated with textiles for wearable applications. MXene composite-based hybrid electrodes in textile biosensors have exhibited the ability to accurately monitor

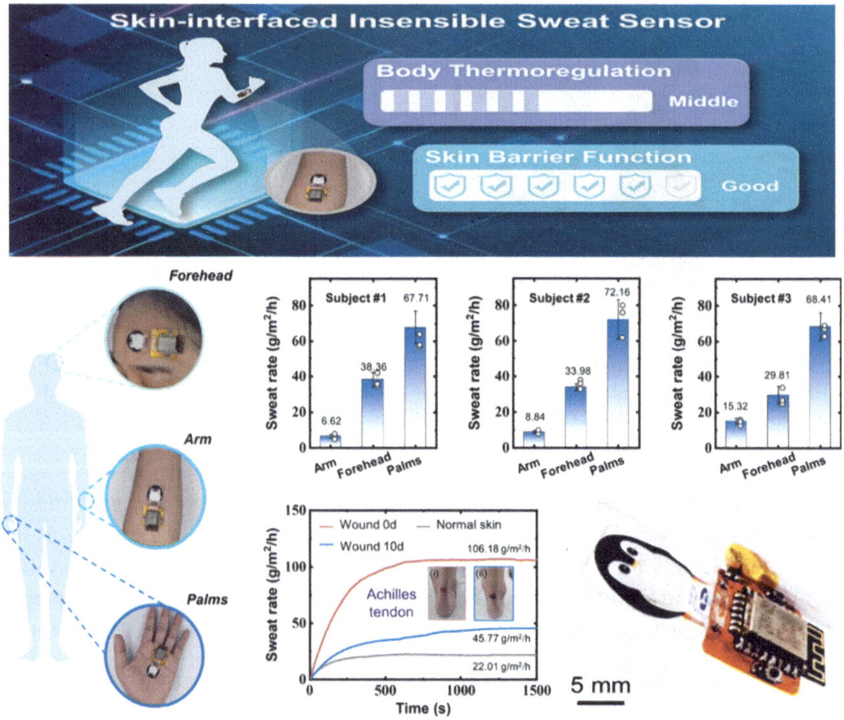

Fig. 5 MXene composite-based textile sweat sensors for evaluating the thermoregulation and skin barrier function of various parts of the human body (arm, forehead, and palm). Reprinted with permission from Liu et al. [51]. Copyright 2023 ACS

the level of sweat glucose. Md Sharifuzzaman et al. [53] developed a nanohybrid-modified flexible, highly sensitive, repeatable, and selective biosensing patch integrated into the textile (see Fig. 7a,f). This hybrid bioelectronics textile (HBeT) sensor can detect sweat glucose with the additional capability of pH adjustment. They fabricated the patch of TiO_2@LCNF (laser-carbonaceous nanofibrous) nanohybrid using electrospun MFNFs (MXene/fluoropolymers nanofibers) by utilizing its advantage of localized fast carbonization process. The HBeT has excellent electrochemical properties with electroconductivity of ~16 Ω/□. The sensitivity of this 2D material-based highly stable textile biosensor was measured to be 77 μA mM^{-1} cm^{-2} (within the physiological concentration of 0.1 μM-2 mM). During the real-time on-body sweat testing analysis for human subjects (2 volunteers), the pH from the HBeT biosensor was found to be similar to the pH value obtained from commercial pH meter devices for both subjects (Fig. 7b,c). This indicates the high applicability of a fabricated textile-based biosensor. Moreover, an increment in both glucose values and current response was observed for the sweat samples collected after the meal in comparison to the sweat samples collected before the meal (Fig. 7d,e). The high

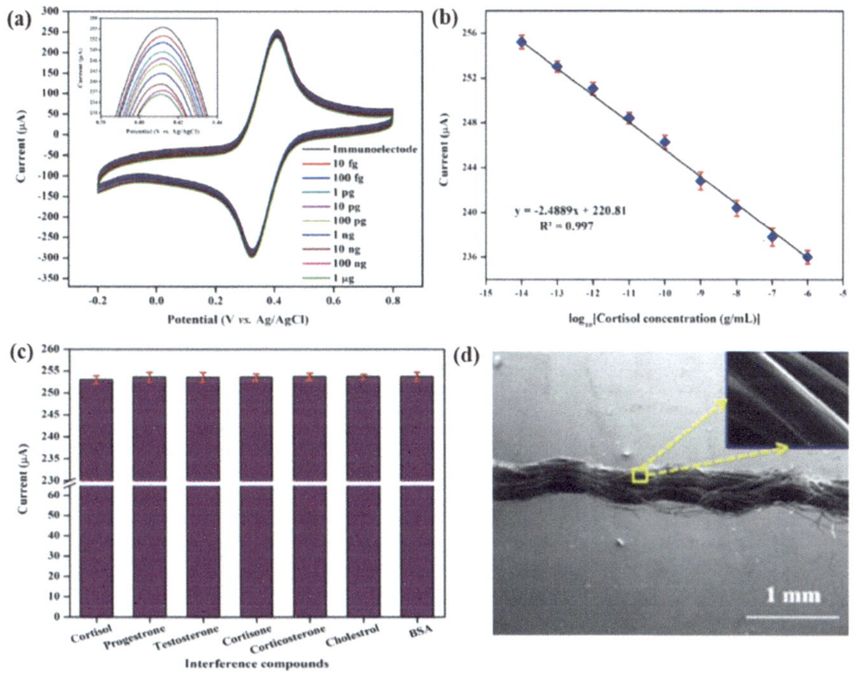

Fig. 6 TMO-CCY-based immunosensor as a highly sensitive cortisol detector: **a** Electrochemical response of immunoelectrode and **b** Calibration curve plotted between electrochemical current response and the logarithm of cortisol concentrations for the CV analysis. **c** Study of interference effect. **d** FESEM image of conductive carbon yarn. Reprinted with permission from Madhu et al. [52]. Copyright 2022 Elsevier

sensitivity of this biosensor is attributed to the synergistic effect of the highly conductive MXene layer, porous disordered carbon, and TiO_2 nanoparticles offering quick heterogeneous electron transmission.

Graphene nanocomposite-based wearable textile biosensors can be extremely useful in assessing the physiological condition of human health by examining the lactate from human sweat. Khan et al. [54] developed an eco-friendly textile electrochemical sensor with graphene nanocomposite material to monitor the lactate from human sweat (Fig. 7 g,h). They discovered the novel composite material of reduced graphene rGO/graphite-polyurethane G- PU/Prussian blue PB for sensor electrodes that offered good conductivity. The highly sensitive sensor showed high performance towards lactate concentration. The limit of detection (LOD) and limit of quantification (LOQ) of the sensor were obtained at 0.4 mM and 1.3 mM, respectively, for lactate from human sweat with a relative standard deviation (RSD) of 3.06%. The G-PU-rGO-PB sensor was also found to be highly reproducible even after several washes using a domestic washing machine.

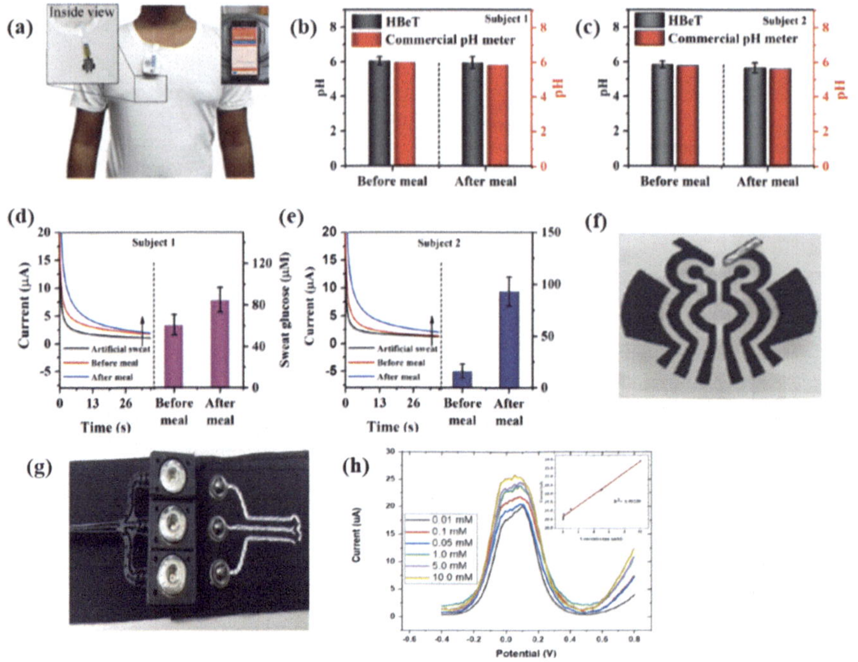

Fig. 7. **a** Subject (volunteer) with HBeT patch integrated sportswear (inset–left side), data recording in the mobile phone (inset—right side). **b** and **c** Comparison of pH of HBeT patch with commercial sensor for two subjects 1 and 2, before and after the meals. **d** and **e** Glucose levels for subjects 1 and 2 were collected from patches before and after the meal. **f** HBeT-based flexible patch and the fabric-containing patch. Reprinted with permission from Md Sharifuzzaman et al. [53]. Copyright 2023 Wiley–VCH GmbH. **g** Design of graphene-nanocomposite-based textile biosensor and **h** differential pulse voltammogram of lactate at various concentrations from 0.01 mM to 10 mM including inset image of the linear calibration curve (right). Reprinted with permission from Khan et al. [54]. Copyright 2022 Elsevier

With advancements in technologies, the simultaneous monitoring of biomarkers with the diagnosis of diseases seems possible in the near future. Recently, a graphene-based transistor has emerged as a noninvasive ultrasensitive sensor for simultaneous biomonitoring. Qing et al. [55] developed graphene-based a high transconductance (g_m ~ 45 mS) fiber–shaped organic electrochemical transistor-based biosensor (FOECTS) which is highly sensitive and has good air permeability as shown in Fig. 8a. Also, it is compatible with textiles suggesting it, is a viable contender for daily healthcare monitoring. They developed a composite material of reduced graphene oxide, porous polypyrrole (PPy) nanowires/flower, and cotton@polyethylene terephthalate (CPET) fiber, showing an excellent tradeoff between large volumetric capacitance and high charge carrier mobility. This makes rGO/CPET/plasma/PPy-FOECTS for simultaneous detection of sweat biomarkers-dopamine (DA), glucose, and lactate (LA). The LOD of this woven biomonitor was achieved up to 1 nM. The sensor exhibited high sensitivity of 235, 85, and 46 mV/dec towards artificial sweat's glucose,

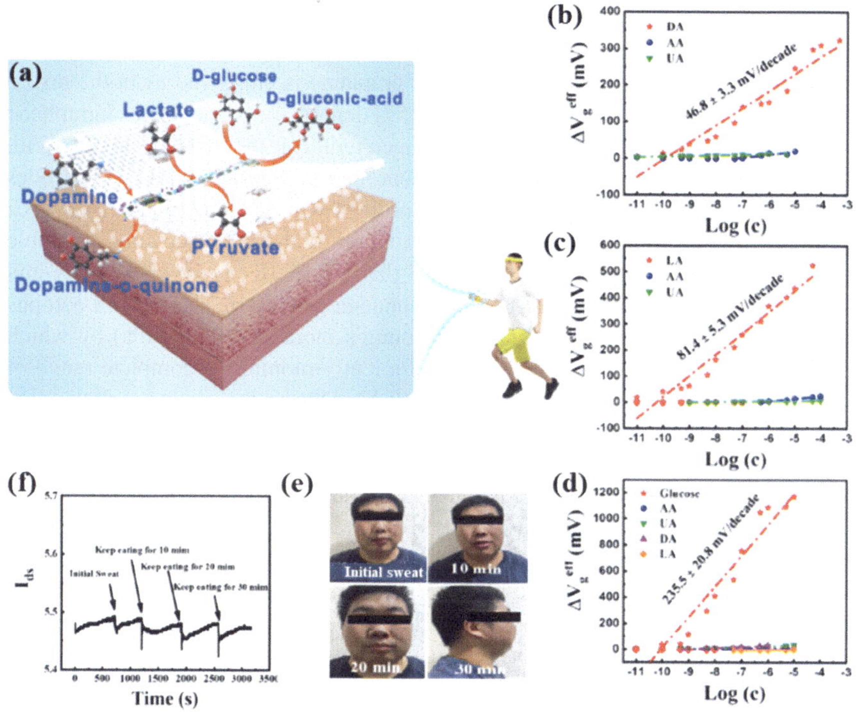

Fig. 8 **a** Graphene-based fiber-shaped transistor bioelectronics sensor. Performance of CPET/plasma/rGO/PPy/Nafion-based **b** DA, **c** LA, and **d** glucose sensors. **e** The optical images and **f** device i–t response to real human sweat before and after feeding. Reprinted with permission from Qing et al. [55]. Copyright 2023 Elsevier

LA, and DA, respectively (see Fig. 8b,c,d) with high reproducibility (RSD ~ 0.33%, 0.74%, and 0.31% for glucose, LA, and DA, respectively). The FOECTS also demonstrated high sensitivity towards real sweat collected at different feeding times (see Fig. 8e,f).

4.2 2D Materials-Based Textile Electrocardiography Sensors

Cardiovascular disease is among the most lethal diseases and an electrocardiogram (ECG) is a potential means to identify it in the form of the heart's electrical activity through repeated cardiac cycles. An electrocardiogram is a graph between voltages versus time for the electrical activity of the heart using an electrode attached to the human skin. Ag/AgCl-based electrodes are typically used in ECG and a gel is used with it in order to improve the conductive contact with human skin. The use of gel has some drawbacks, for instance, in some cases, the user faces skin irritation and allergic reactions. Also, the performance of the standard electrode decreases with time [56]. To resolve these issues, textile-based dry electrodes and sensors can be viable alternatives. In 2D materials, graphene is employed as in the textile electrode for ECG monitoring. Yapici et al. [57] demonstrated the reduced graphene oxide-based textile clad modified with hydrogen iodide as the ECG electrodes. With electroconductivity of 4.5 S/cm, the graphene textile clad-based ECG electrodes displayed excellent correlation in ECG recordings with those of typical Ag/AgCl electrodes in both time–frequency domains. Chun et al. [58] discovered wearable and skin-adherent electronic fabric sensors that work effectively in both dry and wet environments. They developed the graphene-coated fabric sensor with octopus patterns (electroconductivity ~0.22 S/m, Young's modulus ~35.5 MPa) by which it adhered well with the human skin and efficiently monitors a complete range of human activities including ECG (see Fig. 9a,b). During the electrocardiography, ECG signals were monitored upon consistent bending movement of the wrist. The electrodes were attached to the skin without any chemicals. The proposed textile sensor exhibited ECG signals that had the same output up to 120° bending angle in both dry and wet wrist surfaces (Fig. 9c). In another textile-based prototype, Yapici et al. [59] developed a fully wearable electronic textile-functionalized with graphene for ECG monitoring. In the state-of-the-art design, graphene-clad with electronic circuitry was housed within the wristband. The performance of this battery-operated electrocardiography system prototype was found at par with conventional Ag/AgCl electrodes that use the gels (Fig. 9d,e).

4.3 2D Materials-Based Textile Exhaled Breath Sensors

Exhaled breath consists of ample biomarkers and their detection and monitoring is highly helpful in analyzing an individual's physiological level. The developments of user-friendly and handheld monitors are being progressed. The exhaled breath of a human being has a mixture of water vapor and numerous compounds like N_2 (nitrogen), O_2 (Oxygen), CO_2 (carbon dioxide), and many VOCs (volatile organic compounds). There is variation in the markers' level of every human being [60]. By differentiating the VOC level of people, an infected person can be identified. Moreover, the quality of indoor/outdoor environments also significantly and silently affects human health. The presence of toxic gases in the surrounding environment causes detrimental effects on human health, for example, irritations in the eye, nose, and throat; nausea, diseases related to respiratory and pulmonary systems. With the advancement in 2D materials, innovations in wearable textile sensors have been started. The 2D materials contribute tremendously to lifting the capabilities of these sensing monitors and systems which have been proven to be better than the existing sensors [61, 62]. The exhaled breath vapor or other gas sensors based on 2D materials analyze the breath for identifying the particular breath marker for the diagnostics [63]. Bidsorkhi et al. [64] developed a prototype of a graphene-based sensing system for respiratory monitoring. They prepared a low-cost, metal-free, biocompatible smart

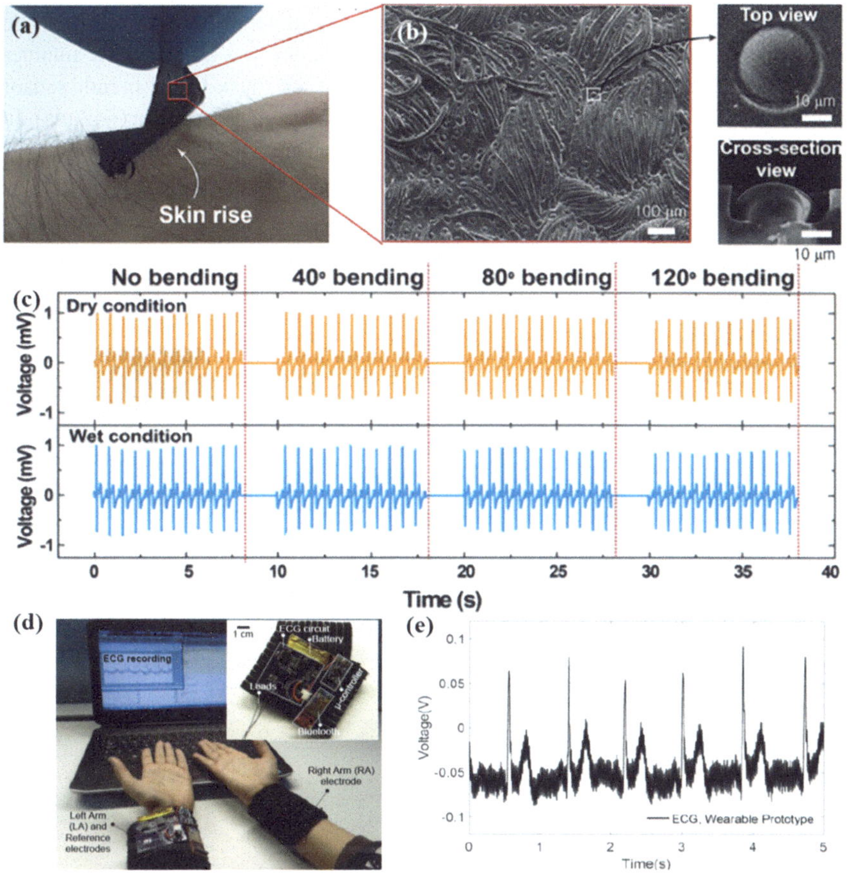

Fig. 9 Graphene Fabric Sensor: **a** graphene-coated fabric on human skin, **b** SEM image of sensor fabric with top and cross-sectional views, **c** Measurements of electrocardiogram (ECG) signals under dry and underwater conditions with bending up to 120° using the GCF attached on a wrist Reprinted with permission from Chun et al. [58]. Copyright 2019 Elsevier. Graphene-functionalized smart-cloth ECG sensors: **d** live demonstration of ECG measurement using the prototype graphene-clad textile embedded wearable wristband with integrated electronics. **e** ECG signal acquired from the wrists. Reprinted with permission from Yapici et al. [59]. Copyright 2018 MDPI

face mask (see Fig. 10a) which can sense the breathing rate of a mask-wearing person, and through the electronic system; the respiration rate can be monitored on a handheld device like a mobile phone. Through this, abnormal respiration can be recognized and generate alert signals during the situation of irregular cough or shortness of breath. In their state-of-the-art prototype, they synthesized the hybrid material of graphene nanoplatelets (GNP) and polycaprolactone (PCL) polymer. They fabricated the piezoresistive strip-lines of graphene/PCL fibrous hybrid on a surgical face mask using a spray coating technique (see SEM image at the top left corner). The piezoresistive stripes of GNP/PCL hybrid detect the resistive variation because of respiration-induced strain. The end user can get and analyze the data on the mobile app through interlinked transmission system in a user-friendly manner (see inset image at top right position). The smart-face mask-based breath sensor demonstrated continuous monitoring with an outstanding fast response (~42 ms). In addition, the graphene-based smart face mask sensor was found to be highly durable (greater than 1000 cycles) and reliable (see bottom inset images).

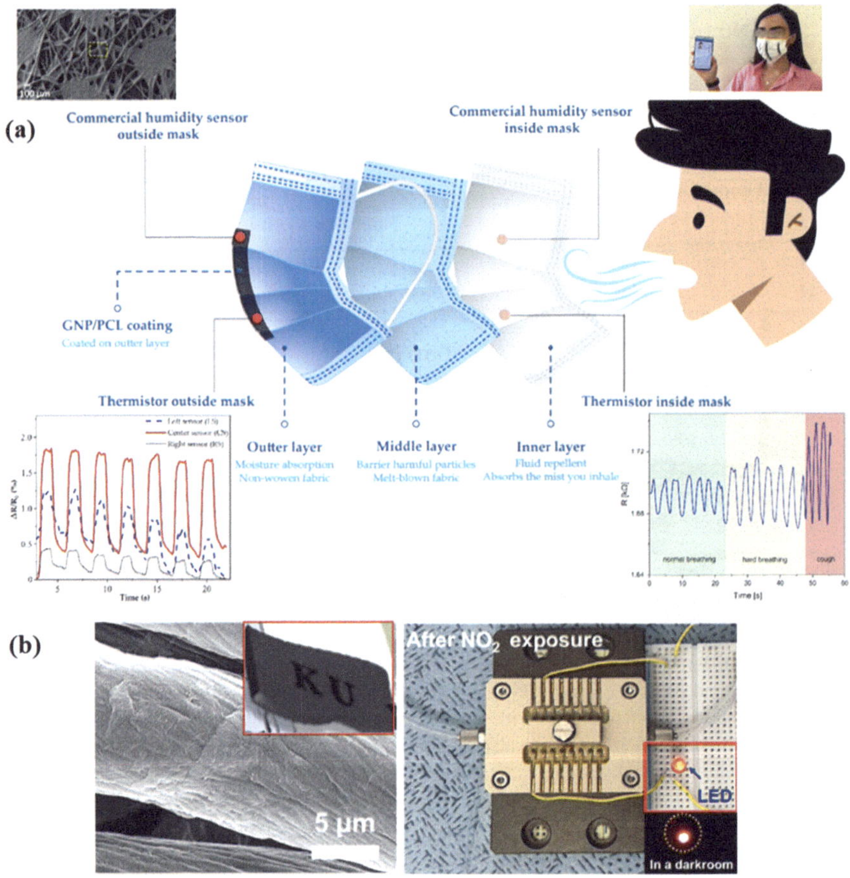

Fig. 10 **a** Smart face mask-based respiratory sensor based on graphene: illustration of the position of the sensor on a three-layer surgical mask. Inset images-SEM image of graphene hybrid coated mask (top left), testing respiration in real time with face mask (top right), resistance variation during testing respiration (artificial) at three different locations—right, center and left on a graphene-coated mask (bottom right). Reprinted with permission from Bidsorkhi et al. [64]. Copyright 2023 Elsevier. **b** Graphene-based textile gas sensor for indoor/outdoor air quality monitor: SEM image of rGO/amyloid nanofibril yarn and gas sensor (inset) and sensing operation (right). The inset shows a glowing LED (red) after NO_2 (10 ppm) exposure for 4 min. Reprinted with permission from Lee et al. [65]. Copyright 2020 ACS

To monitor indoor/outdoor quality, a textile gas sensor based on graphene is found to be very effective for wearable applications. Lee et al. [65] proposed a graphene composite-based highly sensitive and selective e-textile yarn gas sensor for the detection of NO_2 (nitrogen dioxide) which is considered one of the highly toxic gases. They developed textile yarn modified with rGO-amyloid nanofibril that showed high adhesion quality with high electroconductivity (~4 S) as shown in Fig. 10b. The graphene-based e-textile yarn sensor showed very good sensing performance such as sensing efficiency (~0.52 μA/min), sensitivity (~21 nA/ppm), response time (~4 min), and selectivity towards NO_2 gas. With a high-performance graphene/amyloid nanofibril conductive textile yarn sensor, a gas alarming system was developed, that can effectively perform at a 10 ppm threshold concentration of NO_2.

4.4 2D Materials-based Textile Human Motion Sensors

Similarly, graphene fiber (G–F) has emerged as an outstanding choice for wearable sensors. Zhang et al. [66] developed highly flexible and conductive graphene fibers and their sensors to examine human motions. It can detect human motions in a wider bending angle range 0–120°. In addition, the G-F sensor was found to be highly sensitive, accurate, and reproducible along with long cycle stability and exhibited a fast response of ~70 ms. High sensor response (relative resistive variation in transient response) of the G–F sensor under different human motions such as bending and stretching of finger; releasing and clenching of hand; stretching and bending of the arm is presented in Fig. 11a–d along with good cyclic stability.

5 Summary and Prospects

In this chapter, numerous 2D materials–based textile biosensors for human health monitoring applications are discussed. Textile substrates or base of the sensors provide excellent flexibility, stretchability, air/moisture permeability, and wearing comfort making it an excellent choice for wearable e-textile biosensors. The performance of the textile-based health monitors can be enhanced by employing sensing layers/components of 2D materials. Different 2D materials such as graphene, MXenes, TMDs, and their hybrids/composites-based biosensors have displayed excellent performance (high sensitivity, selectivity, faster response) while detecting different biomarkers such as sweat glucose, cortisol, lactate, metal ions; exhaled breath markers and toxic gas molecules. In addition, these textile sensors using 2D materials also exhibited outstanding performance for ECG signal measurements. Portable textile sensors after interfacing with electronic systems consisting of transmission/receiving units, Wi-Fi, and Bluetooth, have produced user-friendly detecting systems that increase the easiness for the user to get physiological information.

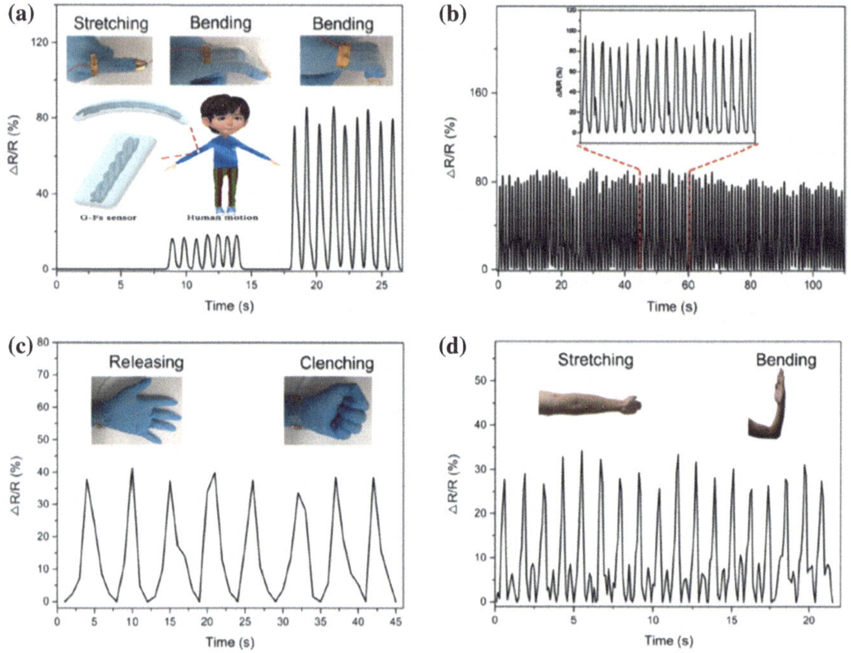

Fig. 11 Wearable G-Fs sensor for human motion detection: **a** Resistance variation during the finger's stretching and bending. The inset illustrates the schematic of a G–Fs sensor, **b** the cyclic stability during finger bending, **c** resistance variation during hand's releasing and clenching, and **d** resistance variation during arm's stretching and bending. Reprinted with permission from Zhang et al. [66]. Copyright 2022 ACS

Although textile biosensors have a lot of potential for monitoring healthcare, there are a few limitations that need to be addressed. The high contact impedance between dry skin and textile electrodes is one of the primary challenges, as it can cause unwanted changes in the resistance and other electric properties of the sensors as a result of washing and wearing [13]. Furthermore, the wearer may experience discomfort if the sensors acquire signals directly on their skin. Another drawback is the lack of a universal standard for biomarker detection due to the variability in the sensitivity and specificity of the sensors, which can vary from person to person [12]. Moreover, the accuracy of the measurements is impacted by motion artifacts and signal noise in textile biosensors [67]. Many wearable devices face challenges related to power consumption, and 2D material-based textile biosensors are no exception. To build self-sustaining systems, it is imperative to investigate novel energy-harvesting technologies and incorporate power-efficient components. For practical, long-term use, power efficiency and the requirement for continuous monitoring must still be balanced [68]. The long-term reproducibility after multiple washing cycles and cost-effective state-of-the-art textile biosensors based on 2D materials are still some focal points that need to be addressed in the near future to thrive in the wearable health monitors market.

Acknowledgements The authors would like to thank the European Union's Horizon 2020, WIDESPREAD-04-2019: ERA Chair 'STRENTEX' (Emerging Technologies and Innovative Research in Stretchable and Textile Electronics: Grant Number 854194).

References

1. Smith, A.A., Li, R., Tse, Z.T.H.: Reshaping healthcare with wearable biosensors. Sci. Rep. **13**(1), 1–16 (2023). https://doi.org/10.1038/s41598-022-26951-z
2. Sulleiro, M.V., Dominguez-Alfaro, A., Alegret, N., Silvestri, A., Gómez, I.J.: 2D Materials towards sensing technology: from fundamentals to applications. Sens. Bio-Sensing Res. **38**, 2022, https://doi.org/10.1016/j.sbsr.2022.100540
3. Ji, W., et al.: Wearable sweat biosensors refresh personalized health/medical diagnostics. Research **2021**(9757126), 1–19 (2021). https://doi.org/10.34133/2021/9757126
4. Khan, S., Ali, S., Bermak, A.: Recent developments in printing flexible and wearable sensing electronics for healthcare applications. Sensors **19**(5), 1–34 (2019). https://doi.org/10.3390/s19051230
5. Yin, J., Li, J., Reddy, V.S., Ji, D., Ramakrishna, S.: Flexible textile-based sweat sensors for wearable applications. Biosensors **13**(127), 1–26 (2023)
6. Pang, Y., Yang, Z., Yang, Y., Ren, T.: Wearable electronics based on 2D materials for human physiological information detection. Small **1901124**, 1–26 (2019). https://doi.org/10.1002/smll.201901124
7. Shi, Y., Zhang, Z., Huang, Q., Lin, Y., Zheng, Z.: Wearable sweat biosensors on textiles for health monitoring. J. Semicond. **44**(021601), 1–14 (2023). https://doi.org/10.1088/1674-4926/44/2/021601
8. Verma et al.: Biosensors and Bioelectronics : X Internet of things (IoT) in nano-integrated wearable biosensor devices for healthcare applications. Biosens. Bioelectron. X **11**, 100153 (2022). https://doi.org/10.1016/j.biosx.2022.100153.
9. Wang, P. et al.: Recent advances in wearable tactile sensors based on electrospun nanofiber platform. Adv. Sens. Res. 2200047 (2023)
10. Yin, J., et al.: Smart textiles for self-powered biomonitoring. Med-X **1**(3), 1–22 (2023). https://doi.org/10.1007/s44258-023-00001-3
11. Li et al.: Advanced textile-based wearable biosensors for healthcare monitoring. Biosensors **13**(10) (2023). https://doi.org/10.3390/bios13100909
12. Hatamie, A., et al.: Review—textile based chemical and physical sensors for healthcare monitoring. J. Electrochem. Soc. **167**(3), 037546 (2020). https://doi.org/10.1149/1945-7111/ab6827
13. Blachowicz, T., Ehrmann, G.: Textile-based sensors for biosignal detection and monitoring. Sensors 21(6042) (2021)
14. Promphet, N. et al.: Non-invasive textile based colorimetric sensor for the simultaneous detection of sweat pH and lactate. Talanta **192**, 424–430 (2019). https://doi.org/10.1016/j.talanta.2018.09.086
15. Simić, M., Stavrakis, A.K., Sinha, A., Premčevski, V., Markoski, B., Stojanović, G.M.: Portable respiration monitoring system with an embroidered capacitive facemask sensor. Biosensors **12**(5) (2022). https://doi.org/10.3390/bios12050339
16. Wang, R., Zhai, Q., An, T., Gong, S., Cheng, W.: Stretchable gold fiber-based wearable textile electrochemical biosensor for lactate monitoring in sweat. Talanta **222**, 121484 (2021). https://doi.org/10.1016/j.talanta.2020.121484

17. Zafar, H., Channa, A., Jeoti, V., Stojanović, G.M.: Comprehensive review on wearable sweat-glucose sensors for continuous glucose monitoring. Sensors **22**(2), 1–35 (2022). https://doi.org/10.3390/s22020638
18. Ali, A.E., Jeoti, V., Stojanović, G.M.: Fabric based printed-distributed battery for wearable e-textiles: a review. Sci. Technol. Adv. Mater. **22**(1), 772–793 (2021). https://doi.org/10.1080/14686996.2021.1962203
19. Arquilla, K., Devendorf, L., Webb, A.K., Anderson, A.P.: Detection of the complete ecg waveform with woven textile electrodes. Biosensors **11**(9), 1–13 (2021). https://doi.org/10.3390/BIOS11090331
20. Stojanović, G. M. et al.: A textile-based microfluidic platform for the detection of cytostatic drug concentration in sweat samples. Appl. Sci. **10**(12) (2020). https://doi.org/10.3390/app10124392
21. Zhang, J., Lei, J., Liu, Z., Chu, Z., Jin, W.: Nanomaterial-based electrochemical enzymatic biosensors for recognizing phenolic compounds in aqueous effluents. Environ. Res. **214**(P3), 113858 (2022). https://doi.org/10.1016/j.envres.2022.113858
22. Malik, P., Gupta, R., Malik, V., Ameta, R.K.: Emerging nanomaterials for improved biosensing. Meas. Sensors **16**, 100050 (2021). https://doi.org/10.1016/j.measen.2021.100050
23. Wang, X., Li, F., Guo, Y.: Recent trends in nanomaterial-based biosensors for point-of-care testing. Front. Chem. **8**(October), 1–6 (2020). https://doi.org/10.3389/fchem.2020.586702
24. Rohaizad, N., Mayorga-Martinez, C.C., Fojtů, M., Latiff, N.M., Pumera, M.: Two-dimensional materials in biomedical, biosensing and sensing applications. Chem. Soc. Rev. **50**(1), 619–657 (2021). https://doi.org/10.1039/d0cs00150c
25. Zhan, B., Li, C., Yang, J., Jenkins, G., Huang, W., Dong, X.: Graphene field-effect transistor and its application for electronic sensing. Small **10**(20), 4042–4065 (2014). https://doi.org/10.1002/smll.201400463
26. Luo et al.: High responsivity graphene photodetectors from visible to near-infrared by photogating effect. AIP Adv. **8**(11) (2018). https://doi.org/10.1063/1.5054760
27. Ahn, S., et al.: A 2D titanium carbide MXene flexible electrode for high-efficiency light-emitting diodes. Adv. Mater. **12**(2000919), 1–7 (2020)
28. Kumar, P., Kang, C.H., Saheed, M.S.M., Irshad, M.I., Mohamed, N.M., Burhanudin, Z.A.: Graphene-based hybrid thin films as transparent conductive electrode for optoelectronic devices. In: IEEE International Conference on Semiconductor Electronics, ICSE, pp. 216–219 (2016)
29. Anasori, B., Lukatskaya, M.R., Gogotsi, Y.: 2D metal carbides and nitrides (MXenes) for energy storage. Nat. Rev. Mater. **2**(16098), 1–17 (2017). https://doi.org/10.1038/natrevmats.2016.98
30. Cai, S., et al.: Pt74Ag26 nanoparticle-decorated ultrathin MoS2 nanosheets as novel peroxidase mimics for highly selective colorimetric detection of H2O2 and glucose. Nanoscale **8**(6), 3685–3693 (2016). https://doi.org/10.1039/c5nr08038j
31. Hassan, J. Z., Raza, A., Din Babar, Z. U., Qumar, U., Kaner, N. T., Cassinese, A.: 2D material-based sensing devices: an update. J. Mater. Chem. A **11**(12), 6016–6063 (2023). https://doi.org/10.1039/d2ta07653e
32. Gosling, J.H. et al.: Universal mobility characteristics of graphene originating from charge scattering by ionised impurities. Commun. Phys. 4(1) (2021). https://doi.org/10.1038/s42005-021-00518-2
33. Kumar, P., et al.: Surface-treatment process related sheet resistance variations in graphene-based thin-film electrodes. Surfaces Interfaces **32**, 1–28 (2022). https://doi.org/10.1016/j.surfin.2022.102161
34. Wang, Y., et al.: Reduced graphene oxide film with record-high conductivity and mobility. Mater. Today **21**(2), 186–192 (2018)
35. Wei, M., et al.: Engineering reduced graphene oxides with enhanced electrochemical properties through multiple-step reductions. Electrochim. Acta **258**, 735–743 (2017). https://doi.org/10.1016/j.electacta.2017.11.120

36. Kumar, P., Woon, K. L., Wong, W. S., Mohamed Saheed, M. S., Burhanudin, Z. A.: Hybrid film of single-layer graphene and carbon nanotube as transparent conductive electrode for organic light emitting diode. Synth. Met. **257**, 1–8 (2019)
37. Huang et al.: Graphene-based sensors for human health monitoring. Front. Chem. **7**, 1–26 (2019). https://doi.org/10.3389/fchem.2019.00399
38. Shen, X., Chen, X., Sun, D., Li, Y.: Fabrication of a magnetite/diazonium functionalized-reduced graphene oxide hybrid as an easily regenerated adsorbent for efficient removal of chlorophenols from aqueous solution. RSC Adv. **8**, 7351–7360 (2018). https://doi.org/10.1039/C8RA00503F
39. Naguib, M., et al.: Two-dimensional nanocrystals produced by exfoliation of Ti_3AlC_2. Adv. Mater. **23**(37), 4248–4253 (2011). https://doi.org/10.1002/adma.201102306
40. Hajian, S., Maddipatla, D., Narakathu, B.B., Atashbar, M.Z.: MXene-based flexible sensors: a review. Front. Sensors **3**(October), 1–24 (2022). https://doi.org/10.3389/fsens.2022.1006749
41. Shahzad, F., Iqbal, A., Kim, H., Koo, C.M.: 2D transition metal carbides (MXenes): applications as an electrically conducting material. Adv. Mater. **32**(51), 1–23 (2020). https://doi.org/10.1002/adma.202002159
42. Riazi, H., Taghizadeh, G., Soroush, M.: MXene-based nanocomposite sensors. ACS Omega **6**, 11103–11112 (2021)
43. Han, S.A., Bhatia, R., Kim, S.: Synthesis , properties and potential applications of two-dimensional transition metal dichalcogenides. Nano Converg. **2**(17), 1–14 (2015). https://doi.org/10.1186/s40580-015-0048-4
44. Terrones, M., Lo´pez-Urı´as, H., Terrones, F.: Novel hetero-layered materials with tunable direct band gaps by sandwiching different metal disulfides and diselenides. Sci. Rep. 3(1549), 1–7 (2013)
45. Bilc, D.I., Benea, D., Pop, V., Ghosez, P., Verstraete, M.J.: Electronic and thermoelectric properties of transition-metal dichalcogenides. J. Phys. Chem. C **125**, 27084–27097 (2021)
46. Kumar, R., Goel, N., Hojamberdiev, M., Kumar, M.: Transition metal dichalcogenides-based flexible gas sensors. Sens. Actuators A. Phys. **303**, 111875 (2020). https://doi.org/10.1016/j.sna.2020.111875
47. Bablich, A., Kataria, S., Lemme, M.C.: Graphene and two-dimensional materials for opto-electronic applications. Electronics **5**(13), 1–16 (2016). https://doi.org/10.3390/electronics5010013
48. Hussain, S., et al.: Synthesis and characterization of large-area and continuous MoS2atomic layers by RF magnetron sputtering. Nanoscale **8**(7), 4340–4347 (2016). https://doi.org/10.1039/c5nr09032f
49. Matte, H.S.S.R., et al.: MoS2 and WS2 analogues of graphene. Angew. Chem. Int. Ed. **49**, 4059–4062 (2010). https://doi.org/10.1002/anie.201000009
50. Kil, M.S., Kim, S.J., Park, H.J., Yoon, J.H., Jeong, J.M., Choi, B.G.: Highly stretchable sensor based on fluid dynamics-assisted graphene inks for real-time monitoring of sweat. ACS Appl. Mater. Interfaces **14**(42), 48072–48080 (2022). https://doi.org/10.1021/acsami.2c10638
51. Liu, Y., et al.: Skin-interfaced superhydrophobic insensible sweat sensors for evaluating body thermoregulation and skin barrier functions. ACS Nano **17**(6), 5588–5599 (2023). https://doi.org/10.1021/acsnano.2c11267
52. Madhu, S., Ramasamy, S., Manickam, P., Nagamony, P., Chinnuswamy, V.: TiO_2 anchored carbon fibers as non-invasive electrochemical sensor platform for the cortisol detection. Mater. Lett. 308(PB), 131238 (2022). https://doi.org/10.1016/j.matlet.2021.131238
53. Sharifuzzaman, M., et al.: MXene/fluoropolymer-derived laser-carbonaceous all-fibrous nanohybrid patch for soft wearable bioelectronics. Adv. Funct. Mater. **33**(21), 1–14 (2023). https://doi.org/10.1002/adfm.202208894
54. Khan, A., Winder, M., Hossain, G.: Modified graphene-based nanocomposite material for smart textile biosensor to detect lactate from human sweat. Biosens. Bioelectron. **10**(100103), 1–9 (2022)
55. Qing, X. et al.: High gain fiber-shaped transistor based on rGO-mediated hierarchical polypyrrole for ultrasensitive sweat sensor. Sens. Actuators A Phys. **354**(February), 114297 (2023). https://doi.org/10.1016/j.sna.2023.114297

56. Shao et al.: Carbon-based textile sensors for physiological-signal monitoring. Materials (Basel) 16(3932) (2023)
57. Yapici, M.K., Alkhidir, T., Samad, Y.A., Liao, K.: Graphene-clad textile electrodes for electrocardiogram monitoring. Sens. Actuators B Chem. **221**, 1469–1474 (2015). https://doi.org/10.1016/j.snb.2015.07.111
58. Chun, S., et al.: Water-resistant and skin-adhesive wearable electronics using graphene fabric sensor with octopus-inspired microsuckers. ACS Appl. Mater. Interfaces **11**(18), 16951–16957 (2019). https://doi.org/10.1021/acsami.9b04206
59. Yapici, M.K., Alkhidir, T.E.: Intelligent medical garments with graphene-functionalized smart-cloth ECG sensors. Sensors **17**(4), 1–12 (2017). https://doi.org/10.3390/s17040875
60. Pham, Y. L., Beauchamp, J.: Breath biomarkers in diagnostic applications. Molecules **26**(18). https://doi.org/10.3390/molecules26185514
61. Gupta, M., Hawari, H.F., Kumar, P., Burhanudin, Z.A., Tansu, N.: Functionalized reduced graphene oxide thin films for ultrahigh co2 gas sensing performance at room temperature. Nanomaterials **11**(3), 1–18 (2021)
62. Gupta, M., Hawari, H.F., Kumar, P., Burhanudin, Z. A.: Copper oxide/functionalized graphene hybrid nanostructures for room temperature gas sensing applications. Crystals **12**(2) (2022). https://doi.org/10.3390/cryst12020264
63. Singh, S. U. et al.: Advanced wearable biosensors for the detection of body fluids and exhaled breath by graphene. Microchim. Acta **189**(236) (2022). https://doi.org/10.1007/s00604-022-05317-2
64. Bidsorkhi, H.C., et al.: Wearable graphene-based smart face mask for real-time human respiration monitoring. Mater. Des. **230**(111970) (2023)
65. Sensors, A.C.S.: Bio-inspired electronic textile yarn-based NO_2 sensor using amyloid–graphene composite. ACS Sensors **6**, 777–786 (2020)
66. Zhang, Z., Zhang, P., Zhang, D., Lin, H., Chen, Y.: Wearable resistive-type sensors based on graphene fibers for monitoring human motions. ACS Appl. Nano Mater. **5**(8), 10912–10921 (2022). https://doi.org/10.1021/acsanm.2c02202
67. Vidhya, C.M., Maithani, Y., Singh, J.P.: Recent advances and challenges in textile electrodes for wearable biopotential signal monitoring: a comprehensive review. Biosensors **13**(7), 1–34 (2023). https://doi.org/10.3390/bios13070679
68. Ali, I., Dulal, M., Karim, N., Afroj, S.: 2D material-based wearable energy harvesting textiles: a review. Small Struct. 2300282 (2023). https://doi.org/10.1002/sstr.202300282

2D Materials—Applications in Photo Sensors

Juri Kalita, Linkon Bharali, and Siddhartha Sankar Dhar

Abstract 2D materials have drawn a lot of interest in the field of photodetection due to their distinct electrical and optical characteristics. Particularly, photoactive materials with considerable promise for application in photo sensors and photodetectors include graphene and transition metal dichalcogenides (TMDs), such as MoS_2, WS_2, $MoSe_2$, and WSe_2. 2D materials have a significant advantage in photodetection due to their low thickness and enormous surface area, which allows for the efficient absorption of photons. Additionally, the high electron mobility of 2D materials makes it easier to transmit charge carriers produced by photoexcitation. 2D materials can be configured in a number of ways in light sensors, including photodiodes, phototransistors, and photovoltaic cells. In comparison to more conventional materials like silicon, the use of 2D materials in optical sensors has a number of benefits. Due to their flexibility and lightweight, 2D materials may be used in wearable and portable electronics. Moreover, 2D materials have a wide spectral response, from ultraviolet to infrared, and can be tuned by changing the thickness or composition of the material. Despite the promise of 2D materials in photodetection, there are still challenges to be addressed, such as improving the efficiency and stability of the devices and developing scalable and cost-effective fabrication methods. This chapter mainly includes the potential applications of 2D materials in photo sensors in the active area of research.

1 Introduction

There is a high demand for new functional materials because of developments in the areas of photonics, electronics, and optoelectronics. These materials must be simple to synthesize, process, and fabricate and allow for the necessary alteration of the chemical and physical features through chemical functionalization as well as the development of hybrid frameworks for potential utilities in optoelectronic and different electronic devices. Two-dimensional (2D) materials are a unique kind of

J. Kalita (✉) · L. Bharali · S. S. Dhar
Department of Chemistry, National Institute of Technology, Silchar, Cachar, Assam 788010, India
e-mail: jurikalita1512@gmail.com

A. Katoch et al. (eds.), *2D Materials*, Engineering Materials,
https://doi.org/10.1007/978-981-97-6258-3_9

nanomaterials employed in various applications and research areas, which include biosensing, chemical sensing, photo sensing, and biomedicine as well as electronics, energy production and storage, etc. Emerging 2D nanomaterials are being investigated for this purpose in response to the rising demand for highly selective, sensitive, economical, low-power, portable, and stable sensors. 2D nanomaterials are an advanced class of atomically thick nanomaterials that comprise one or more layers of atoms. Improved synthesis techniques can provide even longer 2D nanomaterials, with lateral sizes as large as micrometers. 2D materials are sheet-like frameworks that are one atom or several atomic thicknesses, with adjacent sizes ranging from a few nanometers to a hundred nanometers and beyond that [1–4]. Due to their unique characteristics, including phase crystallinity, stability, degree of exfoliation and size, such structures have demonstrated improved physical, chemical, optical, and electrical properties. These characteristics make them potential candidates for biomedical applications, including as photodynamic and photothermal treatments, drug potentiation, and drug transport, in addition to their use as sensors. The ratio of surface-to-volume in 2D materials is higher than that of 0D, 1D, as well as 3D materials [5]. Such a high ratio provides a surface area that is significantly greater for association between the analytic substance and 2D material, which makes such 2D materials more appropriate for sensor-like applications. Low-concentration analytes can be detected using these sensors [6]. Comparing 2D materials to their 1D and 3D analogues, a wide spectrum of conductivities is possible. The bandgaps and conductivities of 2D nanomaterials may vary as a result of fine-tuning their configuration [7]. Additionally, by modifying the morphological characteristics of 2D materials, their electrical characteristics are improved, which increases their signal transduction efficiency as a result of molecule binding events. Variations in their fundamental properties including conductivity, fluorescence, reactivity, and magnetic permeability, are significantly influenced by their atomic-size thickness. The integration of devices is difficult for 0D materials, generating electrical contacts is challenging for 1D materials, and device miniaturization is still an issue for 3D materials [8]. Consequently, 2D materials are ideal for compactly constructed sensors. In contrast to other dimensions, viz., 0D, 1D, and 3D, some 2D materials have remarkable optical transparency and exceptional mechanical strength.

After the successful isolation of the first 2D material, a graphene sheet, from greatly organized pyrolytic graphite in 2004 through micromechanical exfoliation, a revolution in both experimental and theoretical research began to take place and the focus has shifted to other 2D structures, including carbon nitrides (CN), transition metal dichalcogenides (TMDCs), boron nitrides (BN), and transition metal nitrides and carbides (MXenes), metal phosphorus chalcogenides (MPX_3), etc. [9, 10] The scope of this finding has been utilized by other groups with complementary and parallel areas of interest working in material sciences, chemistry, physics, and many other multidisciplinary topics. Other 2D nanomaterials with unique structural, electrical, mechanical, catalytic, optical, and sensing properties, such as TMDCs, black phosphorus (BP), topological insulators (TI), hexagonal boron nitride (h-BN), and MXenes, have also demonstrated promising applicability in diverse areas, including optical devices, optoelectronic devices, batteries, sensors, and biomedicine [11–20]. Surprisingly, TMDCs, which are members of the prime 2D material family, have been extensively employed in the development of optical diagnostic and electrochemical devices due to their good conductivity, wide bandgap, rapid electron transfer and potent fluorescence, electrochemiluminescence, and photoelectrochemical activity. Owing to their reasonable production rate, ease of processing, ease of deposition on various substrates having well-defined thickness controlling of the atomic layers through chemical and mechanical exfoliation, atomic layer etching (ALE), and chemical vapour deposition (CVD) approaches, earth-abundant 2D nanomaterials, including graphene [21–23] and TMDCs [24–26] have appeared as new functional materials of preference. A variety of portable and wearable electronics, such as FETs or field-effect transistors, supercapacitors, sensors, triboelectric nanogenerators, lithium-ion batteries, photodetectors, and solar cells have been studied using adjustable graphene nanosheets [27, 28]. A comparable broad scope of applicability for flexible adjustable atomic layered TMDCs, from wearable electronics to optoelectronics, had been envisaged and is now slowly starting to materialize in response to the achievement achieved through the study of graphene. In the area of robotics surveillance, sports, wearable and portable healthcare, high-speed optical communication, light-emitting diodes (LEDs), flexible photodetectors and biomedical imaging are emerging as a crucial element of optoelectronic technology. Varieties of nanostructures, including atomically layered nanosheets, NWs or nanowires, quantum dots, fibers, and 3D networks of organic and inorganic materials, and as well as their nanohybrids, have been considered for use in flexible photodetectors in wearable optoelectronics.

The usage of these nanostructures as phototransistors, photoconductors, and photodiodes, a range of nanomaterials, such as silicon (Si), selenium (Se), germanium (Ge), and gallium phosphide (GaP), have been used to develop photodetectors along with their hybrids with lead sulfide (PbS), CdO, gold (Au), ZnS, and polymers, $CuInSe_2$, Zn_2GeO_4, $ZnGa_2O_4$, $In_2Ge_2O_7$, In_2S_3, Zn_3P_2, In_2Se_3, Sb_2S_3, Bi_2S_3, Sb_2Se_3, SnS, GaTe, SnS_2, PbI_2, SnO_2, MoO_3, GaS, ZnTe, perovskites, their composites with zinc oxide (ZnO), conjugated polymer of poly(diketopyrrolopyrrole-terthiophene) (PDPP3T), gold, graphene, carbon nanotubes (CNTs), polythiophene, nanocomposites of graphene with zinc oxide quantum dots and with carbon nanotubes, transition-metal trichalcogenides (MX_3, in which M designates a transition metal such as Ta or Nb, Zr or Hf, Ti and X is nothing but a chalcogen, Te, S, or Se), for example., hafnium nanocomposites ($HfSe_3$, HfS_3 etc.) with graphene and also with different graphene-based resources or materials. Applications for optical communications, medical diagnosis, flame detection, radiation and detection of smoke, process control, detection of barcode, motion sensing, sensing in environment, thermography, and astronomy are all developing at the frontiers of device electronics. The incorporation of optics into electrical systems, which has opened up opportunities for multi-functional purposes and improved efficiency, is the most significant advancement during the past decade. The ability to convert light into electric signals more quickly, effectively, and adaptable across a spectrum of wavelengths using photodetectors is one such advancement. The concept of photons absorption for the production of electric response can be achieved by a photodetection process in a variety of ways, including the photo-thermoelectric effect (PTE) [29], photoconductive effect [30], and photovoltaic effect [31]. A photon with enough energy to pass through the junction collides with an atom, removing an electron from the atomic framework. The interior photoelectric effect refers to this mechanism. This process consequences in the formation of hole pairs and free electrons. These unpaired holes and electrons are either joined or remain apart. A photocurrent is created when an external electric field causes the free e^- and h^+ pairs in the depleted layer to move. The quantity of light that enters the intrinsic area has a direct relationship with the photocurrent.

The chapter analyses the photoresponse of various two-dimensional (2D) material architectures, including graphene/TMDC heterostructures, TMDCs, black phosphorous or BP, and others. Due to their optimized characteristics and effective charge trapping, heterostructures have shown to be extremely effective materials for photodetectors and sensors.

The critical photodetector parameters—photoresponsivity, EQE, or external quantum efficiency which refers to the ratio of the total number of charge carriers collected by the device to the number of incident photons of a specific energy, photogain, specific detection activity, response speed, LDR or linear dynamic range that refers to the range of values at which the signals and the analyte concentration in the sample are exactly proportionate, and noise equivalent power (NEP) that signifies the typical measure used to quantify the sensitivity of a photodetector or the power produced by a noise source which have been extensively discussed in the previous literature. Here, such parameters or factors are elaborated in related 2D nanomaterial-based photodetectors, mainly hybrid photodetectors based on graphene, TMDC atomic layers.

2 Characteristics of 2D Materials as Photosensors and in Other Sensing Applications

When employed in sensing protocols, 2D materials have remarkably impressive features such as high strength and lightweight, excellent electrical conductivity, flexibility and transparency, and high thermal conductivity that shows high potential towards sensing. Device manufacturing may benefit greatly from the high-scale construction of such 2D materials with accurate replication of their properties [32]. These 2D materials have a particular thickness range, including lateral dimensions of a few centimetres and atomic-layer lengths of a few microns. Graphene has proven that it is suitable for utilization in photonic devices, electronic devices, sensing, energy conversion, and storage due to its unique chemical, physical, optoelectronic, mechanical, and thermal characteristics [33]. Monolayered 2D materials with higher surface-to-volume ratios than their inert bulk forms, which make them more chemically reactive [5]. Depending on the bandgap, the conductivity of 2D materials varies from the range of an insulator to that of a metal, and it may be altered using a variety of techniques such as adding structural flaws, doping, functionalization, or altering the number of layers. The bandgaps and conductivity of 2D nanomaterials change as a result of fine-tuning their configuration [7]. Introducing a mechanical force or electric field can change the bandgap of molybdenum disulfide monolayers. As a result, they are more effective at transducing signals and are better suited for usage in detectors. Changes in size and shape can affect the proportion of bulk to edges, which can result in spatial confinement in particular dimensions, as a result, modify the mechanical and electrical behaviors. Greater lateral size in 2D materials than in 1D materials allows for the maintenance of reliable electrical connections [34]. They exhibit outstanding compatibility with thin-film production methods due to their atomic-size thickness. Materials used in sensing equipment must be highly tensile strong and flexible. The direct transfer of graphene into the thin-film development mechanism is made possible by properties like its unusually high tensile strength and the adjustable atomic-size thickness [35]. Due to its exceptional flexibility and mechanical strength, graphene is suitable for wearable electrical devices.

Bulk Silicon has enormous applications in solar cells but due to the indirect band gap structure, it cannot emit light. However, it was noticed that some solar cells are characterized by extremely long carrier recombination lifetimes of some milliseconds which is of the order of radiative lifetime. On the other hand, it was observed that a small number of solar cells had extraordinarily long carrier recombination lifetimes, similar to radiative lifetimes, lasting several milliseconds. Bulk silicon has a low photodetector efficiency because of the 1.1 eV indirect bandgap value that restricts its ability to absorb photons to the visible as well as near IR regions of the electromagnetic (EM) spectrum. To detect IR wavelengths, a variety of alternatives, such as 3D-GaAs (Gallium Arsenide) structures, have been put forth. However, these alternatives have industrializing limitations, such as an expansion in device size, packaging, and numerous additive production stages. With the advancement of graphene and also few-layered transition metal dichalcogenides, substitute materials or substances have been discovered that have greater absorption efficacy in the range of visible light and a broad functioning wavelength [36]. These materials have advantages such as improved transparency and absorption efficiency which are associated with thickness on one atomic scale, making them a potential candidate in advanced photovoltaic, photodetection, and photonic applications. TMDs exhibit vertical confinement and quantum confinement effects (out-of-plane), in which the lower thickness limits the excitons, so improves the absorption efficacy [37]. The Van Hove singularities, which are ascribed to the (transition metal) d-orbital localization of electronic bands and result in high-density state peaks at a specific energy, are a defining characteristic of 2D semiconductor crystals. These singularities are seen in 2D semiconductors around the valence and conduction band edges. As a result, the probability that an electron–hole pair will be excited and become responsive to the incident light increases when the photon's energy is close to the bandgap. By altering the optical properties based on strain engineering, this characteristic has expanded their applications to flexible sensors together with a remarkable elastic modulus of $>10\%$ [38]. Depending on the process of converting photons to electrical signals, the

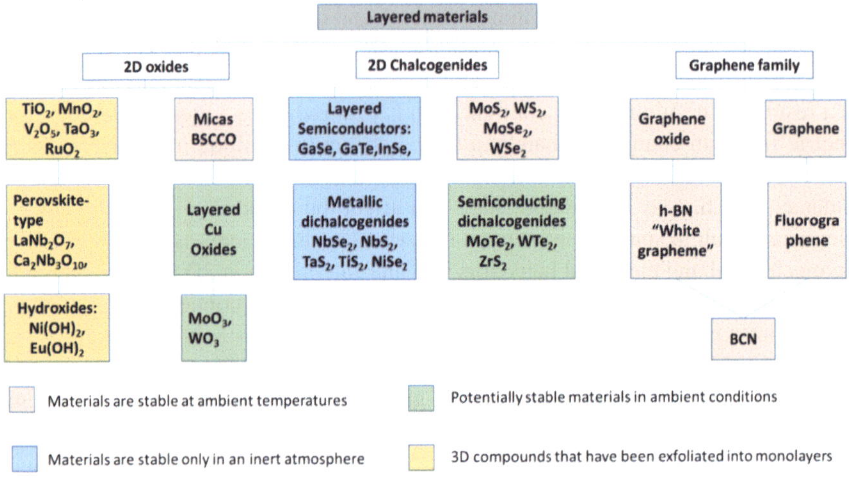

Fig. 1 Classification of layered materials on the basis of stability

photosensing action or effect can be produced by the photoconductive effect, PTE or photovoltaic effect (Fig. 1).

3 Mechanism of Photo Sensing Process

The mechanism of the photo sensing processes is typically divided into three categories of photogating, photovoltaic effect, and photoconduction that are particularly important for FET. The subsequent section includes a discussion of thermal processes such as the photo-thermoelectric effect that can produce/modify photo response in photodetectors.

3.1 Photocurrent Production Caused by e^{-}–h^{+} Partition

3.1.1 Photoconductive Effect

In this effect, more free electron carriers are produced as a result of electromagnetic (EM) radiation absorption, increasing the electrical conductivity of semiconductor substances. A distinctive photoconductor is made up of a semiconductor that acts as the channel and a pair of ohmic contacts attached to its opposite sides serving as source/drain electrodes. The dark current, which is zero when there is no light, flows between two electrodes. Electron–hole pairs that can be separated by applying a voltage are generated by electromagnetic radiation with energies higher than the bandgap. Voltage drift causes these free electrons and holes to travel to their respective dominant sides (against electrodes). As a result, on either edge of the junction, one depletion layer forms. The conductivity of the device is raised as a result of the drifting electrons and holes. The photoconductive effect generates current, which is influenced by the electromagnetic (EM) radiation's intensity, the transistor's channel length, the voltage near the source's drain, and the charge carrier mobility [39].

The band diagram for an FET is shown in Fig. 2a. A tiny source–drain current (I_{dark}) can flow when there is no light present and bias is applied (V_{ds}). Under light, the photon absorption energies greater than the band-gap energies, (more E_{ph} than E_{bg}) produce electron–hole pairs that are detached by the applied V_{ds} (Fig. 2b). Now, the current (I_{photo}) is increased as a result of the photochemically generated (h^+) holes and free (e^-) electrons drifting in reverse ways to the metal. As seen in Fig. 2c, d, such photogenerated current combines with dark current to lower the resistance of the electronic device.

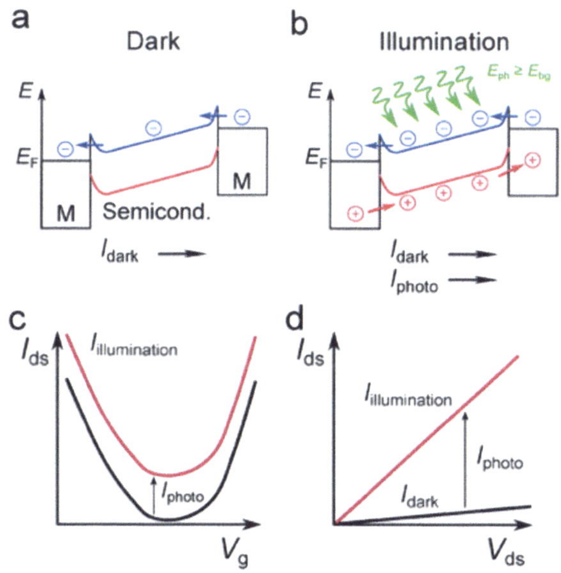

Fig. 2 Schematic diagram of photoconductive effect. **a** Band alignment in the absence of light for a semiconductor channel in contact with two metals (M) with low current (I_{dark}) under external bias. **b** Band alignment in the presence of photons whose energy (E_{ph}) exceeds the E_{bg} (bandgap energy). **c** I_{ds}–V_g marks under illumination and in dark conditions. A positive photocurrent and increased conductivity are produced by illumination throughout the whole range of gate voltage. **d** I_{ds}–V_{ds} curves under illumination and in dark, increasing conductivity and producing a positive photocurrent in light. Reprinted with permission from Late et al. [40] Copyright 2019 Elsevier

3.1.2 Photogating Process

A specific example of the photoconductive phenomenon is photogating. When holes or electrons are confined in localized states, they function as a local gate to modulate the material's resistance. Photogating is often identified as a horizontal change in I_{ds}–V_g lines under light. An e–h pair is produced when a photon is absorbed under illumination. Then, one carrier type is confined in localized states with energy close to the valence band edge.

3.1.3 Photovoltaic Effect

In the photovoltaic effect, specific semiconductor materials convert electromagnetic (EM) radiation energy into electric energy, and an internal electric field separates photogenerated e–h couples. The photovoltaic-based photodetector is made up of p–n kind photodiodes, which are semiconductors made of two different types of doping. (Fig. 3a) Under illumination and no external bias ($V_{ds} = 0$), a significant photocurrent (short-circuit current, I_{sc}) is produced as a result of the inner electric field separating the photo-excited pairs of e^- and h^+. A voltage (known as the open circuit voltage, or V_{oc}) is produced when the circuit is left open because of the formation of carriers or transporters with opposing polarities in various regions of the instrument or device. When these cells are exposed to light of the appropriate wavelength, photon energy is transferred to a single atomic particle of the conducting material or substance at the p–n junction, as well as photodiodes of the Schottky barrier form at the junction of a metal and a semiconductor. The photovoltaic effect requires the built-in electric field differential, which is produced either by electrostatic control through gates or local chemical doping [41, 42]. P–N diodes are frequently used in the arrangement of photodetectors based on this methodology either at zero (0) bias, or with reverse bias, which is consistent with the photovoltaic method.

3.2 Photocurrent Generation Driven by Thermal Mechanisms

3.2.1 Photo-thermoelectric Effect

In this effect, a semiconductor channel experiences a temperature gradient as a result of light-induced heating. The semiconductor channel has a difference in temperature (ΔT) between its ends. As seen in Fig. 4, the Seebeck effect transforms the difference in temperature (ΔT) into voltage differential (ΔV) having a magnitude that is linearly correlated to the Seebeck coefficient (S) and temperature gradient [43].

$$\Delta \mathrm{V} = \mathrm{S} \cdot \Delta \mathrm{T}$$

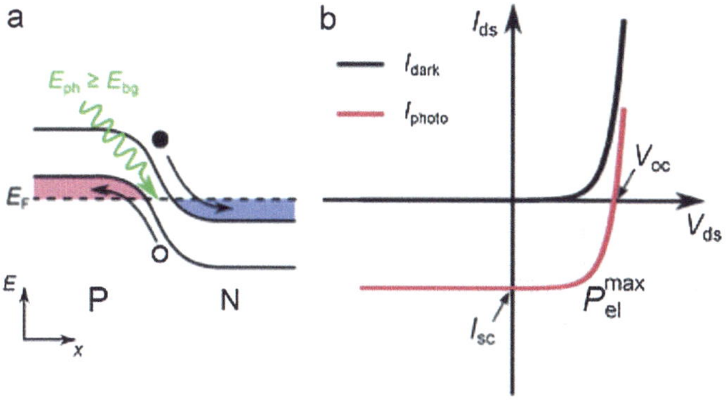

Fig. 3 Schematics of photovoltaic effect. **a** Band orientation in p–n junction. The common E_F (Fermi energy) for P as well as N-doped semiconductors is shown by the dotted line. An electron–hole pair is produced by the photon absorption with $E_{ph} > E_{bg}$, which is subsequently separated and enhanced by the inner electric field at the junction. **b** I_{ds}–V_{ds} graphs under illumination and in dark which causes a short circuit current (I_{sc}) and open-circuit voltage (V_{oc}). When utilized in the fourth quadrant, the apparatus generates electrical power. $P_{el}{}^{max}$ denotes the location of maximal power generation. Reprinted with permission from Late et al. [40] Copyright 2019 Elsevier

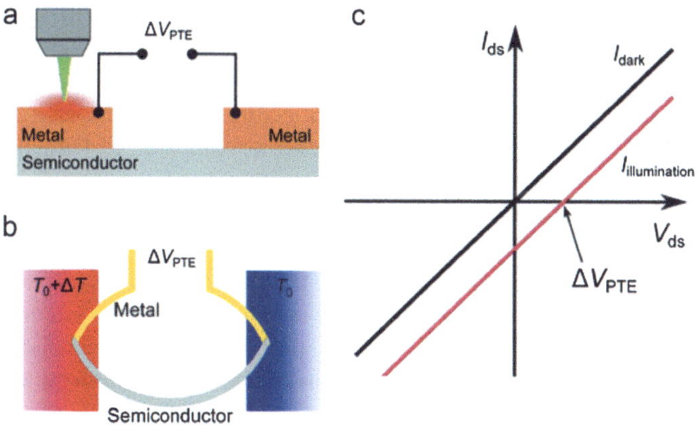

Fig. 4 Photo-thermoelectric effect. **a** Diagram of a FET with a laser light locally illuminating one of the metals interacts with the semiconducting channel. **b** Thermal circuit of the device. **c** I_{ds}–V_{ds} properties under illumination and in the dark of the device in which photoresponse is predominated due to the photo thermoelectric effect. Reprinted with permission from Late et al. [40] Copyright 2019 Elsevier

4 2D Materials Used in Sensing Applications

4.1 Graphene-Based and Graphene-Containing Photodetectors

Because of its structural characteristics, graphene is known to have special physical properties. At low temperatures, loosely suspended monolayers of graphene have carrier mobilities of around (1–2) × 10^6 cm^2 V^{-1} s^{-1}. Graphene layers may have decreased carrier mobility at higher temperatures, which can vary by orders of magnitude in the range of 10^2–10^5 cm^2 V^{-1} s^{-1} depending on the substrate type, synthesis method, and number of layers [44]. A monolayer of graphene absorbs around 2.5% of the incident radiation in the 0.25–2.5 μm range, according to a classic study and its absorption is wavelength-independent [45]. Furthermore, it was shown that reflectivity in monolayer and bilayer graphene can be influenced by an applied field. The investigation of the occurrence of a photo signal in graphene-based photosensitive components of various designs was a significant stage of converting to graphene-based photosensors [46]. Without formally altering the composition or configuration of graphene sheets, a research group, Mueller et al. demonstrated in what way the mere existence of source and metal drain interactions in the design of transistors alters the type of conductivity in the graphene sheets and causes p–n junction generation [47]. Graphene may interact with different types of metals, and depending on the metal, different graphene band structure variations may occur. When metals, such as Co, Ni, and Pd, interact strongly with graphene, the entire band structure of the material is altered. When weakly interacting metallic components such as Cu, Au, Ag, Al and Pt are used, the shifts of Fermi level as compared to Dirac point [48]. The Fermi level rises upward in some metals, such as aluminum, copper, and silver, reflecting transport of electron from the metal atom into sheets of graphene which is analogous to N-doping. For a different set of metals, such as gold, platinum, the Fermi level falls, corresponding to the reverse transfer of electrons (e^-) to the metal atom from grapheme sheets, as a result, a rise in density of hole in graphene, which resembles to p-doping. The variation in work functions of graphene electron and a metal (Schottky barrier analogue) and the interaction between graphene and a metal both contribute to the difference in doping types.

The evaluation of an FET using single-, double-, and triple-layer graphene was reported by Xia et al. [49]. In their experiments using a laser with a = 1.55 m, it was shown that the detector has a photoresponse with multiple light irradiation modulation frequencies, both high (10 MHz to 40 GHz) and low (from unmodulated irradiation to frequencies of as high as 100 Hz). In the area with the strongest field, photogeneration of carriers predominates. The area where graphene and metal are in contact produces the strongest field at low gate voltage. Mueller et al. [50] investigated a bilayer graphene-based phototransistor having an asymmetric drain–source electrode comprised of several metals palladium/gold (20/25 nm) also titanium/gold (20/25 nm) (Fig. 5). For graphene transistor, the development of 50 meV drain–source potential difference is due to varied local doping of graphene, increasing the overall photocurrent. In accordance with a prior publication, the carrier photogeneration is most effective at a distance of around 200 nm from the contacts. The research group [50] constructed an arrangement of electrodes having a width of 250 μm spacings in order to expand the carrier photogeneration area and the entire photosensitive element with a $6 \times 6\ \mu m^2$ area. As a consequence, it was shown that at a wavelength of 1.55 μm, sensitivity might increase up to 6.1 mA/W. In addition, it was suggested to use the band gap (transport) produced in biased double-layer graphene into an electrical field employed vertically to the planes of graphene sheets to lower the bilayer or double-layer graphene-based phototransistor's dark current. This means that in monolayer and bilayer graphene-based phototransistor elements, the photocurrent flow in the source–drain circuit is persuaded under light irradiation as a result of the development of an inherent field in graphene driven by the chemical reaction of a contact material with graphene. By using various metal contact pairings and also an electric field that is in the same direction as the inherent field, it is possible to increase the sensitivity and photocurrent. The photothermoelectric effect that occurs, for instance, at the boundary between single and double layers of graphene, in which the entropy change rises because of the deviations in the densities of states, can also produce the photocurrent in addition to the built-in field [51]. Then, owing to the variance in Fermi levels, the carrier then transports in the photothermoelectric effect in a direction opposite to the inherent field that originates at the junction between the mono- as well as a double layer of graphene. According to [51], the photothermoelectric effect, based on the mono-double layer graphene structure, influences the photocurrent flow in a photosensitive material. In both single- and double-layer graphene, p–n junction zones can be produced by the regional electric field generated under the gate. The p–n junction develops at gate borders when as a result of the field's action, carriers change the sign in the graphene area or region beneath the gate, which is differentiated from the gate substance by a dielectric layer (such as Al_2O_3).

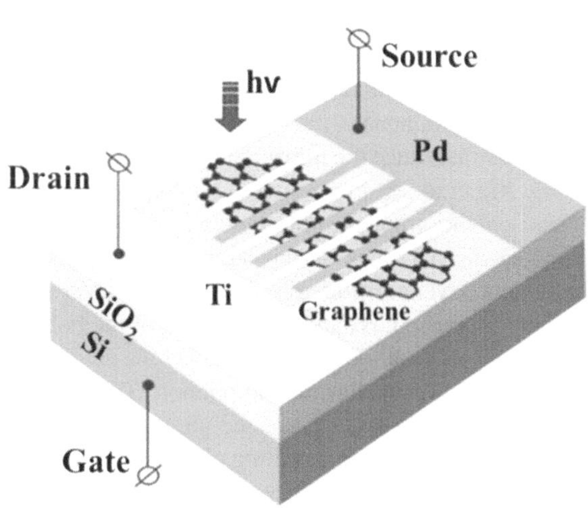

Fig. 5 Illustration of a single layer graphene phototransistor with asymmetric source–drain electrodes. Reprinted with permission from Popov et al. [52] Copyright 2022 Springer Nature

According to Lemme and co-workers [42], photocurrent produced at the p–n junction is the result of photothermoelectric effect, where the production of electric fields by optically induced temperature difference takes place, as well as the direct generation of e^-–h^+ pairs in the vicinity of potential gradient. The ability to manipulate a p–n junction region makes it possible to control a photosignal. It is noteworthy because it paves the way for the development of a long-spectrum radiation detector that neither the photovoltaic nor the photo thermoelectric processes of the emergence of a photoresponse in graphene are wavelength-limited. Additionally, the ability to generate narrow local photosensitive zones makes it possible to produce photodetectors with submicron pixels that have a regulated spectrum sensitivity. A p–n junction develops after chemical doping of the graphene surface. For instance, in a graphene-based FET explained in [53], the p-zones between drain and source were fabricated by covering a portion of the region with poly(methyl methacrylate), an organic polymer whereas the left graphene region between a source and a drain has been modified using an organic amine called PEI or polyethyleneimine to facilitate N-doping. In this instance, the photothermoelectric and photovoltaic effects also take place under irradiation of laser ($\lambda = 0.633$ m), although the photocurrent is mostly produced through the built-in field. Only the top graphene layer was fluorinated by Du et al. [54], who then exploited this to create a graphene/fluorographene heterostructure on the triple-layer graphene surface. This construction demonstrated double-layer graphene with a surface altered (designed) with the fluorographene layer since only the bottom graphene layer of such a device made contact with the chromium/gold electrodes. According to the research group, [54] the photogate effect associated with the electron capture onto fluorographene surface mostly forms the photosignal, which is compatible with the sensor's measured comparatively slow increase and decay times of 0.08–0.2 s, respectively. The photoresponse's long-term stability was improved by a thin aluminum oxide layer that was deposited. The trend of decorating graphene with diverse nanostructures has gained popularity and is discussed in more depth below. The potential to restrict the dispersion of energy from carriers on the lattice, which holds the promise of more effective photodetectors and solar cells, is a crucial consideration for constructing graphene-based photodetectors [55]. Again, it seems that the lattice temperature is substantially lower than the temperature of the electrons in graphene.

In order to reduce the impact of the contacts and substrate on graphene, a research group, Xia et al. [56] presented a phototransistor having polyhydroxystyrene layer placed on the graphene interface and subsequently covered with an HfO_2 layer designed via ALD or atomic layer deposition. In contrast to the inert graphene surface, the organic layer embedded below the HfO_2 layer not only makes it easier to deposit the latter via the $-CH_3$ and –OH groups of the polymer, but also maintains the high carrier mobility and intrinsic features of the active graphene layer by weakening the surface Coulomb scattering and phonon. The speed of a photodetector is a crucial feature. The triple-layer graphene photoresponse time was found to be 1.5 s^{-1} at room temperature using scanning photocurrent microscopy having temporal resolution value at an illumination wavelength (800 nm) [57]. Due to the drag caused by photons on free carriers, one-element or single photodetectors based on thin graphite layers (2–40 layers) and single graphene sheets demonstrate a greater rate of coherent radiation detection over visible to terahertz which is a broad spectral range. High speeds are accomplished at great light intensities that will be highly desired in fiber-optic communication schemes or systems in which speed is crucial. However, owing to the smaller absorption in thin layers, it is challenging for IR viewing systems to identify weak signals, thereby requiring for strong photocurrents, sensitivity, SNR, and specific detectivity.

Graphene-based photodetectors' photocurrent and external quantum efficiency can be greatly improved by employing optical irradiation concentrators that are regular arrangements of metallic structures or particles [58, 59]. Two methods control the photocurrent intensification in the plasmonic antenna instruments or devices. The first of them involves the photon capture by plasmons produced by an EM field of a specific frequency, which leads to a usual rise in the electric field radiation. In the second process, the light scattering properties of metallic nanoparticles take place, which allows for the plasmonic amplification of interactions between light scattered by different particles. In this instance, a larger particle size and, as a result, a larger interaction cross-section, results in a higher amplification efficiency on such concentrators. Here, the particle size is similarly correlated with the spectrum dependence of the maximum wavelength of light amplification. According to Liu et al. demonstration using plasmonic antennas, the light amplification appears directly next to a layer of metallic particles. Whenever tiny particle arrays are utilized as antennas, an average light intensity diminishes very quickly as distance from the nanoparticle plane increases [58]. While plasmonic antennas are positioned between layers of graphene (Fig. 6), another amplification mechanism operates when the architecture incorporates plasmonic antennas and a graphene detector [59]. At this point, the photocurrent is produced due to plasmon decay on the surface of the antenna and hot electron transport generated by plasmons passing through the antenna towards graphene. The higher excitation of inherent electrons of graphene caused by the surrounding field generated by a plasmonic antenna coexists with this mechanism. When compared to a graphene detector with a similar shape without antennas, photocurrent in the visible and infrared bands can be increased by up to 800% with nanoantennas positioned between two graphene layers. In this instance, a system's intrinsic quantum efficiency reaches 20%.

As already known, the low radiation absorption coefficient restricts the sensitivity of ampere–watt (A/W) of the graphene-based photosensors is approximately 10^{-2} A W^{-1} in the lack of photocurrent amplification techniques. The A/W sensitivity and detectivity can be greatly improved by a photocurrent amplification technique in transistor architectures made of 2D materials. Particularly, in 2011, a gain of 10^3 and 10^2 A/W sensitivity was executed in a phototransistor using a field electrode composed of cadmium sulfur and cadmium selenium colloidal quantum dot (CQD), but these values were noticeably lower than the sensitivity attained at the period in the devices based on bulk semiconductor crystals. This resulted from the carriers' inadequately great mobility in the transistor channel that the device discussed was only 15 cm^2 V^{-1} s^{-1} [60]. The union of high electron (e^-) and hole (h^+) mobilities in graphene layers which may reach 6×10^4 cm^2 $V^{-1}s^{-1}$ at ambient temperature

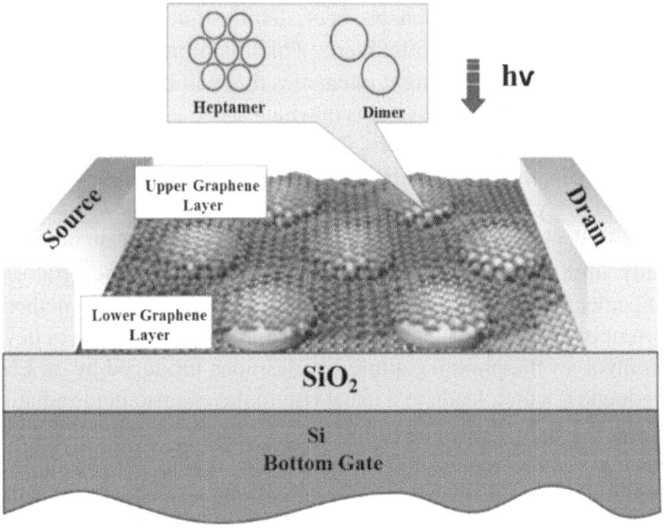

Fig. 6 Structure of a phototransistor with plasmonic antennas positioned in the graphene layers. Reprinted with permission from Popov et al. [52] Copyright 2022 Springer Nature

with photocurrent amplification in transistor designs at the concomitantly greater absorption guaranteed by colloidal quantum dots [61]. In a phototransistor between its source and drain, PbS colloidal quantum dots were deposited on the single-layer graphene surface (Fig. 7). In quasi-0D QD nanocrystals, light is absorbed and electron–hole pairs are produced through photogeneration. At the barrier generated by varied work functions at the quantum dots/graphene interface, electrons and holes are then separated. A voltage applied to the gate causes photogenerated electrons to be captured on QDs and photogenerated holes to recombine with electrons on the graphene surface. The photocurrent is boosted by the recycling of holes in the graphene sheet as long as the QDs or quantum dots are negatively (−ve) charged. Since only graphene is in direct touch with the electrodes in this scenario, the QDs function as decorations on the graphene, changing its properties. Ni et al. were able to achieve the sensitivity in a broad spectrum range, i.e., from 0.375 to 3.9 μm by employing B-doped silicon colloidal QDs placed on the graphene surface in a phototransistor arrangement same as that described previously [62]. At a wavelength of 1.47 μm, this structure's detectivity reached 10^{13} W^{-1} $Hz^{1/2}$ cm. The lengthy signal decay period of this phototransistor (from 3 to 8.5 s) was a severe flaw.

4.2 Photosensors Based on MoS_2

MoS_2 or Molybdenum disulfide is one of the most extensively investigated TMDCs due to its broad availability. To assess the efficiency of MoS_2-based photodetectors

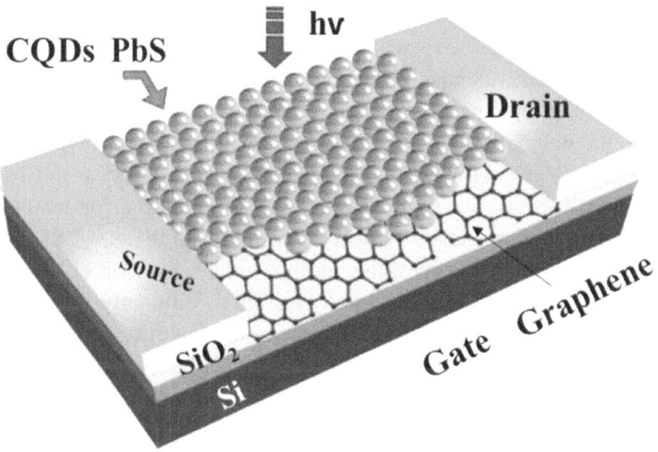

Fig. 7 Phototransistor based on graphene decorated with lead sulfide quantum dots. Reprinted with permission from Popov et al. [52] Copyright 2022 Springer Nature

from the viewpoint of their photodiodes, phototransistors, and photoconductors, it has been discussed how to increase the mobility or movement of carriers and increase the incident light absorption by MoS_2 atomic layers using plasmonic as well as MoS_2 (halide perovskite) heterostructures, 0D–2D and also 2D–2D structures, doping or ICE (interface coupling effect) of MoS_2. Special emphasis is given to the development of layered atom MoS_2-based broadband photodetectors, and from their initial stages to self-sustaining and adaptable photodetectors for wearable optoelectronics. Regarding the development of flexible photodetectors, the activity of pristine MoS_2 atomic layers as well as their hybrid structures with TMDs, graphene, ZnO, CNTs, and MoS_2 atomic layers is explained in terms of their broadband detectivity, photoresponse speed, photoresponsivity, photogain, mechanical flexibility, NEP, EQE, and environmental stability.

To produce a monolayer, bulk MoS_2 is exfoliated using a variety of techniques, including mechanical exfoliation and ultrasonic treatment [63, 64], chemical exfoliation [43]. Researchers would rather develop crystals using procedures like chemical vapor deposition for impurity-free experiments. This approach of home-making crystals also enables the engineering of bandgaps through the introduction of dopants [65]. A photoresponsivity of roughly 7.5×10^{-3} A W^{-1} has been recorded for MoS_2 phototransistors. MoS_2 was deposited on Si/SiO_2 to create a monolayer phototransistor, according to a report. A mono-layer sheet of MoS_2 photoluminescence was detected at ambient temperature utilizing a 488 nm laser and revealed a strong PL spectral peak position at 676 nm. Again, the altitude of single-layer MoS_2 was determined by AFM to be 0.8 nm. Prominent peak results from the direct intraband recombination of the photogenerated e^-–h^+ pairs in the monolayer MoS_2, while a small peak position at 623 nm is thought to be the result of the energy split of MoS_2's valence band (vb) spin-orbital coupling [64, 66, 67]. Unprecedented incident-light control, effective photoresponsivity, and photoswitchability properties of the device made it possible to produce inexpensive and readily available optoelectronic devices. The research for this device also revealed that the photocurrent was influenced by optical effect. The photocurrent was reduced following the emission of dark drain current while the device was exposed to energy beyond 670 nm which is the excitation wavelength. However, when the wavelength was dropped from 670 nm, there was a substantial increase in the photocurrent, satisfying the fundamental idea of input being greater photon energy than the E_{bg} (energy gap), i.e., ~1.83 eV. The electrons with a high incident energy of the photon (greater than 1.83 eV) may produce more photoelectrons, which increases photocurrent. Additionally, this device demonstrated effective photoswitchability and outstanding stability. Whenever the drain current increased to a greater value under photon irradiation and then decreased to a lesser value under darkness, or an off state, the switching behavior was observed [68].

In other instances of MoS_2 that was created by chemical vapor deposition (CVD), a photosensor described by Lopez and co-workers [68], was constructed triangular MoS_2 structures synthesized on Si-SiO_2 substrates by MoO_3, a mild sulfurizing. The metal connections, apparently 4 nm thick in Ti and 100 nm thick in Au, were made using electron beam lithography at 107 Torr pressure. The photoluminescence spectrum with 670 nm maximum intensity and photon energy of 1.86 eV was visible in the Raman spectra acquired at 488 nm laser excitation. Because of D_{3h} point group symmetry, a feature of noncentrosymmetric monolayer MoS_2, the first-order Raman modes are visible. In the space between the cathode and the anode, PL and Raman spectra were recorded, indicating the existence of single-layer MoS_2 and also its bandgap (direct). When the photosensitivity of the sensor was tested using two light wavelengths between 488 and 514.5 nm, the sensor response was detected. A current versus time plot, which includes values every 200 ms, complements the observation of the bias voltage and laser excitation between 0 and 90 s. The dark current hardly flickered when the employed bias voltage grew from 1.5 to 2.0 V for about 52 s, but the excitation laser lighted the device for the duration of 5 s beginning at $t = 14$ s and 64 s.

4.3 Molybdenum Diselenide ($MoSe_2$)-Based Photosensor

Due to its similarity to MoS_2 in terms of direct bandgap ($E_g = 1.6$ eV) as well as strong exciton BE or binding energy, the monolayer $MoSe_2$ has become one of the most widely used substances to study both electrical and optical features. The observation of significant photoluminescence in $MoSe_2$ strengthens the case even more [69]. Furthermore, few-layered $MoSe_2$ have good anti-photocorrosion stability and almost degenerate direct and indirect bandgaps [70, 71]. Larentis et al. [72] revealed the development of the first back-gated FET made from a few layered $MoSe_2$ that demonstrated a large I_{on}/I_{off} ratio, which is crucial for digital devices. A single-layer $MoSe_2$-based photodetector was recently suggested to have been developed by Xia et al. [73], although the photoresponsivity values were low. In the solar spectrum range, it is expected to have an improved photoresponse, and it has good antiphotocorrosion stability. Titanium (Ti) electrodes with a 50 nm thickness were deposited on silicon substrates using electron beam evaporation to make a phototransistor [74]. Polydimethylsiloxane (PDMS) transfer technique was used to deposit $MoSe_2$ onto the Ti electrodes. In comparison with multilayered $MoSe_2$, the phototransistor structure exhibits a higher saturation current. The relationship between the drain current (I_d) and drain-source voltage (V_{ds}) at various gate voltage values is clearly described by Abderrahmane et al. [74] Linear dependence has been observed for lower source–drain voltages, and transfer features for a variety of source–drain voltages. Thus, this demonstrates the ohmic nature of the contacts. At the junction of the gate oxide and $MoSe_2$, the phototransistor displayed n-type behavior, collecting electrons. While a large threshold voltage is undesirable for small power consumption utilities, the current on/off ratio and threshold voltage (V_{th}) at $V_{ds} = 10$ V were reported to be approximately 3×10^4 and -25 V, respectively. The threshold voltage will be lower and the current on/off ratio will be improved if high dielectric materials are employed in place of SiO_2 as the gate oxides. The back gate sweep was used to estimate the effective mobility using the formula, $\mu_{eff} = [dI_d/dV_g] \times [L/(WC_iV_{ds})]$, where C_i represents the gate capacitance determined using the formula, $C_i = \varepsilon_0\varepsilon_r/d$. For the SiO_2 in this equation, $\varepsilon_r = 3.9$. The value of gate capacitance is 1.15×10^{-4} Fm^{-2}, the oxide is 300 nm thick, and W and L stand for the channel breadth and length, respectively. The device had an effective mobility value of 19.7 cm^2 V^{-1} s^{-1}. When the laser power (P) was 0.015 μW, the maximum photoresponsivity (R_λ), which is determined by $R_\lambda = I_{ph}/P$, was around 97.1 A W^{-1}. As the laser power increased, this value decreased. From the investigations, it was found that the photoresponsivity value of $MoSe_2$ was greater than in other studies on photodetectors with multilayer including GaSe [76], GaS [75], InSe [78], and MoS_2 [77].

4.4 Tungsten Disulfide-Based Photosensor (WS_2)

For the fabrication of future ultrathin devices, tungsten disulfide (WS_2) is one of the transition metal dichalcogenide (TMDC) materials that have received the most attention. Due to the similarity of its two-dimensional (2D) structure to that of graphene nanosheets, it offers a wide range of possible applications including field effect transistors (FETs), optoelectronic devices for use in telecommunications, and electroluminescent devices [79–81]. When WS_2 is a thin film or when it is in bulk form, its indirect and direct bandgaps, which range in value from 1.4 to 2.1 eV, give rise to its semiconducting properties [82, 83]. Additionally, they can be employed as high-temperature activity devices due to their extra thermal stability in an inert atmosphere (up to the temperature 1100 °C).[84] WS_2 has a tunable bandgap, enabling absorption which extends from the UV (ultraviolet) range to the visible range. Functionalizing such materials could result in the alteration of bandgap, which is essential in optics as well as transistor technology. Additionally, WS_2 has a considerable 5–10% absorption of sunlight. In comparison to gallium arsenide (GaAs) and silicon (Si) materials, WS_2 is a factor of an order of magnitude higher. As a result, WS_2 is a crucial component of transistor and optics technology for highly photoresponsive ultraviolet (UV) and visible photodetectors [85]. Due to the fact that the sheets have different lattice constants and are only joined by a weak van der Waals force, as in layered materials, it is also conceivable to construct vertical hetero-structures using distinct 2D materials without the typical lattice mismatch. As a result, even atomically thin films substantially interact with light, such as monolayer tungsten disulfide (WS_2), which can absorb around 10% at excitonic resonances and cover a very wide electromagnetic spectrum due to its diverse electronic characteristics [86]. The highly significant absorption of WS_2 has also been demonstrated in recent research [87]. As a sandwiching material between metal heterojunction contacts for the manufacture of high-performance devices [88], WS_2 offers significant potential for high-responsivity photo-detecting applications [89, 90]. In recent years, there has been a notable advancement in the synthesis and various uses of WS_2. Nevertheless, several challenges still need to be addressed to encourage the commercial application of WS_2. For example, sophisticated synthesis techniques should be developed to produce 2D WS_2 materials with excellent uniformity and quality at least manufacturing costs. It is still difficult to synthesize high-quality WS_2 materials at the wafer size with control over thickness.

It has been reported that WS_2 materials, which are produced via a combination of thermal annealing and magnetron sputtering in an argon gas environment, are used for photodetection [91]. With thicknesses ranging from 100 to 200 nm, this showed the first photoconductivity characteristics for WS_2 thin films. As the voltage decreased in proportion to the energy of the photons, photoconductivity was achieved under constant current conditions. There have been several reports of evaluation of photocurrent behavior of WS_2 grown-up by CVD on quartz substrates. The spectral responsivity was found at 2.0 $\mu A\ W^{-1}$ in case of lower excitation (wavelength, $\lambda = 647$ nm; energy, $E = 1.91$ eV), whereas greater value at 21.2 $\mu A\ W^{-1}$ for higher energy (wavelength, $\lambda = 568$ nm; energy, $E = 2.71$ eV). The spectral responsivity was proportional to the excitation source wavelength. The sensitivity at various excitation intensity levels was discovered to validate the result that the I_P (photocurrent) is not linearly proportional to the laser's excitation intensity [92]. A substance that is photosensitive in this way is sensitive well at lower intensities and sensitive much better at higher intensities. A vacuum chamber with a pressure of 1×10^{-6} Torr was used for the photocurrent measurements, which were connected to a micro-Raman spectrometer. The time-varying component of all photocurrent measurements is the photocurrent generated by turning on and off the laser beam. As the laser excitation wavelength changed, a significant photocurrent has been observed by measuring the spectral responsiveness of the WS_2 electronic device. With regular or periodic laser beam illumination, the photocurrent's temporal dependency was observed, demonstrating the photocurrent's substantial reliance on both illumination power and wavelength. The quick measured response produced response and recovery times of 5.3 ms with good responsiveness, stability, and comparison to other transition metal dichalcogenides, suggesting WS_2 as a solid alternative for optoelectronics [90]. In relation to WS_2, Zheng and co-workers reported a special photoelectrical alteration feature having 53.3 A W^{-1} high responsivity and 1.22×10^{11} Jones high detectivity at the wavelength of 365 nm, also a method for the large-scale synthesis of WS_2 film that may be applied to the development of photoelectrochemical cells, photosensors, and solar cells [93].

According to reports, a nickel (Ni) thin layer is employed to encourage the formation of WS_2 films with properties resembling those of WS_2 single crystals. According to the Perea-Lopez research group and Zeng research group [93, 94], the annealing procedure allows for the development of WS_2 crystallization layers that are stacked in the c direction. Here, the observed visible photoresponsivity that is in tens of A/W is significantly greater than WS_2 created with a lack of annealing procedure. Heterostructure design can improve the photoresponsivity as well as additional physical features of a thin film that has been prepared. However, this method has a drawback because samples processed at low temperatures have been found to have slow rise and fall periods. A single-layer WS_2 photodetector supported by gold nanoparticles having 1050 A/W responsivity was demonstrated by Liu et al. so as to improve the effect of light trapping in WS_2.[95] This is due to the Au nanoparticles' confined surface plasmon resonance, which amplified the electromagnetic field locally. Additionally, a multilayered WS_2 photodetector with ~0.27 A/W responsivity under 370 nm illumination was displayed for photodetection broadband in the range from 370 to 1064 nm [96].

Ahmad et al. studied the photoconduction behavior of a fabricated Ag/WS_2/SiO_2/p-Si/Ag heterostructure (Schottky) photodetector under UV-irradiation using a LED or light-emitting diode at a wavelength of 380 nm (Fig. 8) [97]. The process is carried out at various light intensities and bias voltages for direct current (DC). In order to develop an affordable UV photodetector, a unique technique called drop casting is used in this study to build a thin film of WS_2 on the substrate SiO_2/p-Si. The solution of WS_2 nanoflake is ultrasonically sonicated along with iron (III) O powders, which serves as sub-electrode to create a worthy Schottky contact with tungsten disulfide nanoflakes so as to obtain desirable electrical connections and adhesion with the substrate. In early experiments, the drop-casted tungsten disulfide solution showed weak and unreliable photoconduction. Different materials were added to the tungsten disulfide solution to enhance its performance and offer improved responsiveness. It was found that adding Fe powder to the WS_2 solution produced a solution with superior conductivity, greater stability, and responsiveness and that ions of iron strongly formed a Schottky contact between the silver conducting paste and tungsten disulfide film. For this experiment, 1 mL iron solution was introduced to the tungsten disulfide (WS_2) solution. Again, photoconduction of the constructed heterostructure photodetector WS_2/SiO_2/p-Si varies when the ultraviolet power source changes, which may be a method for the improvement of UV (ultraviolet) photoconduction.

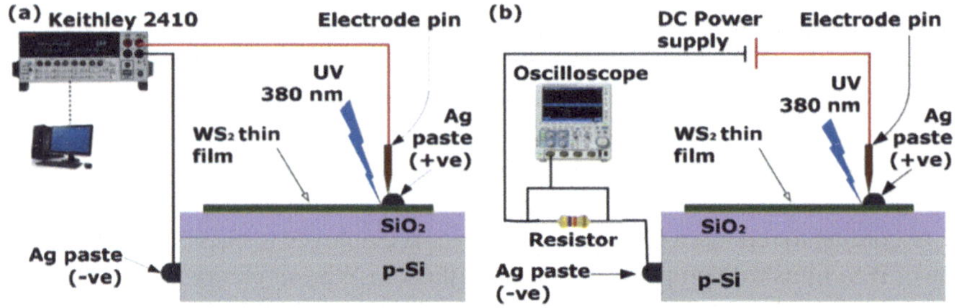

Fig. 8 **a** *I–V* characteristics setup and **b** temporal measurement setup of Ag/WS_2/SiO_2/p-Si/Ag heterostructure Schottky photodetector. Reprinted with permission from Ahmad et al. [97] Copyright 2019 Elsevier

4.5 Black Phosphorous-Based Photosensor

A multilayer crystal of phosphorus that is stable at ambient temperature is known as black phosphorus (BP). It has recently sparked interest when studies revealed that monolayer and few-layer flakes can be exfoliated similarly to grapheme [98–100]. But unlike graphene, BP has a direct bandgap that is expected to be between 1.8 and 2.0 eV for the monolayer. The bandgap narrows and eventually reaches the bulk value at 0.3 eV when the layer number rises [101]. The elemental 2D semiconducting material black phosphorus (BP) has ~1.5 eV direct bandgap for single-layers [102]. Black phosphorus is an interesting contender for broadband optoelectronic utilities because of its extremely small bandgap, which spans the visible to near-infrared spectrum. High-performance photodetection requires a number of key components, including a direct bandgap, excellent light absorption efficiency across the broad spectrum, and high carrier mobility, all of which few-layer BP possesses. Additionally, BP has demonstrated superior electrical characteristics, including high hole mobility of up to 1,000 $cm^2\ V^{-1}\ s^{-1}$ at ambient temperature, an on–off ratio up to 10^5, and excellent current saturation in field-effect devices [103].

The rectangular in-plane lattice of the layered BP crystal has a highly anisotropic along both x and y directions, and every two phosphorus atom rows puckered up and down in an alternate manner to formulate an 'armchair-like' geometry along the x-direction only. This is in contrast to the in-plane hexagonal lattice in other 2D materials like MoS_2 and graphene that are not sensitive to linear polarization of light of incident radiation. Because of this, the electrons and photons in BP can exhibit highly anisotropic behavior inside the layer plane [98, 104]. BP offers a fascinating possibility for incorporating intrinsic crystal anisotropy due to the mirror reflection symmetry only in the direction of the y-axis for linear-polarization-sensitive photodetection, contrary to existing photodetectors for linear dichroism detection that is based on extrinsic geometric effects (for example, wire-grid patterns 115–118). In the article reported by Hongtao Yuan et al., they showed how a linear dichroic photodetector based on black phosphorus transistors works. The absorption of polarised light in dual in-plane crystal axes in this device in various ways throughout a wide spectral

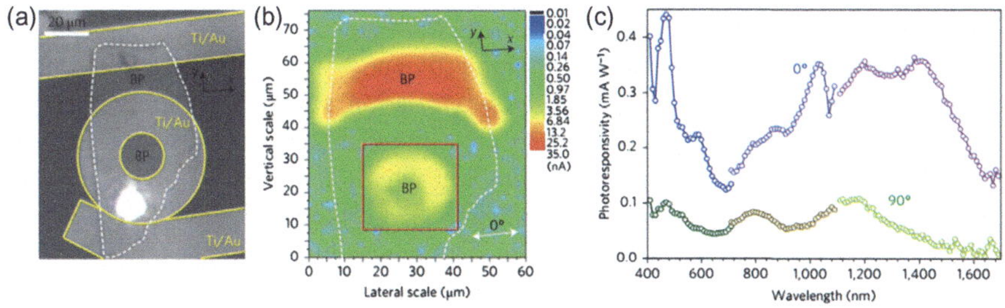

Fig. 9 **a** Optical representation of a black phosphorous (BP) photodetector with a photocurrent collector in the form of a ring. Ti/Au electrodes are shown by yellow lines, and the BP flake is contained by a white line. **b** Corresponding photocurrent microscopy photograph of the apparatus illustrated in (**a**), with illumination at 1,500 nm and polarization along the x direction (white arrow). **c** Polarization dependence of photoresponsivity with illumination from 400 to 1,700 nm. Reprinted with permission from Hongtao Yuan et al. [106] Copyright 2015 Springer Nature

range, from around 400–3,750 nm. A new level of freedom in manipulating the electrical and optoelectronic properties of multilayer BP is made possible by the optical choice criteria in black phosphorus, which permit the broadband light absorption polarised preferentially in the direction of the x-axis. We specifically spatially isolate the photogenerated e^{-}–h^{+} pairs in the channel, reducing the recombination rate and increasing the photodetector's photoresponsivity, employing the perpendicular p–n junction caused via ionic gel-gated transistor.

A photodetector based on the BP layer exhibits high responsiveness in near-infrared region (wavelength 900 nm) with values of around 7×10^6 A/W at 20 K and 10^3 A/W at 300 K. (Fig. 9) Because of strong ohmic contact with black phosphorus, the photogenerated carriers are successfully captured [105]. To achieve high responsiveness photodetection, Black phosphorus was also well combined with silicon (Si) waveguide. When doping levels are low the intrinsic responsiveness approaches the value of 135 mA/W for 11.5 nm of approximate thickness and also 657 mA/W for 100 nm thickness at ambient temperature and has bandwidth with high response surpassing 3 GHz, photovoltaic current far exceeds photocurrent.

5 Conclusion and Recent Development

Beyond graphene, 2D materials have rapidly developed in recent years. These materials provide new possibilities for a wide range of biomedical and environmental monitoring applications, including medication delivery systems and bio/ sensors. Since several of the newly discovered 2D materials, such phosphorene, germanene, and silicene, are biocompatible and/or biodegradable, they present new opportunities. Additionally, the distinctive optical, chemical, and physical features of 2D materials show advancements in the capabilities of the produced gadgets as well as discussed applications in the field of biomedicine. These characteristics comprised high photothermal conversion efficiency, strong photoluminescence, electrical conductivity, metallicity, degree of exfoliation, strong photoluminescence, and surface functionalization. One of the most popular 2D materials, MXene are appropriate NIR or near-infrared preoccupation which is employed for synergetic cancer treatment and photothermic diagnosis [1]. Furthermore, FRET or Förster resonance energy transfer sensors may be used to precisely measure distances on the nanoscale and detect changes both in vivo and in vitro in biological applications. As a consequence, nanobiosensors with remarkable selectivity, enhanced biostability, and enhanced sensitivity have been developed [2]. Although good research work has been already done, still there is room for the discovery of novel 2D materials, which means that there will be more potential for their use in biochemical and medical research. Due to the unique optical and electrical properties, newer chemical characteristics, and complex physics of 2D materials investigation or research into these have increased recently. As a result of their atomically tiny dimensions, they also possess very intriguing and exceptional mechanical properties. New opportunities for layer-by-layer materials and hybrid/heterostructure device engineering are opened up by the large-area production and isolation of different novel 2D thin atomic materials. These materials allow for investigating and exploiting superior or previously unexplored features, which assurance a variety of new device utilities and technologies. Despite the fact that many technologies and applications are still in their early years, it is important to continue researching the many features of 2D materials because of their remarkable potential and wide range of uses. This book chapter discusses the most recent findings on the uses of sensor devices, unsolved issues, and 2D material properties. Graphene, metal dichalcogenides (WS_2, $MoSe_2$, MoS_2, etc.), black phosphorous, are some examples of 2D materials in this chapter.

The most extensive knowledge on the mechanisms governing the production of photosignals has recently been gathered, especially for the material graphene. The application of graphene in photosensors was particularly explored in systems where monolayers of graphene are utilized for photocarriers as a transport layer produced in different areas of the device. As a result of this characteristic, industrial businesses are already considering integrating graphene into their photodetector systems. The decorated graphene/fluorographene heterostructure working in the middle wave IR range, in which detectivity is the same as an uncooled bolometer, has made the most advancement in the direct usage of graphene as one of the photosensitive materials. Additionally, one remarkable outcome for disoriented graphene-like structures (3D graphene) was demonstrated, also their application allowed for a significant expansion of silicon's spectral capabilities as well as the development of a strategy that will eventually be able to compete with InGaAs in the fabrication of economic SWIR detectors.

It is still difficult to integrate MoS_2 or molybdenum disulfide photodetectors with varied flexible electrical, optoelectronic functional, and also photonic devices. Investigation on MoS_2-based photodetectors is progressing quickly [107–109]. It is possible to investigate the broad array of 2D materials for creating photodetection technology. A novel high-performance photodetector can be developed using 2D

TMDs, such as PtS_2, $MoTe_2$, $PtSe_2$, PdS_2, $PtTe_2$, PdS_2, $GeSe_2$, GeS_2, TiS_2, HfS_2, $PdTe_2$, FeS_2, CoS_2, SnS_2, NiS_2, $TaSe_2$, and $NbSe_2$. There are a lot of possibilities for transition metal dichalcogenides-based materials with different nanomaterials, such as graphene, black phosphorus, h-BN, single-chalcogenide layered substances or materials (GaTe, GaS, GaSe, InTe, InSe, SnSe, SnS, InS, SnTe, etc.), 2D transition-metal nitrides, carbonitrides and also carbides. Certain characteristics, including responsiveness, internal quantum efficiency, EQE, temporal response, noise equivalent, and wavelength contribute to identifying photosensor systems. Even though they were developed on the same substrate, photodetectors made of various 2D materials, such as MoS_2, $MoSe_2$, and WS_2, exhibit a great deal of responsive diversity. When compared to $MoSe_2$ or WS_2 devices, the photodetectors based on MoS_2 exhibit significantly faster response times. Being sensitive to environmental changes is one characteristic that unites them, and that makes them a strong alternative to gas detectors which are light-sensitive. Due to the semiconducting di as well as tri chalcogenides' lack of photoresponse at telecommunication wavelengths, the BP was investigated. The BP exhibits significant responsivity (approximately 0.1 A/W) with a response speed of f3dB $\approx$ 3 GHz under the wavelength $\lambda = 1550$ nm illumination. In the IR region, BP has emerged as a viable contender for quick and broad detection, which might be utilized for infrared light energy harvesting. This is clear that semiconducting layered materials based on photodetectors exhibit significant variance in their responsiveness (approximately 10 orders of magnitude). The innovative characteristics of layout changes of transition metal dichalcogenides-based photodetectors have been the limited dark current and improved device performance. The greater photon energy than the bandgaps of the distinct TMDs materials in the case of 2D hybrid structures for visible range, and each stacked material produces an adequate quantity of photogenerated free carriers, leading to a noticeably increased photoresponse. In addition to these, the gallium, tin chalcogenides, and also indium are quickly holding up since the compounds made of Ga, Sn, and In have responsivities that are greater than equal to those observed in TMDC-based photo-FETs. Since the process may be expanded to the ultraviolet region, they can also be employed as UV detectors. The emergent features of 2D materials, which result in ultrafast response time, carrier production, and improved switchability control, are what give them their superior performance. They are now a superior option for optoelectronic and photonic devices as a result of all these aspects.

Acknowledgements The authors would like to express their gratitude to all individuals and organizations who allowed us to reprint figures and other important information in order to complete the chapter. The authors would also like to thank the Department of Chemistry, National Institute of Technology, Silchar, Assam.

Conflict of Interest The authors declare no competing interests.

References

1. Miró, P., Audiffred, M., Heine, T.: An atlas of two-dimensional materials. Chem. Soc. Rev. **43**, 6537–6554 (2014)
2. Chia, X., Pumera, M.: Characteristics and performance of two-dimensional materials for electrocatalysis. Nat. Catal. **1**, 909–921 (2018)
3. Tan, C., Cao, X., Wu, X.-J., He, Q., Yang, J., Zhang, X., Chen, J., Zhao, W., Han, S., Nam, G.-H., Sindoro, M., Zhang, H.: Recent advances in ultrathin two-dimensional nanomaterials. Chem. Rev. **117**, 6225–6331 (2017)
4. Zhai, S., Wei, L., Karahan, H.E., Chen, X., Wang, C., Zhang, X., Chen, J., Wang, X., Chen, Y.: 2D materials for 1D electrochemical energy storage devices. Energy Storage Mater. **19**, 102–123 (2019)
5. Rao, C.N.R., Sood, A.K., Subrahmanyam, K.S., Govindaraj, A.: Graphene: the new two-dimensional nanomaterial. Angew. Chem. Int. Ed. **48**, 7752–7777 (2009)
6. Sun, Y., Lei, F., Gao, S., Pan, B., Zhou, J., Xie, Y.: Atomically thin tin dioxide sheets for efficient catalytic oxidation of carbon monoxide. Angew. Chem. **125**, 10763–10766 (2013)
7. Xia, F., Wang, H., Xiao, D., Dubey, M., Ramasubramaniam, A.: Two-dimensional material nanophotonics. Nat. Photon **8**, 899–907 (2014)
8. Xu, S., Yan, Z., Jang, K.-I., Huang, W., Fu, H., Kim, J., Wei, Z., Flavin, M., McCracken, J., Wang, R., Badea, A., Liu, Y., Xiao, D., Zhou, G., Lee, J., Chung, H.U., Cheng, H., Ren, W., Banks, A., Li, X., Paik, U., Nuzzo, R.G., Huang, Y., Zhang, Y., Rogers, J.A.: Assembly of micro/nanomaterials into complex, three-dimensional architectures by compressive buckling. Science **347**, 154–159 (2015)
9. Chang, H., Wu, H.: Graphene-based nanocomposites: preparation, functionalization, and energy and environmental applications. Energy Environ. Sci. **6**, 3483 (2013)
10. Perreault, F., Fonseca de Faria, A., Elimelech, M.: Environmental applications of graphene-based nanomaterials. Chem. Soc. Rev. **44**, 5861–5896 (2015)
11. Zhu, J., Xiao, P., Li, H., Carabineiro, S.A.C.: Graphitic carbon nitride: synthesis, properties, and applications in catalysis. ACS Appl. Mater. Interfaces **6**, 16449–16465 (2014)
12. Zhang, K., Feng, Y., Wang, F., Yang, Z., Wang, J.: Two dimensional hexagonal boron nitride (2D-hBN): synthesis, properties and applications. J. Mater. Chem. C **5**, 11992–12022 (2017)
13. Yankowitz, M., Ma, Q., Jarillo-Herrero, P., LeRoy, B.J.: van der Waals heterostructures combining graphene and hexagonal boron nitride. Nat. Rev. Phys. **1**, 112–125 (2019)
14. Pumera, M., Sofer, Z.: 2D Monoelemental arsenene, antimonene, and bismuthene: beyond black phosphorus. Adv. Mater. **29**, 1605299 (2017)
15. Naguib, M., Kurtoglu, M., Presser, V., Lu, J., Niu, J., Heon, M., Hultman, L., Gogotsi, Y., Barsoum, M.W.: Two-dimensional nanocrystals produced by exfoliation of Ti_3AlC_2. Adv. Mater. **23**, 4248–4253 (2011)
16. Lloret, V., Rivero-Crespo, M.Á., Vidal-Moya, J.A., Wild, S., Doménech-Carbó, A., Heller, B.S.J., Shin, S., Steinrück, H.-P., Maier, F., Hauke, F., Varela, M., Hirsch, A., Leyva-Pérez, A., Abellán, G.: Few layer 2D pnictogens catalyze the alkylation of soft nucleophiles with esters. Nat. Commun. **10**, 509 (2019)
17. Hart, J.L., Hantanasirisakul, K., Lang, A.C., Anasori, B., Pinto, D., Pivak, Y., van Omme, J.T., May, S.J., Gogotsi, Y., Taheri, M.L.: Control of MXenes' electronic properties through termination and intercalation. Nat. Commun. **10**, 522 (2019)
18. Chia, X., Pumera, M.: Layered transition metal dichalcogenide electrochemistry: journey across the periodic table. Chem. Soc. Rev. **47**, 5602–5613 (2018)
19. Chhowalla, M., Shin, H.S., Eda, G., Li, L.-J., Loh, K.P., Zhang, H.: The chemistry of two-dimensional layered transition metal dichalcogenide nanosheets. Nat. Chem. **5**, 263–275 (2013)
20. Akinwande, D., Petrone, N., Hone, J.: Two-dimensional flexible nanoelectronics. Nat. Commun. **5**, 5678 (2014)
21. Zhang, Z., Huang, H., Yang, X., Zang, L.: Tailoring electronic properties of graphene by π–π stacking with aromatic molecules. J. Phys. Chem. Lett. **2**, 2897–2905 (2011)

22. Wang, D., Wang, L., Dong, X., Shi, Z., Jin, J.: Chemically tailoring graphene oxides into fluorescent nanosheets for Fe^{3+} ion detection. Carbon **50**, 2147–2154 (2012)
23. Ma, T., Liu, Z., Wen, J., Gao, Y., Ren, X., Chen, H., Jin, C., Ma, X.-L., Xu, N., Cheng, H.-M., Ren, W.: Tailoring the thermal and electrical transport properties of graphene films by grain size engineering. Nat. Commun. **8**, 14486 (2017)
24. Wang, H., Yuan, H., Sae Hong, S., Li, Y., Cui, Y.: Physical and chemical tuning of two-dimensional transition metal dichalcogenides. Chem. Soc. Rev. **44**, 2664–2680 (2015)
25. Ryder, C.R., Wood, J.D., Wells, S.A., Hersam, M.C.: Chemically tailoring semiconducting two-dimensional transition metal dichalcogenides and black phosphorus. ACS Nano **10**, 3900–3917 (2016)
26. Najmaei, S., Zou, X., Er, D., Li, J., Jin, Z., Gao, W., Zhang, Q., Park, S., Ge, L., Lei, S., Kono, J., Shenoy, V.B., Yakobson, B.I., George, A., Ajayan, P.M., Lou, J.: Tailoring the physical properties of molybdenum disulfide monolayers by control of interfacial chemistry. Nano Lett. **14**, 1354–1361 (2014)
27. Tetsuka, H., Nagoya, A., Tamura, S.: Graphene/nitrogen-functionalized graphene quantum dot hybrid broadband photodetectors with a buffer layer of boron nitride nanosheets. Nanoscale **8**, 19677–19683 (2016)
28. Liu, N., Tian, H., Schwartz, G., Tok, J.B.-H., Ren, T.-L., Bao, Z.: Large-area, transparent, and flexible infrared photodetector fabricated using P–N junctions formed by N-doping chemical vapor deposition grown graphene. Nano Lett. **14**, 3702–3708 (2014)
29. Buscema, M., Barkelid, M., Zwiller, V., van der Zant, H.S.J., Steele, G.A., Castellanos-Gomez, A.: Large and tunable photothermoelectric effect in single-layer MoS_2. Nano Lett. **13**, 358–363 (2013)
30. Konstantatos, G., Sargent, E.H.: Nanostructured materials for photon detection. Nat. Nanotech. **5**, 391–400 (2010)
31. Furchi, M.M., Pospischil, A., Libisch, F., Burgdörfer, J., Mueller, T.: Photovoltaic effect in an electrically tunable van der Waals heterojunction. Nano Lett. **14**, 4785–4791 (2014)
32. Vlassiouk, I.V., Stehle, Y., Pudasaini, P.R., Unocic, R.R., Rack, P.D., Baddorf, A.P., Ivanov, I.N., Lavrik, N.V., List, F., Gupta, N., Bets, K.V., Yakobson, B.I., Smirnov, S.N.: Evolutionary selection growth of two-dimensional materials on polycrystalline substrates. Nat. Mater. **17**, 318–322 (2018)
33. Castro Neto, A.H., Guinea, F., Peres, N.M.R., Novoselov, K.S., Geim, A.K.: The electronic properties of graphene. Rev. Mod. Phys. **81**, 109–162 (2009)
34. Zeng, M., Xiao, Y., Liu, J., Yang, K., Fu, L.: Exploring two-dimensional materials toward the next-generation circuits: from monomer design to assembly control. Chem. Rev. **118**, 6236–6296 (2018)
35. Gao, L.: Flexible device applications of 2D semiconductors. Small **13**, 1603994 (2017)
36. Novoselov, K.S., Geim, A.K., Morozov, S.V., Jiang, D., Zhang, Y., Dubonos, S.V., Grigorieva, I.V., Firsov, A.A.: Electric field effect in atomically thin carbon films. Science **306**, 666–669 (2004)
37. Britnell, L., Ribeiro, R.M., Eckmann, A., Jalil, R., Belle, B.D., Mishchenko, A., Kim, Y.-J., Gorbachev, R.V., Georgiou, T., Morozov, S.V., Grigorenko, A.N., Geim, A.K., Casiraghi, C., Neto, A.H.C., Novoselov, K.S.: Strong light-matter interactions in heterostructures of atomically thin films. Science **340**, 1311–1314 (2013)
38. Castellanos-Gomez, A., Roldán, R., Cappelluti, E., Buscema, M., Guinea, F., van der Zant, H.S.J., Steele, G.A.: Local strain engineering in atomically thin MoS_2. Nano Lett. **13**, 5361–5366 (2013)
39. Koppens, F.H.L., Mueller, T., Avouris, P., Ferrari, A.C., Vitiello, M.S., Polini, M.: Photodetectors based on graphene, other two-dimensional materials and hybrid systems Nat. Nanotech. **9**, 80–93 (2014)
40. Late, D.J., Bhat, A., Rout, C.S.: Photo sensor based on 2D materials. In: Fundamentals and Sensing Applications of 2D Materials, pp. 465–479. Elsevier (2019)
41. Farmer, D.B., Golizadeh-Mojarad, R., Perebeinos, V., Lin, Y.-M., Tulevski, G.S., Tsang, J.C., Avouris, P.: Chemical doping and electron−hole conduction asymmetry in graphene devices. Nano Lett. **9** 388–392 (2009)

42. Lemme, M.C., Koppens, F.H.L., Falk, A.L., Rudner, M.S., Park, H., Levitov, L.S., Marcus, C.M.: Gate-activated photoresponse in a graphene p–n junction. Nano Lett. **11**, 4134–4137 (2011)
43. Tsymbal, E.Y., Pettifor, D.G.: Perspectives of giant magnetoresistance. Solid State Phys. **56**, 113–237 (2001)
44. Ponomarenko, V.P., Popov, V.S., Popov, S.V., Chepurnov, E.L.: Photo- and nanoelectronics based on two-dimensional materials. Part I. Two-dimensional materials: properties and synthesis. J. Commun. Technol. Electron. **65**, 1062–1104 (2020)
45. Nair, R.R., Blake, P., Grigorenko, A.N., Novoselov, K.S., Booth, T.J., Stauber, T., Peres, N.M.R., Geim, A.K.: Fine structure constant defines visual transparency of graphene. Science **320**, 1308–1308 (2008)
46. Wang, F., Zhang, Y., Tian, C., Girit, C., Zettl, A., Crommie, M., Shen, Y.R.: Gate-variable optical transitions in graphene. Science **320**, 206–209 (2008)
47. Mueller, T., Xia, F., Freitag, M., Tsang, J., Avouris, P.: Role of contacts in graphene transistors: a scanning photocurrent study. Phys. Rev. B **79**, 245430 (2009)
48. Giovannetti, G., Khomyakov, P.A., Brocks, G., Karpan, V.M., van den Brink, J., Kelly, P.J.: Doping graphene with metal contacts. Phys. Rev. Lett. **101**, 026803 (2008)
49. Xia, F., Mueller, T., Lin, Y., Valdes-Garcia, A., Avouris, P.: Ultrafast graphene photodetector. Nat. Nanotech. **4**, 839–843 (2009)
50. Mueller, T., Xia, F., Avouris, P.: Graphene photodetectors for high-speed optical communications. Nat. Photon **4**, 297–301 (2010)
51. Xu, X., Gabor, N.M., Alden, J.S., van der Zande, A.M., McEuen, P.L.: Photo-thermoelectric effect at a graphene interface junction. Nano Lett. **10**, 562–566 (2010)
52. Popov, V.S., Ponomarenko, V.P., Popov, S.V.: 2D material-based photo- and nanoelectronics. Part III. Photosensors based on graphene, graphene-like, and related 2D nanomaterials. J. Commun. Technol. Electron. **67**, 1152–1174 (2022)
53. Guo, B., Fang, L., Zhang, B., Gong, J.R.: Graphene doping: a review. Insciences J. 80–89 (2011)
54. Gao, Y., Lei, S., Kang, T., Fei, L., Mak, C.-L., Yuan, J., Zhang, M., Li, S., Bao, Q., Zeng, Z., Wang, Z., Gu, H., Zhang, K.: Bias-switchable negative and positive photoconductivity in 2D $FePS_3$ ultraviolet photodetectors. Nanotechnology **29**, 244001 (2018)
55. Bistritzer, R., MacDonald, A.H.: Electronic cooling in graphene. Phys. Rev. Lett. **102**, 206410 (2009)
56. Xia, F., Farmer, D.B., Lin, Y., Avouris, P.: Graphene field-effect transistors with high on/off current ratio and large transport band gap at room temperature. Nano Lett. **10**, 715–718 (2010)
57. Sun, D., Aivazian, G., Jones, A.M., Ross, J.S., Yao, W., Cobden, D., Xu, X.: Ultrafast hot-carrier-dominated photocurrent in graphene. Nat. Nanotech. **7**, 114–118 (2012)
58. Liu, Y., Cheng, R., Liao, L., Zhou, H., Bai, J., Liu, G., Liu, L., Huang, Y., Duan, X.: Plasmon resonance enhanced multicolour photodetection by graphene. Nat. Commun. **2**, 579 (2011)
59. Fang, Z., Liu, Z., Wang, Y., Ajayan, P.M., Nordlander, P., Halas, N.J.: Graphene-antenna sandwich photodetector. Nano Lett. **12**, 3808–3813 (2012)
60. Lee, J.-S., Kovalenko, M.V., Huang, J., Chung, D.S., Talapin, D.V.: Band-like transport, high electron mobility and high photoconductivity in all-inorganic nanocrystal arrays. Nat. Nanotech. **6**, 348–352 (2011)
61. Konstantatos, G., Badioli, M., Gaudreau, L., Osmond, J., Bernechea, M., de Arquer, F.P.G., Gatti, F., Koppens, F.H.L.: Hybrid graphene–quantum dot phototransistors with ultrahigh gain. Nat. Nanotech. **7**, 363–368 (2012)
62. Ni, Z., Ma, L., Du, S., Xu, Y., Yuan, M., Fang, H., Wang, Z., Xu, M., Li, D., Yang, J., Hu, W., Pi, X., Yang, D.: Plasmonic silicon quantum dots enabled high-sensitivity ultrabroadband photodetection of graphene-based hybrid phototransistors. ACS Nano **11**, 9854–9862 (2017)
63. Splendiani, A., Sun, L., Zhang, Y., Li, T., Kim, J., Chim, C.-Y., Galli, G., Wang, F.: Emerging photoluminescence in monolayer MoS_2. Nano Lett. **10**, 1271–1275 (2010)
64. Yin, Z., Li, H., Li, H., Jiang, L., Shi, Y., Sun, Y., Lu, G., Zhang, Q., Chen, X., Zhang, H.: Single-layer MoS_2 phototransistors. ACS Nano **6**, 74–80 (2012)

65. Zhang, W., Huang, J.-K., Chen, C.-H., Chang, Y.-H., Cheng, Y.-J., Li, L.-J.: High-gain phototransistors based on a CVD MoS_2 monolayer. Adv. Mater. **25**, 3456–3461 (2013)
66. Weng, W.Y., Chang, S.J., Hsu, C.L., Hsueh, T.J.: A ZnO-nanowire phototransistor prepared on glass substrates. ACS Appl. Mater. Interfaces **3**, 162–166 (2011)
67. Terrones, H., Corro, E.D., Feng, S., Poumirol, J.M., Rhodes, D., Smirnov, D., Pradhan, N.R., Lin, Z., Nguyen, M.A.T., Elías, A.L., Mallouk, T.E., Balicas, L., Pimenta, M.A., Terrones, M.: New first order Raman-active modes in few layered transition metal dichalcogenides. Sci. Rep. **4**, 4215 (2014)
68. Perea-López, N., Lin, Z., Pradhan, N.R., Iñiguez-Rábago, A., Laura Elías, A., McCreary, A., Lou, J., Ajayan, P.M., Terrones, H., Balicas, L., Terrones, M.: CVD-grown monolayered MoS_2 as an effective photosensor operating at low-voltage. 2D Mater. **1**, 011004 (2014)
69. Ugeda, M.M., Bradley, A.J., Shi, S.-F., da Jornada, F.H., Zhang, Y., Qiu, D.Y., Ruan, W., Mo, S.-K., Hussain, Z., Shen, Z.-X., Wang, F., Louie, S.G., Crommie, M.F.: Giant bandgap renormalization and excitonic effects in a monolayer transition metal dichalcogenide semiconductor. Nat. Mater. **13**, 1091–1095 (2014)
70. Tongay, S., Zhou, J., Ataca, C., Lo, K., Matthews, T.S., Li, J., Grossman, J.C., Wu, J.: Thermally driven crossover from indirect toward direct bandgap in 2D semiconductors: $MoSe_2$ versus MoS_2. Nano Lett. **12**, 5576–5580 (2012)
71. Coehoorn, R., Haas, C., de Groot, R.A.: Electronic structure of $MoSe_2$, MoS_2, and WSe_2. II. The nature of the optical band gaps. Phys. Rev. B **35**, 6203–6206 (1987)
72. Larentis, S., Fallahazad, B., Tutuc, E.: Field-effect transistors and intrinsic mobility in ultra-thin $MoSe_2$ layers. Appl. Phys. Lett. **101**, 223104 (2012)
73. Xia, J., Huang, X., Liu, L.-Z., Wang, M., Wang, L., Huang, B., Zhu, D.-D., Li, J.-J., Gu, C.-Z., Meng, X.-M.: CVD synthesis of large-area, highly crystalline $MoSe_2$ atomic layers on diverse substrates and application to photodetectors. Nanoscale **6**, 8949 (2014)
74. Abderrahmane, A., Ko, P.J., Thu, T.V., Ishizawa, S., Takamura, T., Sandhu, A.: High photosensitivity few-layered $MoSe_2$ back-gated field-effect phototransistors. Nanotechnology **25**, 365202 (2014)
75. Yang, S., Li, Y., Wang, X., Huo, N., Xia, J.-B., Li, S.-S., Li, J.: High performance few-layer GaS photodetector and its unique photo-response in different gas environments. Nanoscale **6**, 2582–2587 (2014)
76. Hu, P., Wen, Z., Wang, L., Tan, P., Xiao, K.: Synthesis of few-layer GaSe nanosheets for high performance photodetectors. ACS Nano **6**, 5988–5994 (2012)
77. Choi, W., Cho, M.Y., Konar, A., Lee, J.H., Cha, G.-B., Hong, S.C., Kim, S., Kim, J., Jena, D., Joo, J., Kim, S.: High-detectivity multilayer MoS_2 phototransistors with spectral response from ultraviolet to infrared. Adv. Mater. **24**, 5832–5836 (2012)
78. Tamalampudi, S.R., Lu, Y.-Y., U. R K, Sankar R, Liao C-D, B. K M, Cheng C-H, Chou F C and Chen Y-T 2014 High Performance and Bendable Few-Layered InSe Photodetectors with Broad Spectral Response *Nano Lett.* **14** 2800–2806
79. Carozo, V., Wang, Y., Fujisawa, K., Carvalho, B.R., McCreary, A., Feng, S., Lin, Z., Zhou, C., Perea-López, N., Elías, A.L., Kabius, B., Crespi, V.H., Terrones, M.: Optical identification of sulfur vacancies: bound excitons at the edges of monolayer tungsten disulfide. Sci. Adv. **3**, e1602813 (2017)
80. Ataca, C., Şahin, H., Ciraci, S.: Stable, single-layer MX_2 transition-metal oxides and dichalcogenides in a honeycomb-like structure. J. Phys. Chem. C **116**, 8983–8999 (2012)
81. Georgiou, T., Jalil, R., Belle, B.D., Britnell, L., Gorbachev, R.V., Morozov, S.V., Kim, Y.-J., Gholinia, A., Haigh, S.J., Makarovsky, O., Eaves, L., Ponomarenko, L.A., Geim, A.K., Novoselov, K.S., Mishchenko, A.: Vertical field-effect transistor based on graphene–WS_2 heterostructures for flexible and transparent electronics. Nat. Nanotech. **8**, 100–103 (2013)
82. Long, H., Shi, Y., Wen, Q., Tsang, Y.H.: Ultrafast laser pulse (115 fs) generation by using direct bandgap ultrasmall 2D GaTe quantum dots. J. Mater. Chem. C **7**, 5937–5944 (2019)
83. Zhou, Y., Tan, H., Sheng, Y., Fan, Y., Xu, W., Warner, J.H.: Utilizing interlayer excitons in bilayer WS_2 for increased photovoltaic response in ultrathin graphene vertical cross-bar photodetecting tunneling transistors. ACS Nano **12**, 4669–4677 (2018)

84. Kumar, A., Ahluwalia, P.K.: Electronic structure of transition metal dichalcogenides monolayers ^{1}H-MX_2 (M = Mo, W; X = S, Se, Te) from ab-initio theory: new direct band gap semiconductors. Eur. Phys. J. B **85**, 186 (2012)
85. Shahar, C., Zbaida, D., Rapoport, L., Cohen, H., Bendikov, T., Tannous, J., Dassenoy, F., Tenne, R.: Surface functionalization of WS_2 fullerene-like nanoparticles. Langmuir **26**, 4409–4414 (2010)
86. Xu, Z.-Q., Zhang, Y., Wang, Z., Shen, Y., Huang, W., Xia, X., Yu, W., Xue, Y., Sun, L., Zheng, C., Lu, Y., Liao, L., Bao, Q.: Atomically thin lateral p–n junction photodetector with large effective detection area. 2D Mater. **3**, 041001 (2016)
87. Bernardi, M., Palummo, M., Grossman, J.C.: Extraordinary sunlight absorption and one nanometer thick photovoltaics using two-dimensional monolayer materials. Nano Lett. **13**, 3664–3670 (2013)
88. Lv, Q., Yan, F., Wei, X., Wang, K.: High-performance, self-driven photodetector based on graphene sandwiched GaSe/WS_2 heterojunction. Adv. Opt. Mater. **6**, 1700490 (2018)
89. Shelke, N.T., Karche, B.R.: Hydrothermal synthesis of WS_2/RGO sheet and their application in UV photodetector. J. Alloy. Compd. **653**, 298–303 (2015)
90. Perea-López, N., Elías, A.L., Berkdemir, A., Castro-Beltran, A., Gutiérrez, H.R., Feng, S., Lv, R., Hayashi, T., López-Urías, F., Ghosh, S., Muchharla, B., Talapatra, S., Terrones, H., Terrones, M.: Photosensor device based on few-layered WS_2 films. Adv. Funct. Mater. **23**, 5511–5517 (2013)
91. Ballif, C., Regula, M., Schmid, P.E., Remkar, M., Sanjin, R.: Preparation and characterization of highly oriented, photoconducting thin films
92. Zeng, H., Liu, G.-B., Dai, J., Yan, Y., Zhu, B., He, R., Xie, L., Xu, S., Chen, X., Yao, W., Cui, X.: Optical signature of symmetry variations and spin-valley coupling in atomically thin tungsten dichalcogenides. Sci. Rep. **3**, 1608 (2013)
93. Zeng, L., Tao, L., Tang, C., Zhou, B., Long, H., Chai, Y., Lau, S.P., Tsang, Y.H.: High-responsivity UV-Vis photodetector based on transferable WS_2 film deposited by magnetron sputtering. Sci. Rep. **6**, 20343 (2016)
94. Fan, Y., Hao, G., Luo, S., Qi, X., Li, H., Ren, L., Zhong, J.: Synthesis, characterization and electrostatic properties of WS_2 nanostructures AIP Adv. **4**, 057105 (2014)
95. Chen, W.-H., Chen, Q.-W., Chen, Q., Cui, C., Duan, S., Kang, Y., Liu, Y., Liu, Y., Muhammad, W., Shao, S., Tang, C., Wang, J., Wang, L., Xiong, M.-H., Yin, L., Zhang, K., Zhang, Z., Zhen, X., Feng, J., Gao, C., Gu, Z., He, C., Ji, J., Jiang, X., Liu, W., Liu, Z., Peng, H., Shen, Y., Shi, L., Sun, X., Wang, H., Wang, J., Xiao, H., Xu, F.-J., Zhong, Z., Zhang, X.-Z., Chen, X.: Biomedical polymers: synthesis, properties, and applications. Sci. China Chem. **65**, 1010–1075 (2022)
96. Yao, J.D., Zheng, Z.Q., Shao, J.M., Yang, G.W.: Stable, highly-responsive and broadband photodetection based on large-area multilayered WS_2 films grown by pulsed-laser deposition. Nanoscale **7**, 14974–14981 (2015)
97. Ahmad, H., Thandavan, T.M.K.: Ultraviolet photoconduction in tungsten disulphide based Schottky heterostructure photodetector. Opt. Mater. **92**, 255–261 (2019)
98. Xia, F., Wang, H., Jia, Y.: Rediscovering black phosphorus as an anisotropic layered material for optoelectronics and electronics. Nat. Commun. **5**, 4458 (2014)
99. Liu, H., Neal, A.T., Zhu, Z., Luo, Z., Xu, X., Tománek, D., Ye, P.D.: Phosphorene: an unexplored 2D semiconductor with a high hole mobility. ACS Nano **8**, 4033–4041 (2014)
100. Li, L., Yu, Y., Ye, G.J., Ge, Q., Ou, X., Wu, H., Feng, D., Chen, X.H., Zhang, Y.: Black phosphorus field-effect transistors. Nat. Nanotech. **9**, 372–377 (2014)
101. Das, S., Zhang, W., Demarteau, M., Hoffmann, A., Dubey, M., Roelofs, A.: Tunable transport gap in phosphorene. Nano Lett. **14**, 5733–5739 (2014)
102. Qiao, X.-F., Wu, J.-B., Zhou, L., Qiao, J., Shi, W., Chen, T., Zhang, X., Zhang, J., Ji, W., Tan, P.-H.: Polytypism and unexpected strong interlayer coupling in two-dimensional layered ReS_2. Nanoscale **8**, 8324–8332 (2016)
103. Wang, H., Wang, X., Xia, F., Wang, L., Jiang, H., Xia, Q., Chin, M.L., Dubey, M., Han, S.: Black phosphorus radio-frequency transistors. Nano Lett. **14**, 6424–6429 (2014)

104. Qiao, J., Kong, X., Hu, Z.-X., Yang, F., Ji, W.: High-mobility transport anisotropy and linear dichroism in few-layer black phosphorus. Nat. Commun. **5**, 4475 (2014)
105. Guillaumée, M., Dunbar, L.A., Santschi, C., Grenet, E., Eckert, R., Martin, O.J.F., Stanley, R.P.: Polarization sensitive silicon photodiodes using nanostructured metallic grids. Appl. Phys. Lett. **94**, 193503 (2009)
106. Yuan, H., Liu, X., Afshinmanesh, F., Li, W., Xu, G., Sun, J., Lian, B., Curto, A.G., Ye, G., Hikita, Y., Shen, Z., Zhang, S.-C., Chen, X., Brongersma, M., Hwang, H.Y., Cui, Y.: Polarization-sensitive broadband photodetector using a black phosphorus vertical p–n junction. Nat. Nanotech **10**, 707–713 (2015)
107. Mutz, N., Park, S., Schultz, T., Sadofev, S., Dalgleish, S., Reissig, L., Koch, N., List-Kratochvil, E.J.W., Blumstengel, S.: Excited-state charge transfer enabling MoS_2/phthalocyanine photodetectors with extended spectral sensitivity. J. Phys. Chem. C **124**, 2837–2843 (2020)
108. Li, J., Nie, C., Sun, F., Tang, L., Zhang, Z., Zhang, J., Zhao, Y., Shen, J., Feng, S., Shi, H., Wei, X.: Enhancement of the photoresponse of monolayer MoS_2 photodetectors induced by a nanoparticle grating. ACS Appl. Mater. Interfaces **12**, 8429–8436 (2020)
109. Jang, H.Y., Nam, J.H., Yoon, J., Kim, Y., Park, W., Cho, B.: One-step H_2S reactive sputtering for 2D MoS_2/Si heterojunction photodetector. Nanotechnology **31**, 225205 (2020)

2D Materials' Sensing Mechanisms

Yadagiri L N K Mallela

Abstract The amazing class of two-dimensional (2D) materials has evolved, offering new possibilities in a variety of domains, including sensing applications. In order to take advantage of their properties and advance sensing technologies, it is essential to comprehend the principles of sensing mechanisms designed particularly for 2D materials. The possibilities and difficulties of 2D materials for sensing applications are discussed with fundamental concepts and mechanisms that underlie such sensing. It underlines how these materials are great candidates for sensing technologies due to the high surface-to-volume ratio, adjustable electrical characteristics with superior mechanical flexibility. Their usage in biosensing for medical applications, chemical sensing for industrial processes, strain sensing for structural health monitoring, and gas sensing for environmental monitoring is well known in scientific communities.

1 Introduction

Two-dimensional (2D) materials have become crucial for many applications, particularly for the real-time detection of dangers including physical, chemical, and biological exposures [1, 2]. The demand for simple, quick, cost-effective sensors is rising. It has become difficult to monitor and sense dangerous microorganisms, radiation, chemicals, biomarkers, and temperatures on a global scale. As a result, 2D materials with distinctive features that show the sensing capabilities in fabricated devices ranging from environmental monitoring to health care are receiving more attention [3, 4]. Most of the materials in focus for the 2D sensors are inspired by the initial research done on graphene [5]. From then onwards, research had extended to other chemically different 2D layered materials that include transition metal dichalcogenides [6] (TMDC), Mxenes [7] (transition metal carbon nitrides, nitrides, carbides), boron nitrides [8], and phosphorenes [9, 10]. These 2D layered materials are widely adopted

Yadagiri L N K Mallela (✉)
The University of Texas at Austin, Austin, TX 78712, USA
e-mail: mylnkishore@gmail.com

A. Katoch et al. (eds.), *2D Materials*, Engineering Materials,
https://doi.org/10.1007/978-981-97-6258-3_10

for various applications, especially for the sensing applications [11]. Examples in this chapter give the idea about only the 2D material-based sensors.

The main reason to use 2D materials for sensing is that these materials:

- Yield a long operating life with consistent conductivity;
- Manifest a high volume-to-surface ratio;
- Have transparent optics;
- Show stable mechanical performance;
- Enable easy processing;.
- Work with current technologies like ultra-thin silicon channel technology [12–15], various printing methods [16], wearable flexible electronics [17], metal electrodes [18].

The 2D materials have some drawbacks that include:

- Lack of large-area, high-quality, homogeneous mass production of material;
- Limited stability of dome forms under ambient circumstances due to the lack of simple, dependable, and effective solutions for device integration [14, 15].

2 Sensing Mechanisms

Understanding the fundamental characteristics of these materials and their sensing methods is crucial given the development of new types of 2D layered materials beyond graphene. Interestingly, 2D materials themselves lack the inherent ability to interact with target molecules in a selective manner. As a result, by altering their fundamental characteristics, 2D nanosheets and molecules/ions can interact in ways that can be controlled. The initial stage in sensing is detection, and the change that the 2D materials undergo yields a sensing response. The detection mechanism of the 2D materials can be broadly related to two types of interactions, namely, chemisorption and physisorption [5]. To manage sensing, chemisorption primarily entails the creation of covalent bonds between 2D materials and analytes. As opposed to chemisorption-based sensors, physisorption—which takes place through non-covalent interactions—offers several advantages with regard to the reusability of the sensors and is preferred over chemisorption. Noncovalent interactions have excellent recovery rates and hold promise for on-site analysis/real-time monitoring. π-π interactions, Van der Waals, electrostatic, and are additional non-covalent physical interactions for 2D materials that can be utilized to controllably adjust doping levels. Due to its reliance on the properties of the molecules and the interaction surface, physisorption is subject to various limitations. Ineffective and unstable molecular immobilization is involved in this, which may result in false-positive and false-negative findings. In addition to these ways of detection, the electronic measurements of a material's properties and the desired response determine a material's capacity for sensing. The general design of the sensors based on various 2D materials and their sensing mechanisms will be covered in this chapter.

2D materials can also be used as sensitive elements for detecting mechanical stimuli, such as pressure, strain, vibration, sound, and force, as well as chemical/biological analytes, such as gases, molecules/ions, biomarkers. In this chapter, we will focus on the sensing mechanisms of 2D materials for sensors that give physically measurable or observable responses and can be categorized mostly into four categories: electrochemical, electrical, piezoelectric, and optical. Most of the sensing platforms fall into one of these four categories. It's important to note that these sensing mechanisms often overlap, and many 2D materials can exhibit multiple sensing modalities simultaneously. Additionally, the specific sensor's mechanism depends on the properties of 2D material and the design of the sensing device or platform. The following section will discuss the sensing of mechanisms in which transducers leverage one or combinations of the four mechanisms.

1. Electrochemical

   When this method involves scanning, electrodes are used as transducer elements. Electrochemical sensors are a popular choice in analysis due to their precision, quick outputs, responsiveness, and ease of installation [19]. Electrochemical sensors monitor reactions that occur between analytes and the electrode surfaces. They often transform the chemicals into valuable electrical signals. Amperometric, potentiometric, and conductometric methods form the basis of electrochemical sensors [20]. Thus, quantitative investigation of electrochemical signals (e.g., electron transfer, oxidation, or reduction-specific measurements) can be enhanced using the custom design of 2D objects—to replace working electrodes. When the analyzer does not respond, an external device as an intermediary must provide an electrochemical signal equivalent to the measuring instrument [21].

   Most of the electrochemical sensors are based on voltammetric and amperometric measurements [22, 23]. The voltammetric sensor in voltammetry finds a current proportional to the change in potential during the redox reaction of an electroactive analyte or medium [24]. Main applications are based on voltammetry, including human health monitoring via small molecule detection. Biomolecules important for maintaining health include ascorbic acid, dopamine, and uric acid [25, 26]. Apart from biomolecules, currently, sensors are used to monitor food, environment, health, and agriculture [26]. In amperometry, measurements are made at a constant potential. Through the redox process, electroactive molecules produce faradaic currents that are continuously recorded and used to calculate chemical concentrations [27].

2. Electrical

   Electrical sensors are designed using measurements of conducting and semiconducting materials. When it comes to semiconducting devices, the behavior of electrons and holes is employed to produce the appropriate response for detecting analytes via changes in the target systems' resistance. In the case of conductors, when analytes interact with electrodes, changes in the electric current can be used for designing the sensors. For electrical sensing, 2D materials can be utilized in conjunction with other components like field effect transistors (FETs). The transport channels that enable the electron-hole mobility in FETs can be made of 2D materials. In this view, FETs employing 2D materials are the most promising electrical sensor materials because they can offer enhanced sensitivity, robustness, quick detection, and simple operability [19, 20]. 2D materials-based FETs have the same behavior as semiconductors in terms of electron-hole mobility. The performance of the FETs also depends on both the positive and negative species that migrate between electrodes. In this case, the charge transfer among the adsorbed molecules in 2D materials and semiconductor materials causes fluctuation in the resistance [21, 22]. Applications in gas, ion, protein, optical, mechanical, and bioelectric sensing have been observed for FETs and electrical sensors made of 2D materials [23]. In the case of conductometric sensors, 2D materials that are mostly used include materials like graphene, TMDs, and black phosphorus (BP) and they play a significant role in room temperature gas sensors [24, 25]. Multiple workspaces, large areas, hot and cold areas, and adjustable optics make them more useful than current sensors, and the results are good.

3. Piezoelectric

   More importantly, there are many piezoelectric sensors incorporated into 3D scaffolds, polymer microstructures, or (micro) fibers [35]. The examples in this section only give an idea of 2D objects as sensors. However, vertical and in-plane piezoelectricity of TMDs was proposed and reported for graphene (d33). Difficulties of controlled syntheses with centimeter-scale sized, good crystalline, and high reproducibility limits TMDs' use in widely amalgamated, flexible sensor arrays [36]. The best example for piezoelectric 2D materials mainly includes graphene due to its linear change in resistance with strain. Three specific mechanisms, including structural deformation, excessive graphene sheet connection, and the tunneling effect between nearby sheets, are thought to be the causes of graphene piezo-resistance.

   Piezo electric mechanism is mainly applied for various strain-related physical sensor applications. High-sensitivity physical sensors with a variety of uses include piezoelectric sensors. The resistance changes in piezoelectric sensors under pressure depend on geometric elements and the effects of resistivity. The semiconductor resistance under stress can be determined using total resistance, length, and average area of cross-section area. The Gauge Factor (GF) determines the sensitivity of strain sensors that transform external inputs (changes in physical parameters such as force and twists in materials) to electrical pulses. Here, sensitivity is defined as the ratio of the normalized resistance change compared to the reference mode.

4. Optical

   In this section, we go through the three primary optical processes that optical sensors use to convey light signals: fluorescence resonance energy transfer (FRET), surface plasmon resonance (SPR), and surface-enhanced Raman spectroscopy (SERS) [26, 27]. Optical sensors are frequently straightforward and reasonably priced tools that can measure several aspects of light, including wavelength, frequency, intensity, scattering, and polarization. Due to the chemical inertness, immunity to electromagnetic interferences, and the ability for remote sensing, optical approaches are advantageous over electrical sensors. Additionally, optical sensors may be extremely sensitive, perhaps being able to find single molecules. The described 2D materials shone as the optimal platform for optical sensors with unique optical characteristics, desired conductivity, excellent electron mobility, and high surface-to-volume ratio [26]. Additionally, dopants or metallic nanostructures may help to improve the target signal [28, 29]. In general, the optical sensors built with 2D materials are expanding quickly to a number of critical industries, including biomedicine, environmental monitoring, and food analysis [27].

4.1. Surface Plasmon Resonance

In reality, the direct use of 2D nanomaterials in plasmonic sensors has been constrained by their low absorption wavelengths that fall in the visible to NIR range of the spectrum. Therefore, to improve the performance of plasmonic sensors, Au or Ag is usually utilized in place of 2D nanomaterials. Despite the drawbacks, a small number of researchers have used 2D materials' plasmonic properties directly in the creation of SPR sensors [30, 31]. As a result, the use of plasmonic sensors for biochemical and environmental monitoring has been increasing significantly. The SPR method depends on the conduction electrons' velocity on material surfaces and the oscillation of electromagnetic surface waves at certain inputs such as photons, phonons, and electrons, which is the theoretical basis of plasmonic sensors [32]. Some 2D materials, such as TMD and graphene, are often used in SPR sensors because of their attractive physicochemical properties [30].

4.2. Surface-Enhanced Raman spectroscopy

Due to inexpensive, simple synthesis, excellent optical characteristics, and strong biocompatibility, 2D materials provide effective surface-enhanced Raman spectroscopy (SERS) substrates [33]. These factors make 2D materials an intriguing substitute for noble metals, the cutting-edge components of SERS. Noble metal really has a number of disadvantages while being quite effective, including expensive cost, strong metal-adsorbate interaction, instability with age, photobleaching, and catalytic effects [34]. Combining noble metals and 2D materials to create nanocomposites that can overcome the shortcomings of both is a common tactic.

A common analytical technique called Raman spectroscopy enables non-destructive structural fingerprinting of molecules. Raman scattering's ability to be used in sensing is frequently constrained by the weak signals obtained with low analyte concentrations. SERS, which was developed by Fleischmann et al., has generated a great deal of attention since it can detect analyte traces that are multiplied by a factor of up to 10^{10} with extreme sensitivity [35, 36]. The electromagnetic/chemical (EM/CM) enhancement phenomena serve as a foundation for the mechanisms underlying SERS.

In more detail, charge transfer resulting from analyte to substrate dipole-dipole interactions gives EM, which is derived from the localized surface plasmon resonance (LSPR), and thus causes CM.

4.3. Fluorescence Resonance Energy Transfer (FRET)

The FRET technique involves a fluorescent donor with a light-absorbing acceptor. When the donor's emission spectrum overlaps about 10 nm of the acceptor's absorption spectrum, this event occurs [37]. 2D materials typically exhibit good quenching properties [38]. TMDs have also been used as acceptors in FRET sensors in a variety of applications, such as single molecule spectroscopy, cellular imaging, environmental monitoring, and biochemical analysis [39–41]. The 2D materials can be employed as a quencher or fluorophore in fluorescence-based chemical sensors because of their photoluminescent capabilities [42]. As a result, fluorescence sensors offer increased sensitivity, simple, affordable designs, and portability [43].

Figure 1 shows the various mechanisms with which the 2D materials work for sensing applications.

3 Applications

Most of the sensing devices are made based on the above-discussed mechanisms or a combination of these often involved to improve the precision of the devices. The sensing devices can be broadly classified into physical and chemical sensors which are used to detect particular physical changes (light, fluorescence, strain, color, vibrations) and analytes (biomolecules, pollutants, toxins, etc.) respectively. The following section will discuss the various mechanisms used in the particular devices aimed at detecting particular responses and analytes. Accordingly, the sensing applications are divided into the following sections.

1. Gas Sensors

1.1. Conductometric: Due to their flexible manufacturing methods and usefulness, conductometry-based gas sensors are frequently applied for specific gas detections, especially under normal atmospheric circumstances [44–46]. The interaction rate can be determined using modifications in the optical characteristics, capacitance, conductivity, mass, work function, and/or reaction kinetics [45]. The concept of reverse interaction between the surface material and the gas molecules is essential to the operation of the fuel cell [47]. The physical makeup of materials, such as their surface area, additives, as well as some external characteristics like temperature and humidity, all have an impact on this interaction. These surface interactions can change and enhance the conductometric semiconducting gas sensors' key sensitivity. The MoS_2 monolayer' interaction with physisorbed gases acting with electron-donating or accepting roles resulted in the proposed device demonstrating great selectivity towards diverse gases/vapors [48].

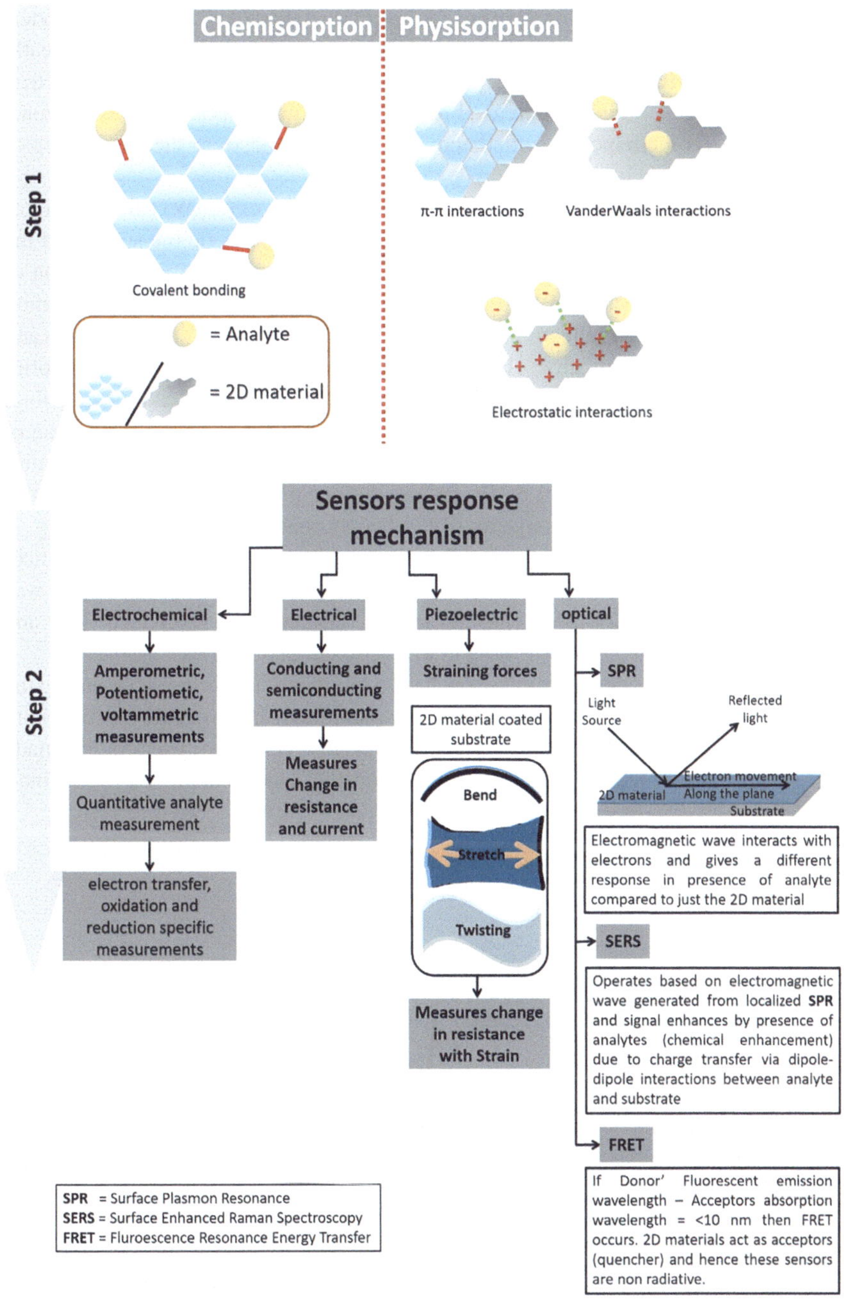

Fig. 1 Schematic sketch showing the different sensing mechanisms. Most of the sensors developed so far either work on one single mechanism or on the combination of the mechanisms for effective sensing application of 2D materials

By examining the fluctuations in the conductance for a variety of analytes, it is possible to assess the sensing kinetics. The conductance varies greatly for different kinds of gases; for instance, triethylamine was an electron donor gas where it increased quickly, but water vapor, dichlorobenzene, dichloromethane, nitromethane, and nitrotoluene were electron acceptor gases where it climbed slowly.

1.2. Surface Acoustic Sensors: Surface Acoustic Waves (SAW) can be used for sensing applications. Physical and chemical changes on the surface of the SAW detection platform have an effect on the propagation/attenuation of the acoustic waves [49, 50]. SAW sensors' response/selectivity modulation is controlled by adjusting the sensing materials' physical characteristics and by adding chemically selective compounds to them [51, 52]. The majority of SAW gas sensors identify gas molecules by mass loading and fluctuating conductivity characteristics of sensing materials [53]. Because of their compact size, outstanding sensitivity, resilience, and ability to identify a wide range of gases, SAW devices are one of the cutting-edge technologies of sensors that are primarily utilized as gas sensors [51, 54]. In Fig. 2, various designs of thin film piezoelectric ZnO SAW sensors are depicted. By coating a glass substrate with a sensing layer, higher performance was attained via GO [55]. The GO layers' surface area, resonance (R_f), and sensitivity of SAW sensors all have inverse relationships.

1.3. Chemiresistors: Chemiresistors have a number of advantages, including the capacity for further downsizing, easy manufacturing procedures, simplicity of use, low cost, and eco-friendliness [56, 61]. Chemiresistors are made of binary metallic transducers (or modified electronically) with an inert

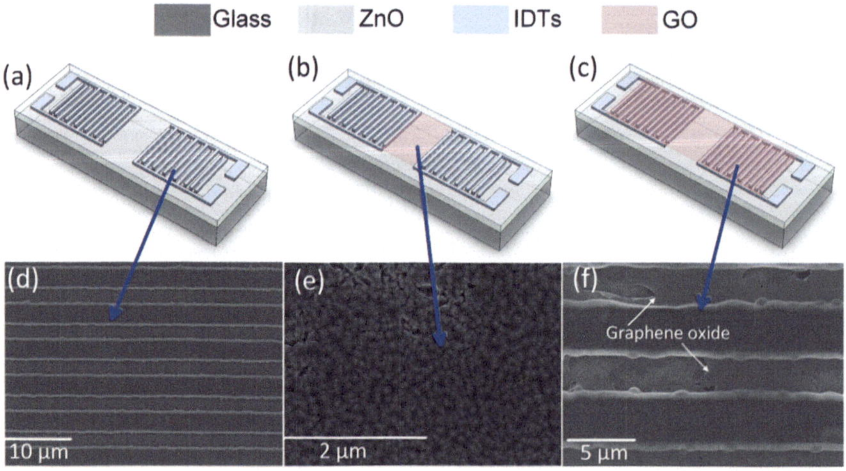

Fig. 2 Image showing SAW devices **a** GO layer coated on a clean surface **b** middle and **c** entire area. SEM images of the IDTs **d** GO layer coated on a clean surface **e** middle and **f** entire area. Taken from Ref. 67. Copyright 2014 The Author, Springer Nature

substrate forming the sensing materials. Chemiresistors have been the most thoroughly explored among the gas-sensing devices. Surface adsorbed gas molecules on the detecting substrate in this kind of sensor alter the electrical resistance, enabling a sensitive detection [57]. Along with other organic gases, they can detect CO, O_2, Cl_2, H_2, SO_x, and NO_x [58, 59]. Chemical resistors are made using a layer of chemically formed graphene sheet (with spin coating) over dual sensor substrates made of various materials, including four-point electrodes with small hot plates. Design of micro-hot plates of Si_3N_4 using medium pressure with simultaneous four-point electrode integration of thermally generated SiO_2. It was possible to identify NH_3, NO_2, 2,4-dinitrotoluene using the suggested sensors. The temperature dependency of the NO_2 reaction could also be studied using the micro-hot plate sensors concurrently [60]. Cho et al. used molybdenum trioxide (MoO_3) and sulfur powder as precursors in chemical vapor deposition (CVD) to create multilayer MoS_2 [61]. They created interdigitated electrodes with a 100 mm interval by depositing the material on a sapphire substrate (see Fig. 3a). NH_3 and NO_2 are used to control the change in temperature of the sensor. NO_2 increases resistance at ambient temperature and 100 °C (very sensitive). Here, the electron acceptor is NO_2, which lowers the Fermi level and brings it closer to the valence band, causing p-doping. MoS_2, which drives the Fermi level to the conduction band edge, exhibits negative characteristics at both temperatures because NH_3 operates as an electron donor (e.g., n-doped) (see Fig. 3b). Regardless of operating temperatures, the product surface is saturated at about 20 ppm as shown in Fig. 3c. The authors also confirmed the charge transitions involving gas molecules and MoS_2 and their best positions in the prepared process using DFT (see Fig. 3d).

1.4. Schottky Diodes: By applying the same techniques used to create electronic devices, Schottky devices are created with stacked semiconductor heterojunctions. For example, three-terminal devices with good communication between atomically thin graphene and substrates such as hydrogenated silicon have been built using layer-by-layer techniques. This device with this variable barrier made of graphene was called "barristor" [62]. Gas molecules adsorption on the sensor surface can vary the Schottky barrier height (SBH) of the heterojunction as well as the Fermi energy. The main factor determining the SBH, a variation that results in reverse currents, is the adsorbed gas molecule density. These characteristics enable quick and simple manufacturing of high selectivity Schottky diode gas sensors [63, 64]. It is simple to comprehend, and modify the sensitivity of devices because reverse bias influences the actual work function of SBHs and graphene. Schottky diodes at indistinguishable bias voltage/environmental conditions exhibit lower power with high sensitivity rates to NH_3 and NO_2 identification compared to graphene chemiresistor sensors.

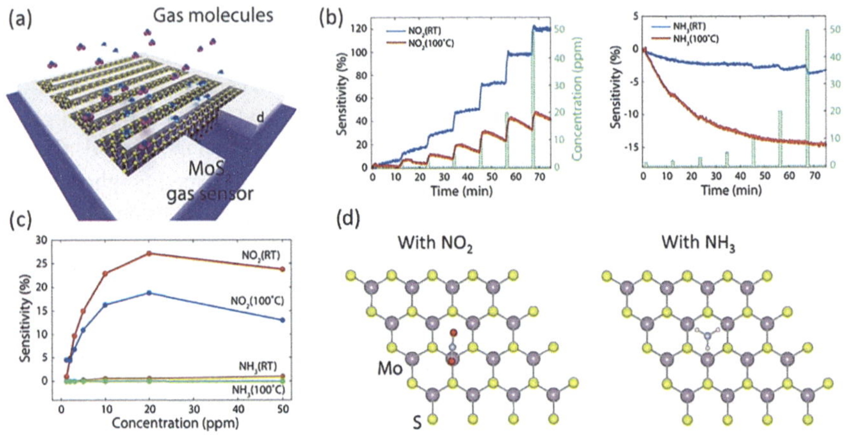

Fig. 3 **a** 3-D schematic of MoS2 nested electrodes on sapphire. **b** Temporary curing of NH_3 (right) and NO_2 (left) gases from running at 100 °C and room temperature, 1.5 to 50 ppm. Resistance increases (very sensitive) in NO_2 gas mode. **c** Comparison of the sensitivity of NO_2 and NH_3 gases at various concentrations and operation temperatures. **d** MoS2's best NO_2 (left) and NH_3 (right) configuration top view. With permission, modified from Ref. 5 Copyright 2023, RSC Publishers Creative Commons License

2. Electrochemical Devices: The development of diverse sensors with various sensing processes has resulted from the rapid advancement of sensor devices. Electrochemical sensors (there are electrodes that act as sensors in the analyte presence) are their first choice due to well-documented accuracy. These sensor devices are a desirable option for analytic platforms because of sensitivity, and fast response using a simple setup [65].

   These sensors depend on the detection of reactions occurring electrochemically between the analytes and electrode surface in order to function. They are made to transform chemical concentrations into electric signals, giving the levels of concentration a numerical measurement. These sensors frequently use amperometry, potentiometry, and conductometry as measurement methods. In some cases, the use of external mediators is required to produce signals electrochemically proportionate to the analyte quantity as some analytes do not readily elicit redox reactions at the electrode surface [66]. A particular working electrode design can be used with 2D materials, to enhance the precision of monitoring electrochemical signals of analyte-antigen interactions. The surface area is increased, the electron transport characteristics are improved, and the sensitivity and selectivity toward the target analyte (antigen-antibody interaction) are all improved [67, 68].

   The basic research method of communication in electronic devices is the transfer of electrons between the functional electrodes immersed in the electrolyte and active material placed/coated on the surface of the working electrode. In the case of gas sensors, gases diffuse via a membrane, which oxidizes them or reduces them at the electrode surface. Electrochemically designed sensors can use cyclic voltammetry (CV), differential pulse voltammetry, amperometry, three unique techniques, to produce various kinds of signals. Large surface area, intrinsic redox potential, and remarkable electrical conductivity are only a few of the crucial factors that determine how effective electrode materials are in electrochemical sensors. Extensive research has concentrated on using nanoparticles as active sensor materials to improve the performance of electrode materials. Nanomaterials are suited for effective mass transfer as a result of their small size providing vast surface area relative to their bulk counterparts. Furthermore, nanoparticles have improved signal-to-noise ratios, an important property used in electrochemical sensors. Because they have the potential to enhance sensor performance and sensitivity in comparison to traditional bulk materials, nanoparticles have been intensively studied and acknowledged for their use in electrochemical sensors [69, 70]. Due to their excellent electrical conductivity, excellent electrochemical properties, and excellent surface-to-volume aspect ratio, 2D materials are applied as electronic materials for electrochemical sensors [71, 72]. In the realm of electrochemical sensors, 2D graphene and its derivatives [73], including reduced graphene oxide (rGO), have gained significant attention. However, it is important to explore other 2D materials for their potential in electrochemical sensing. Several diverse 2D materials, apart from graphene derivatives, have been investigated for this purpose. Among them, materials such as MoS_2, WS_2, SnS_2, FeS_2, and Ni_3S_2 are potential candidates for sensing electrodes in detecting various analytes [74]. For instance, Zhang et al. conducted experiments for qualitative/quantitative detection of glucose utilizing a monolayer MoS_2 as an electrode. The usage of MoS_2 in electrochemical applications, such as sensors, demonstrates the expanding landscape of 2D materials beyond graphene derivatives in this field. MoS_2 has displayed notable redox activity, and the electrochemical reduction of MoS_2 (r-MoS_2) has proven effective in facilitating electron transport and improving conductivity across various electrolytes, including the $Fe[(CN)_6]^{3-/4-}$ and $[Ru(NH_3)_6]^{2+/3+}$ redox systems. Utilizing these properties, a sensor incorporating MoS_2 was developed for dopamine (DA) identification and quantitation. The sensing device demonstrated excellent selectivity for DA, even with interfering agents like ascorbic acid, uric acid [75]. This highlights the capability of MoS_2-based sensors to effectively detect and discriminate DA from complex sample matrices, making them highly promising for reliable electrochemical detection applications. It is worth noting that pristine transition metal dichalcogenide (TMDC) nanosheets typically do not demonstrate exceptional sensor performance because of limited electrical conductivity. Consequently, an approach to enhance the sensitivity involves the synthesis of hybrid nanosheets by incorporating rare earth metal nanoparticles (NPs), for example, Pd, Au, Ag, and Pt. The combination of TMDCs and NPs exhibits synergistic effects [74, 76], leading to improved electrocatalytic performance and enhanced sensing capabilities with respect to pure TMDC nanosheets. This strategy allows for the optimization of sensing performance by harnessing the advantageous characteristics of both TMDCs and metal NPs.

A similar strategy is applied to various other 2D materials with inferior electrical conductivity for electrochemical sensing. In pursuit of developing cutting-edge electrochemical sensing materials, researchers have explored the utilization of MXenes for electrode materials. MXenes offer electrical conductivity, as metals provide, redox-active sites, intrinsic surface terminations, high mass loading, and a vast surface area, enabling quick chemical signal transduction. For example, Zhu et al. made a mediator-free Ti_3C_2-MXene [77] modified with immobilized hemoglobin (Hb). This sensor exhibited exceptional performance in nitrite detection, achieving a low limit of detection (LOD) (0.12 mM), and wide linear range detection, i.e., 0.5–1800 mM. In addition, the application of 2D metal-organic framework (MOF) nanosheets for electrodes to detect hydrogen peroxide (H_2O_2) was described by Zhang et al. These MOF nanosheet-based electrodes demonstrated their potential for sensitive and selective H_2O_2 detection [78]. Thus, the integration of 2D materials, such as MXenes and MOF nanosheets, as electrode materials in electrochemical sensors expands the range of possibilities for developing advanced sensing platforms with enhanced performance and capabilities.

Despite the significant achievements of 2D materials with respect to electrochemical sensors, it is important to acknowledge their limitations. One major challenge is the insufficient electrical conductivity observed in several 2D materials, resulting in a low detection limit. To overcome this limitation, the researchers investigated that the incorporation of rare earth metal nanoparticles (NPs) into 2D materials improves redox activity and electrical conductivity, thus improving performance. However, the use of noble metals in NPs sizes, can be costly and increase fabrication expenses. Alternatively, modifying the crystalline phases of 2D transition metal dichalcogenide (TMDC) materials presents an alternative to mitigate the issue of below-par electrical conductivity. For instance, transforming TMDC metal nanosheets from the 2H phase to the 1T phase can significantly enhance their electrical conductivity. This phase transformation offers a comprehensive solution to overcome the conductivity limitation associated with 2D TMDC materials. Another drawback of 2D materials in electrochemical sensors is their finite stability. It is crucial to address this issue to ensure sustained and reliable sensor performance over extended periods. In summary, while 2D materials' electrochemical sensors have demonstrated remarkable potential, challenges such as poor electrical conductivity, high fabrication costs, and limited long-term stability need to be addressed to fully harness their capabilities for practical applications.

The incorporation of nanoparticles (NPs) into the surface of 2D materials has been shown to improve performance, but the use of noble metallic NPs is expensive and increases manufacturing costs. Alternatively, modifying the crystalline phases of 2D transition metal dichalcogenide (TMDC) materials presents an approach to address their poor electrical properties. For instance, transforming TMDC metal nanosheets from the 2H phase to the 1 T phase can comprehensively overcome this limitation, as 1 T phase nanosheets exhibit higher electrical conductivity.

3. Wearables: Moving on to wearable sensors, they have gained significant interest in recent years for real-time monitoring of human health. These sensors enable the analysis of biomarkers present in sweat, providing valuable information for physiological health monitoring. Meng et al.'s work focuses on creating an electrochemical sensor with a highly integrated sensing paper for in-the-moment sweat monitoring. The gadget was made using a straightforward printing process that combined sensor paper, a hydrophobic barrier wax, and an array of conductive electrodes. $Ti_3C_2T_x$/MB, sometimes known as MXene/methylene blue, was the active substance used. Chitosan-containing glucose oxidase (GOx) solution immobilized after dripping $Ti_3C_2T_X$-MXene aqueous dispersion onto a carbon electrode made by screen printing method during manufacture. The sensor demonstrates high sensitivity and selectivity by simultaneously measuring lactate and glucose with a limit of detection (LOD) of 0.49 μA mM^{-1} and 4 nA mM^{-1}, respectively. In summary, the integration of NPs, phase modification, and development of wearable electrochemical sensors offer potential solutions to address challenges such as electrical conductivity, fabrication costs, and long-term stability, facilitating advancements in the field of electrochemical sensing for various applications, including real-time human health monitoring.

The remarkable adsorption capability of $Ti_3C_2T_x$, attributed to its large surface area, was further enhanced by the synergistic effect of methylene blue modification, resulting in improved electrochemical response of the device as reported in several studies [79–81]. The highly integrated application of the paper-based sensor with promising properties to be used in wearable bioelectronics provides an opportunity to develop modern technology. In a recent development, Lu et al. successfully fabricated a flexible pressure sensor utilizing carbonized silk (Csilk) fabric with vertically aligned MoS_2 nanosheets, referred to as MoS_2/Csilk [82]. This unique design and combination of materials enabled the creation of a sensitive, flexible pressure sensor that has the ability to detect and measure pressure variations. The pyrolysis process of commercial silk fiber products leads to the production of carbonized fibers with excellent sp^2 hybrid graphite structure. The hierarchical structure of MoS_2/Csilk was achieved through a solvothermal reaction in two steps. In the first step, a mixture of urea, MoO_3, and thioacetamide was added to a solution containing 0.3 g, 60 mg, and 70 mg of ethanol (25 mL) and deionized water (15 mL), respectively. Agitation of the mixture at high speed for approximately 1 h promoted the formation of MoS_2 on the fibers. The second stage involved soaking the carbonized silk fiber in the aforementioned combination and hydrothermally treating the entire mixture at 220 °C for 24 h. The resultant MoS_2/Csilk fiber was carefully cleaned with ethanol and water before being allowed to dry at 60 °C overnight [90]. In summary, the production of highly conductive carbonized fibers and the formation of the MoS_2/Csilk hierarchical structure involved a pyrolytic reaction and a two-step solvothermal process, resulting in a well-prepared material with promising properties for various applications. To construct the pressure sensor, two thin films of MoS_2/Csilk were utilized and arranged in a face-to-face configuration. The film was wrapped securely with 3M VHB 9469 double-sided tape. This configuration causes the two MoS_2/Csilk films to change their contact surfaces with applied pressure, resulting in a change in contact resistance. Therefore, pressure can be measured by observing the change in electrical signal (as shown in Fig. 4). This sensor's adaptability makes it appropriate for incorporation into self-powered or wearable devices. Its outstanding performance makes it well-suited for real-time physiological parameter monitoring and potential application in clinical diagnostics. By leveraging the unique properties of MoS_2/C silk, this pressure sensor holds potential for various health care applications and beyond. Overall, these advancements in highly integrated sensing paper-based sensors and flexible pressure sensors utilizing 2D materials like $Ti_3C_2T_x$ and MoS_2 provide new insights for the development of bioelectronic products and smart technologies.

Researchers led by Wang developed an electrochemical sweat-sensing platform using a unique combination of materials. They used a free-standing graphene (GP) sheet coated with a MoS_2 nanocrystalline monolayer and submicron copper buds to perform the two detection tasks of both lactate and glucose. Here, the fabrication of the electrode sensor involved a sequential process. First, they used a hydrothermal method to grow 0D-MoS_2 nanocrystals on the 2D freestanding GP.

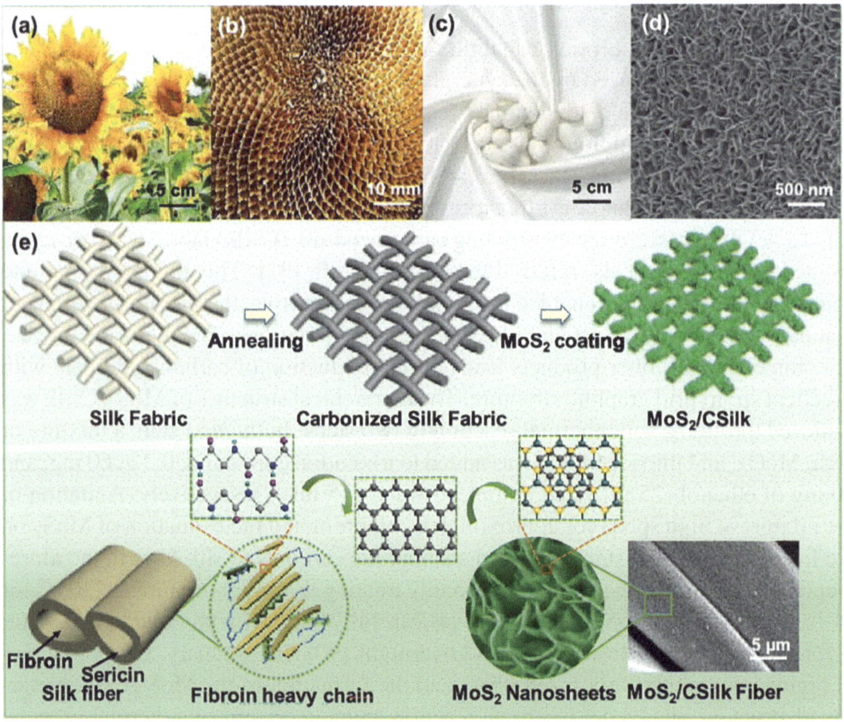

Fig. 4 Design, manufacturing method, and structural alteration of MoS_2/Csilk. **a** and **b** images of flower petals, **c** commercial products containing silk, **d** MoS_2/Csilk SEM images, **e** Method demonstration and modification of the structure. Photo courtesy of Ref. 91 Copyright 2020, American Chemical Society

Next, they electrodeposited 3D Cu submicron buds onto the structure. The addition of GP enhances charge transfer to the working electrode when lactate oxidase is coated on the triad (Cu-MoS_2-GP) structure, which creates an ideal environment for enzyme immobilization and enhanced transfer.

The final lactate oxidase-Cu-MoS_2-GP electrode exhibits excellent electrocatalytic performance, long-term stability, and repeatability. Lactate concentrations range from 0.01 to 18.4 mM and glucose concentrations range from 5 to 1775 μM, suggesting wide range detection capability exists. Glucose and lactate sensitivity were determined as 3.38 mA cm^{-2} μM^{-1} and 83 mA cm^{-2} M^{-1} with respective detection limits of 500 nM and 0.1 mM. The integration of 0D transition metal sulfides with 2D graphene and 3D transition metals offers a promising strategy for the development of lactate and glucose detection platforms for sweat analysis [83]. Wang et al. also suggested a flexible, non-enzymatic sensor with a metal-organic framework (MOF) electrode. They successfully combined $Cu_3(btc)_2$-NH_2-GP ($Cu_3(btc)_2$-NH_2-functionalized freestanding GP) with self-assembled and orientated $Cu_3(btc)_2$ nanocubes to form the sensor [84]. The proposed sensors for sweat glucose and lactate checking gave good results when compared with similar enzymes. Although only small amounts of lactate oxidase were used (5 mM for lactate and 30 nM for glucose), they gave a wide detection range of 0.05 mM to 22.6 mM for lactate and 0.05 to 1775.5 μM for Glucose. These sensors were found to be more efficient with 0.029 mA cm^{-2} μM^{-1} for lactate and 5.36 mA cm^{-2} μM^{-1} for glucose, respectively. Tightly packed $Cu_3(btc)_2$ nanocubes and functionalized graphene nanosheets enhance the pathway for conductivity and are responsible for this sensitivity [91].

A similar approach was used for electrochemical blood glucose detection using GOx immobilized polymer electrodes made from Kapton® polyimide films containing gold, MoS_2, and gold nanofilms (Au/MoS_2/Au-nanofilm). An Au layer was coated on the pretreated polymer substrate using the Au sputtering technique, and then MoS_2 nanoparticles (MoS_2/Au-nanofilm) were spin-coated onto this Au layer. A second gold layer is then added to form the Au/MoS_2/Au nanofilm polymer electrode. Binder (6-MHA) was added to immobilize GO_x on the electrode surface (GO_x/Au/MoS_2/Au-nanofilm) [93] (as shown in Figure 5) to improve Au-thiol binding properties. For the quantitative determination of glucose, amperometric measurements (I-T) were performed showing LODs as high as 10 nM in the range of 500 nM to 10 nM.

In particular, these sensors revealed better sugar detection even with inhibitory substances, for instance, uric acid and ascorbic acid. The mutually beneficial effects resulting from the integration of Au and MoS_2 nanoparticles contributed to enhanced electrocatalytic activity, while the polymer substrate added increased flexibility to the sensor design.

To detect HIV-1, particularly the glycoprotein GP120 (gp120), Shin et al. prepared a light sensor consisting of MoS_2 stacked along two gold layers (Au/MoS_2/Au) supported by polyethylene terephthalate (PET) substrate [94]. HIV virus has an envelope of glycoproteins, GP120 and GP41 bind non-covalently, and GP120 plays an important role in the HIV virus by binding to the host cell group. To construct this biosensor, first, a single gold layer (Au/PET) was sputtered onto a PET substrate,

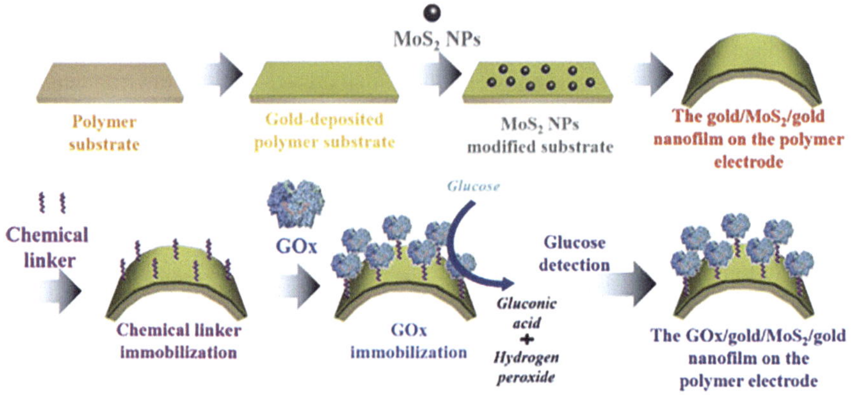

Fig. 5 Schematic representation of the manufacturing protocol of the glucose sensing electrochemical biosensor. Reproduced with permission from Ref. 241 Copyright 2019 by Elsevier BV

and then MoS_2 nanoparticles were coated onto a gold-coated PET substrate (MoS_2/Au/PET) by spin coating. Ultimately, Au was recoated on the substrate using sputtering (Au/MoS_2/Au/PET) [94]. The immobilization of the GP120 antibody onto the electrode was achieved through the cysteamine-modified Au/MoS_2/Au surface. The fabrication of the sensor was confirmed through spectroscopic analysis as well as scanning electron microscopy (SEM), atomic force microscopy (AFM), and cyclic voltammetry (CV) measurements. Using square-wave voltammetry to detect analytes, the flexible sensor exhibits impressively high sensitivity and selectivity showing a 0.1 pg mL^{-1} limit of detection (LOD) [94]. Overall, physiological signals are created by the body's natural processes and offer important information on the condition of various physiological systems. Electrophysiology is a field of science that studies the voltage/currents fluctuations at various scales that depict the electrical properties of cells and tissues, from individual protein-based ion channels and ultimately the whole organs. Analyzing and monitoring these signals are of great significance in diagnosing and monitoring a wide range of diseases. By studying and interpreting physiological signals, healthcare practitioners gain insights into the underlying physiological processes and make informed decisions regarding patient care. Wearable sensors can be briefly represented for monitoring living functions as shown in Figure 6 [85].

4. Breath Analysis: Monitoring respiration, including the quantity and quality of breaths, is crucial for assessing the well-being of hospitalized patients in real time. In general, wearable masks equipped with strain sensors are used to monitor changes in airflow pressure as you breathe. Graphene materials have demonstrated excellent capabilities in this application, offering a wide operating range and high sensitivity, which are essential requirements. Due to their accessibility, affordability, simplicity of preparation, softness, and environmental friendliness,

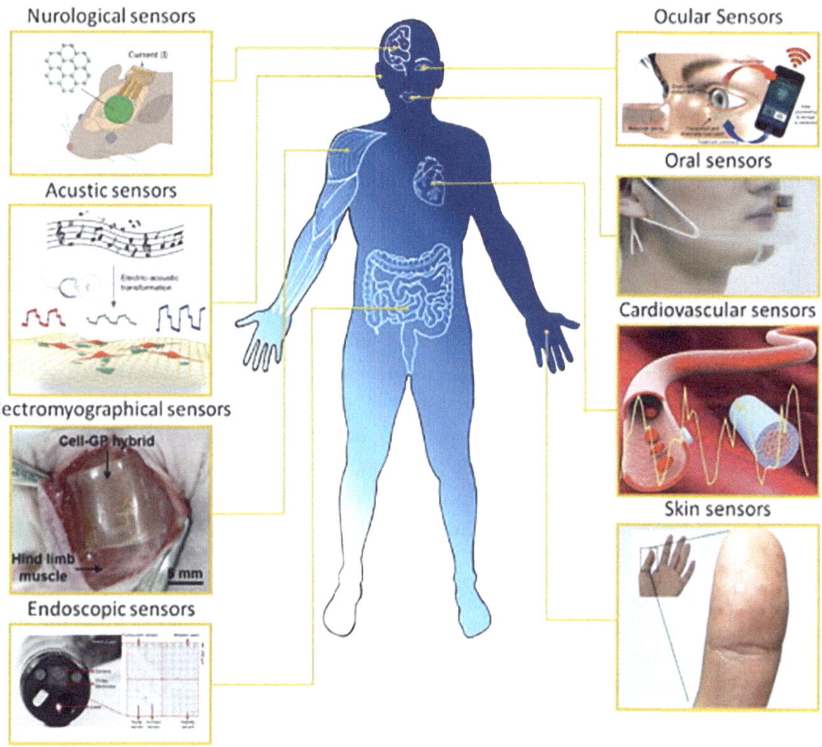

Fig. 6 Picture depicting the brief idea of monitoring the organs with wearables, implantable 2D materials-based sensors with an example of each organ. Courtesy of Ref. 96 Copyright 2022 by Elsevier BV

graphene-based textile and paper sensors have become more popular for respiration monitoring among the numerous substrates available [86, 87]. For example, thermally reduced thin sheets soaked in graphene oxide (GO) exhibit excellent sensitivity (17.2 kPa) and wide operating range (0–20 kPa), enabling applications in motion, impact detection, speech recognition, and voice detection [88]. Alternatively, the patient's chest attached with strain sensors served the dual purpose. Ramirez et al. created a graphene/palladium device supported by PEDOT:PSS that concurrently monitored breathing and heartbeat as a prominent example [89]. This sensor's low strain and high sensitivity allowed the detection of both subtle heartbeat variations and more pronounced respiratory movements (Fig. 7). Humidity sensors provide an alternative approach to breath monitoring by measuring relative humidity during exhalation and inhalation. Graphene oxide (GO) is widely utilized because of water permeability and large surface area, enabling proton generation upon water contact. For instance, He et al. created a wireless GO/polydopamine-based humidity sensor that functioned by continuously forming and breaking hydrogen bonds with water molecules in the polymer [90]. These sensors can detect slow, medium, and fast breathing in real time with low impedance, high sensitivity, low hysteresis, and ultra-fast response. It has also been shown that WS_2 films exhibit excellent moisture sensing performance with fast response and are satisfactory for mask-free analyses [101]. Asthma, smoking-related diseases, and cystic fibrosis are among the various respiratory disorders in humans that can be evaluated through the measurement of air humidity and volatile organic compounds (VOCs) [91, 92]. By continuously monitoring and recording the respiratory profile of patients, these sensors provide valuable insights. Additionally, patients' state of dehydration may be detected and evaluated using wearable humidity sensors [93]. A growing body of evidence suggests that variations in air humidity can have an impact on asthma symptoms [94]. Additionally, the precise detection of volatile organic compounds (VOCs) at ultra-low levels, even in parts per billion (ppb), is crucial for studying contamination. Also, accurate detection of volatile organic compounds (VOCs) at extremely low concentrations, even in parts per billion (ppb), is essential for analyzing pollution, performing breath testing and identification of contaminants. This ability is especially important for early detection [106].

Respiratory gas acetone levels may signal the early development of diabetes [95]. Currently, the analysis of moisture and VOCs in human breath is the least invasive and best diagnostic method. Wearable biosensors; allow continuous monitoring of certain biomarkers, including water, NH_3, sulfur dioxide, and

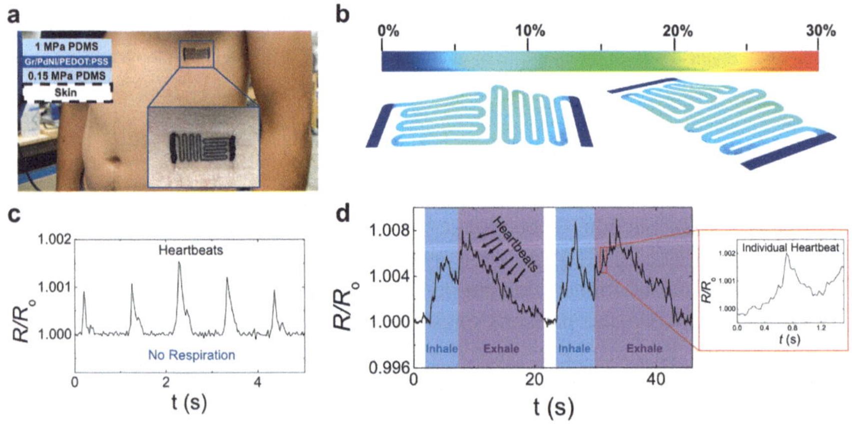

Fig. 7 Field test of Gr/Pd/PEDOT:PSS materials using 1 MPa PDMS encapsulator and 0.15 MPa PDMS substrate was modeled in the serpentine strain gauge model. **a** Image and side view diagram of the wearable Gr/Pd/PEDOT:PSS sensor in the human body with 1 MPa PDMS encapsulant and 0.15 MPa PDMS substrate. **b** Finite element analysis (FEA) of the prototype Gr/Pd/PEDOT:PSS sensor simulating the forces caused by heartbeat and/or breathing as the material is subjected to stress caused by normal forces. Piezoresistive response of the exemplary Gr/Pd/PEDOT:PSS sensor for human subjects simulating **c** obstructive sleep apnea and **d** normal sleep. From Ref. 100 with permission. Copyright 2019, American Chemical Society

other VOCs, without immersing sensors in body fluids [104]. Graphene-based materials, including pure graphene, GO (graphene oxide), and rGO (reduced graphene oxide), are attractive options for developing wearable electronics for respiration monitoring in view of the flexibility, and vast effective surface area with humidity-sensing capabilities. These sensors demonstrate excellent flexibility and bending capability when applied to curved skin surfaces, indicating their remarkable conformal ability. Key performance parameters for humidity sensors include response/recovery time, linearity, and sensitivity. PEDOT:PSS (poly(3,4-ethylene dioxythiophene)-poly(styrene sulfonate)) has been used to enhance the sensing characteristics of porous graphene, resulting in considerable improvements in linearity, sensitivity, and response/recovery times [104]. Wearable humidity sensors integrated into medical masks can monitor changes in breath humidity, aiding in the detection of symptoms such as apnea and irregular breathing patterns. Additionally, humidity sensors can measure the degree of dehydration before and after drinking water, which is advantageous for those who have trouble speaking. Water molecules can stick to graphene oxide (GO) layers' hydrophilic functional groups, which change the humidity sensor's electrical characteristics [104]. Ultra-thin GO films on interdigital electrodes have demonstrated rapid response to moisture, allowing real-time recording of subtle human speech and respiration. Principal component analysis (PCA) and the rapid Fourier transform have made it possible to classify and differentiate different whistled tunes with high accuracy, providing a useful validation approach. Other humidity sensors, like as rGO-based sensors, may assess physiological signals in the human body and offer real-time information on humidity variations [90]. For example, a MoS_2-coated etched single-mode fiber (ESMF) humidity sensor has demonstrated a considerable increase in light transmission power in response to changes in relative humidity (RH), with quick reaction and recovery times [96]. To increase sensitivity, detection limit, and reaction time, MoS_2 has been widely investigated in hybridization with materials such as Ag NPs, S NO_2, Cu_2S, and graphene [97–100]. These developments make it possible to measure humidity changes caused by breathing at various speech frequency patterns. It is essential to create high-efficiency analyzers with significant selectivity and detection limits as low as parts per billion (ppb) in complicated gas mixtures in order to identify volatile organic compounds (VOCs) in exhaled breath [101, 102]. Graphene-based sensors modified with ssDNA (single-stranded DNA) have shown improved selectivity and sensitivity, facilitating discrimination of various VOCs and biomarkers related to halitosis, kidney disorders, and cancer [103]. Additionally, graphene-based gas sensors, when equipped with heaters, can achieve rapid reaction and recovery times for detecting gases such as NO_2 [114]. Multiple functionalized reduced graphene oxide (rGO) sensors can simultaneously monitor VOCs, and physiological signals void of signal interference, enlarging their application scope in deterrent medicine [113]. Other 2D materials, such as MoS_2, WS_2, and their hybrids, offer semiconducting characteristics and adjustable bandgaps, making them suitable for gas sensor fabrication [104–107].

MXene materials, known for their exceptional surface chemistry, conductivity, and biocompatibility, have also been successfully employed in solid state.

5. Voice and sound intensity sensors: Lately, increment of an interest in speech recognition and restoration of speech abilities for human–computer interaction. For the purpose of capturing the minute muscle movements involved in speech and achieving quick reaction times, several studies have concentrated on enhancing pressure sensors with increased sensitivity [85]. One particularly ground-breaking and cutting-edge work introduced an artificial throat wearable constructed with laser-induced graphene (LIG) onto a polyimide substrate. This remarkable technology not only enabled simultaneous detection but also the generation of sounds (as depicted in Fig. 8) [108]. The artificial throat had exceptional qualities, including great sensitivity, a low detection threshold, strong thermal conductivity, and a low heat capacity, which made it the perfect choice for thermoacoustic sound creation. The scientists also showed how it could distinguish between distinct hums, coughs, and screams of varied intensities and volumes. They were able to distinguish between spoken words and phrases by converting ambiguous throat vibrations into regulated noises. This ground-breaking study is regarded as a crucial step in the creation of particular phonation identification systems, providing aid and training in speech therapy for people with impairments.

So far the 2D materials for sensing applications are discussed and sufficient to give a broad outlook for how the main mechanisms, and their combinations, are used for sensing applications. Apart from above discussed applications, various sensing platforms exist and are noteworthy. These include neurotransmitter [109–112], tear [113, 114], sweat [115–122], saliva [123, 124], human motion [125–129], photodetector [130–134], fluorescent sensory devices. At the time of writing this chapter, many journals are in the process of publication or some are left out for the brevity of the content. This proves that 2D materials can occupy the major part of the sensing platforms for future sensor applications.

Setbacks and Outlook of 2D Materials

Due to their various properties, 2D materials have many advantages for detecting applications. The high surface region-to-volume proportion and phenomenal physical and chemical capacities are essential for sensing/detecting applications. This unique feature allows for the design of highly sensitive sensing devices by introducing modifications to the framework of 2D materials. However, a major limitation lies in the oxidative instability of different 2D nanostructures when exposed to ambient conditions. This instability can lead to structural changes and material decomposition in severe cases, severely limiting their practical application in functional devices. Innovative methods are required to increase the oxidative and thermal stability of 2D materials during their production, storage, and practical use in order to solve this obstacle. In addition, the chemically tunable ability of 2D materials allows for the creation of multiple components, allowing multiple or multiple analytes to be detected in a single instrument. Sensing capacities can be greatly improved by the presence of free radicals and by the increased reactivity of 2D materials. Non-selective binding

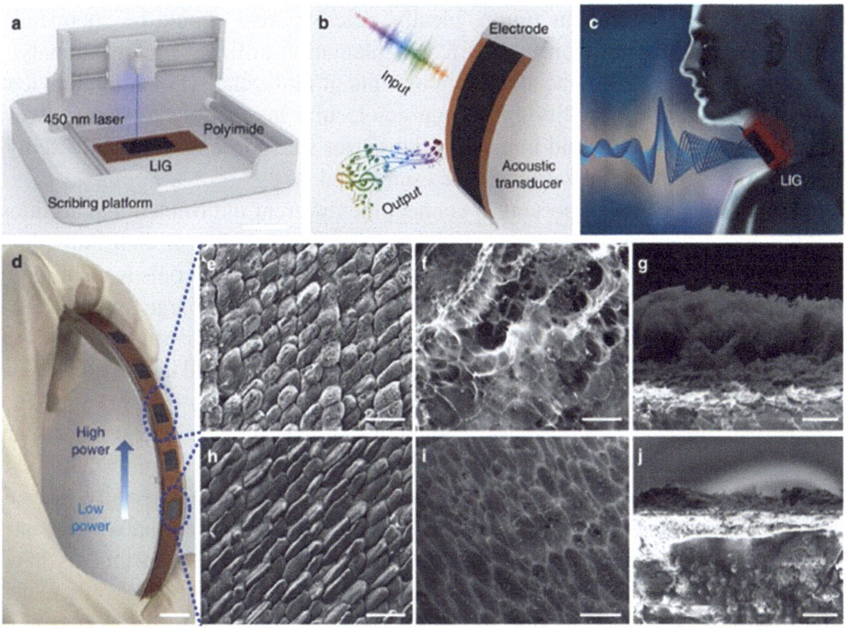

Fig. 8 Schematic representation of the production process and morphology of LIG. **a** LIG's one-step manufacturing process. PI was directly converted to LIG by irradiation with a 450 nm laser. Scale bar, 2.5 cm. **b** LIG can transmit and record audio on a single device. **c** The artificial throat can detect the movement of the throat and produce controlled sounds accordingly. **d** Six samples of LIG produced by a 450 nm laser with a power of 20 to 350 mW. Scale bar, 5 mm. **e** Morphology of the LIG structure produced at 290 mW in the electron microscope. Scale bar, 150 μm. **f** Morphology of a LIG sample produced in large size at 290 mW. Scale bar, 5 μm. **g** Sectional view of a typical LIG design at 290 mW. scale bar, 12.5 μm. **h** Morphology of the LIG structure produced at 125 mW in the electron microscope. Scale bar, 150 μm. **i** Morphology of a LIG sample generated at 125 mW at high magnification. Scale bar, 5 μm. **j** Sectional view of the LIG design at 125 mW, Scale bar, 12.5 μm. copyright 2017, the author, Springer Nature. Ref. 120

of analytes, however, may lead to insufficient recovery, slow reaction times, or poor selectivity. Biological and metal-relevant molecules have been detected with success using optical sensors that make use of the photoluminescence (PL) properties of 2D materials after antigen-analyte interactions. Additionally, electrochemical sensors show promise due to their practical operation and rapid detection capabilities.

Strong interactions with target analytes might not occur in the presence of perfect or defect-free sheets. However, defects are important in 2D materials because they provide bandgap modularity and optoelectronic/electronic control of devices. Defects also make 2D materials more chemically reactive, allowing for further functionalization of their reactive spots. This leads to enhanced sensitivity and a lower limit of detection (LOD) for target analytes. Nevertheless, uncontrolled defect engineering can have long-lasting effects on 2D materials. Ad hoc flaws in 2D materials will need to be included in a well-ordered way with exact control over their concentration and spatial resolution in the near future. Ideally, reactive areas should be chemically functionalized in conjunction with defect implantation utilizing preferred analyte receptors. By better comprehending the concepts of molecular recognition this feat may be achievable. By modifying the supramolecular interactions associated with ultra-sensitive 2D devices and improving the sensor's response time and selectivity, a clear path to commercialization can be established.

In addition, it is also important to connect the different multifunctional abilities of 2D materials and electronics together through the integration of their individual components. For example, the creation of porous electronic materials leads to integration with surfaces, which increases hydrophobicity, reduces charge transfer resistance, and increases capacitance. This feature can be used to increase the efficiency of potentiometric devices. The combination of magnetic and electrical interactions in 2D materials, on the other hand, can enhance transport resistance in antigen analysis. The development of this technology has the potential to replace the scanning process with advantages such as cost-effectiveness, ease of use, and enhanced cognitive abilities. New 2D materials are still being explored, and given their amazing diversity, there is still plenty of time to establish them as leaders in the disciplines of nanoscience and electrical transduction sensors. More importantly, observable advancements have been achieved in the development of two-dimensional materials such as the materials based on graphene, the MXene family, and transition metal dichalcogenides (TMDCs), which are used widely in many equipment such as temperature sensors, gas sensors, filters, and pressure sensors. Furthermore, there is a growing likelihood of the introduction of these materials in industrial and commercial technologies in the coming years. In general, flexible electronic devices, particularly optoelectronics, offer promise in the processing of human physiological data, particularly in wearable devices used for diagnosis and therapy. In addition, research continues on integrated electronic devices with multiple signal detection and provide various physiological information.

Graphene along with other 2D materials has remarkable mechanical and electrical qualities including rigidity and flexibility. Utilizing these characteristics, flexible sheets may incorporate material-based 2D sensors to construct smart wearables for monitoring the environment and human interaction. Additionally, electrical contacts, material stability, and device performance may all be enhanced by mixing several 2D materials to create van der Waals heterostructures and integrating them with electrodes. These heterostructure interfaces might be used to improve the stability, sensitivity, and selectivity of sensing devices. Additionally, the creation of multifunctional devices that combine chemical and physical elements is a potential way to precisely evaluate physical conditions. Signal crosstalk, however, makes it difficult to detect chemical signals and huge strain changes at the same time. There is a demand for wearable chemical systems with remarkable selectivity, better sensitivity, and lower limits of detection in complicated human body settings. Additionally, combining diagnostic and therapeutic functions provides an alternative pathway to expand the functionality and applications of these devices. This is where medical treatment technologies that combine physical therapy and medicine administration, such as electrical stimulation, show promise. Although several studies have shown that human physiological signals may be recovered, usually, laboratory research on commercial goods usually takes long time periods. Practical collaboration with electronic engineers should be used to produce signal processing that uses flexible circuit boards, low power consumption, and compact sizes. Existing circuit technologies, which are normally put together using ordinary chips, might not be appropriate for some uses, including chips incorporated inside contact lenses for intraocular pressure monitoring. In addition, it is important to create long-lasting and continuously working small equipment in order to monitor human health information 24 h a day. Medical specialists should evaluate the signals that these gadgets pick up in order to ensure the security of wearable systems. Finding more useful information can be accomplished by comparing the findings to those of similar medical tests. Human physiological signals are evaluated and continuously monitored to provide data consistency, which is essential for efficient individualized treatment. Although impactful progress was made in the evaluation of nanoindentation-based mechanical devices, issues such as structural damage during preparation and evaluation still need to be addressed. The creation of large-scale production of high-quality novel 2D materials, and characterizations/measurements using more computational and experimental methods should be the main goals of the next generation of 2D ultrafine nanomaterials research. Recent research points to the potential for "on-the-skin" uses of 2D materials, providing opportunities for the identification of clinically important biomarkers. Thus, tattoos based on 2D materials can work as dry sensors of the skin, reducing the motion artifacts regarding skin's sensitivity, thus having a possibility of mitigating the disadvantages of traditional electrodes and dry sensors for monitoring health.

Designing Sensor Based on Computational Methods

The development of stimulus-sensitive 2D materials and their important applications in chemical sensing has a great deal of promise thanks to the collaboration of computational assessment and curiosity-motivated research. A difficult long-term goal in this area is the development of computer models and their incorporation into materials design and synthesis. Consequently, two important approaches might appear later on. First, use simulations to construct a growth chamber design and synthesis parameters that adequately represent the experimental thermophysical environment, expected growth, and reaction kinetics. To provide a strong design, this method demands a thorough sensitivity analysis and excellent model accuracy. Second, the model, which analyzes the system and sends them to the controller, can constantly monitor the growth in real time, which can create the basis for the creation of the processes for particular applications. The computing power of the models used will determine how effective this strategy is. The chemical programmability of 2D materials can be used to enable future developments in multi-analyte detection by assembling multiple components. Current mathematical models are very comprehensive and frequently fail to capture entire physical details that are needed in generating 2D data, limiting their potential in management. Experimental and computational data are infused into Machine learning models, trained, for developments such as 2D material models, and those in the 2D Crystals Society's Innovative Materials promise to solve this problem. To the end, methods/models used to deduce the evolution of 2D materials, may be applied, to recognize the evolution of other materials via synthesis methods including thin-film chemical vapor deposition (CVD) development. A vast number of possible sensors may be developed in a shorter amount of time thanks to computational designing.

Importance and Methods to Understand Mechanisms of 2D Materials

Spectroscopic methods used in situ and in operation are extremely sophisticated instruments for understanding the concepts and procedures related to gas sensing [5]. The sensing processes of chemiresistive gases can be better understood by integrating models for ion sorption, oxygen vacancy, and charge transfer. The creation of high-performance intelligent sensor devices can be aided by a thorough grasp of the precise sensing processes made possible by the employment of spectroscopic technologies in a variety of operational settings. Additionally, in order to confirm the actual results gained from operando spectroscopic investigations, comprehensive theoretical analyses must be conducted. The creation of high-performance gas sensors is made easier by the incorporation of various operando spectroscopic methods and theoretical studies, which can offer new perspectives on sensing technology. The 2D luminescent materials are viable candidates for imaging and biosensing applications. The vast field of 2D material data offers an invisible potential to be used as a platform for many experiences. Although research and development have made great advances, a successful solution for sensor devices has remained elusive. The development of chemical sensors that are reliable, inexpensive, and environmentally friendly with superior sensitivity and selectivity should be the target of future research. Additionally, the development of portable sensor technology promises an intriguing new direction for growth.

In conclusion, the promise of 2D materials for creating physical, chemical, biological, and wearable sensors has thus far received a lot of attention. The main benefit is that they have a large contact surface area compared to volume, which allows for high sensitivity and limit of detection (LOD). The choice of a 2D material-based sensing platform is crucial for applications requiring chemical and physical sensors. As pristine materials may alter selectivity, anchor groups must be present on the surface to improve analyte absorption. Gas sensing, particularly for NO_2, NH_3, and humidity, has been the focus in chemiresistive devices, but exploring other toxic gases and alternative approaches is necessary. Decorating 2D materials with metals has been found to enhance electron transfer, selectivity, sensitivity, and LOD in optical-based devices. Although there are still many manufacturing issues, graphene has potential in piezoresistive materials such as electrified skin, human performance, healthcare, and wearables. The performance of 2D TMD materials for electronic devices requires improvements in bonding, fabrication processes, and material-substrate interactions. Although graphene-based skin sensors have been very successful, limitations of nanomaterials still exist, such as the need for modification with biomolecules, high production costs, and lack of stability. In vitro biocompatibility of biomedical devices based on implantable 2D materials has been found, but in vivo biocompatibility is still a problem that can be solved with the aid of biomimetic properties and biocompatible layers. The commercialization of wearable and implantable devices depends on their durability and performance. Further study is necessary due to the possibility of skin irritability and allergic reaction in graphene-based or TMDs-based sensors. Reproducibility and multiplatform sensing using different materials are the next horizons, including cross-linked heterostructures. Although 2D materials have shown significant sensing properties, realizing their full potential requires multidisciplinary advanced research. Although the creation and discovery of 2D materials are moving forward quickly, some materials still need to be thoroughly investigated for chemical sensing. For a product to be successful on the market, trustworthy product design, scalability, and affordable production are essential industrial factors. For 2D materials, scalable synthesis and sensor fabrication techniques have been developed, yet contamination and flaws still fall short of industrial requirements. As 2D materials, in particular graphene, TMDs, and their heterostructures, are seen as feasible possibilities for industrial manufacturing, overcoming these challenges is a fundamental objective. Thoroughly analyzing the challenges and research in 2D-based sensing devices establishes a better understanding and strengthens the academia-industry connection.

Acknowledgements This is the independent work of the corresponding author of this chapter.

Conflict of Interest The author affirms that he has no known financial or interpersonal conflicts that may have looked to have influenced the research presented in this study.

References

1. Arduini, F., Cinti, S., Scognamiglio, V., Moscone, D.: How cutting-edge technologies impact the design of electrochemical (bio)sensors for environmental analysis. A Review. Anal. Chim. Acta (2017). https://doi.org/10.1016/j.aca.2016.12.035
2. Chen, X., Leishman, M., Bagnall, D., Nasiri, N.: Nanostructured gas sensors : from air quality and environmental monitoring to healthcare and medical applications
3. Article, R.: An atlas of two-dimensional materials. Chem. Soc. Rev. **43**, 6537–6554 (2014). https://doi.org/10.1039/b000000x
4. Tyagi, D., Wang, H., Huang, W., Hu, L., Tang, Y., Guo, Z., Ouyang, Z., Zhang, H.: Recent advances in two-dimensional-material-based sensing technology toward health and environmental monitoring applications. Nanoscale **12**(6), 3535–3559 (2020). https://doi.org/10.1039/c9nr10178k
5. Hassan, J. Z., Raza, A., Din Babar, Z. U., Qumar, U., Kaner, N. T., Cassinese, A.: 2D material-based sensing devices: an update. J. Mater. Chem. A 6016–6063 (2023). https://doi.org/10.1039/d2ta07653e.
6. Wang, Q.H., Kalantar-zadeh, K., Kis, A., Coleman, J.N., Strano, M.S.: Electronics and optoelectronics of two-dimensional transition metal dichalcogenides. Nat. Publ. Gr. **7**(November), 699–712 (2012). https://doi.org/10.1038/nnano.2012.193
7. Barsoum, M.W.: The MN+1AXN phases: a new class of solids; thermodynamically stable nanolaminates. Prog. Solid St. Chem. **28**, 201–281 (2000)

8. Molaei, M. J., Younas, M., Rezakazemi, M.: A comprehensive review on recent advances in two-dimensional (2D) hexagonal boron nitride (2021). https://doi.org/10.1021/acsaelm.1c00720
9. Li, L., Yu, Y., Ye, G. J., Ge, Q., Ou, X., Wu, H., Feng, D., Chen, X. H., Zhang, Y.: Black phosphorus field-effect transistors. 9 (March) (2014). https://doi.org/10.1038/nnano.2014.35
10. Ye, P.D., Liu, H., Neal, A.T., Zhu, Z., Luo, Z., Xu, X., Engineering, C., Lafayette, W., States, U., Lansing, E., States, U., Engineering, M., Lafayette, W., States, U.: Phosphorene : an unexplored 2D semiconductor with a high hole. ACS Nano **4**, 4033–4041 (2014)
11. Radhakrishnan, S., Mathew, M., Rout, C.S.: Microfluidic sensors based on two-dimensional materials for chemical and biological assessments. Mater. Adv. **3**, 1874–1904 (2022). https://doi.org/10.1039/d1ma00929j
12. Liu, X., Ma, T., Pinna, N., Zhang, J.: Two-dimensional nanostructured materials for gas sensing **1702168**, 1–30 (2017). https://doi.org/10.1002/adfm.201702168
13. Neri, G., Neri, G.: Thin 2D: The New Dimensionality in Gas Sensing Thin 2D: The New Dimensionality in Gas Sensing. https://doi.org/10.3390/chemosensors5030021
14. Tan, C., Cao, X., Wu, X., He, Q., Yang, J., Zhang, X., Chen, J., Zhao, W., Han, S., Nam, G., Sindoro, M., Zhang, H.: Recent advances in ultrathin two-dimensional nanomaterials. Chem. Rev. **117**, 6225–6331 (2017). https://doi.org/10.1021/acs.chemrev.6b00558
15. Varghese, S.S., Varghese, S.H., Swaminathan, S., Singh, K.K.: Two-dimensional materials for sensing: graphene and beyond. https://doi.org/10.3390/electronics4030651
16. Gómez, I.J., Alegret, N., Dominguez-alfaro, A., Sulleiro, M.V.: Recent advances on 2D materials towards 3D printing
17. Eric Singh, M. Meyyappan, H.S.N.: Flexible graphene-based wearable gas and chemical sensors. ACS Appl. Mater. Interfaces 40, 34544–34586 (2017). https://doi.org/10.1021/acsami.7b07063
18. Zeng, M., Xiao, Y., Liu, J., Yang, K., Fu, L.: Exploring two-dimensional materials toward the next-generation circuits: from monomer design to assembly control. Chem. Rev. **118**(13), 6236–6296 (2017). https://doi.org/10.1021/acs.chemrev.7b00633
19. Islam, N., Channon, R. B.: Electrochemical sensors. Elsevier Ltd. (2020). https://doi.org/10.1016/B978-0-12-813886-1.00004-8.
20. Lee, C.W., Suh, J.M., Jang, H.W.: Chemical sensors based on two-dimensional (2D) materials for selective detection of ions and molecules in liquid. Front. Chem. **7**(November), 1–21 (2019). https://doi.org/10.3389/fchem.2019.00708
21. Chowdhury, A., Biswas, S., Singh, T., Chandra, A.: Redox mediator induced electrochemical reactions at the electrode-electrolyte interface: making sodium-ion supercapacitors a competitive technology. Electrochem. Sci. Adv. **2**(1), 1–14 (2022). https://doi.org/10.1002/elsa.202100030
22. Labib, M., Sargent, E.H., Kelley, S.O.: Electrochemical methods for the analysis of clinically relevant biomolecules. Chem. Rev. **116**(16), 9001–9090 (2016). https://doi.org/10.1021/acs.chemrev.6b00220
23. Thévenot, D.R., Toth, K., Durst, R.A., Wilson, G.S.: Electrochemical biosensors: recommended definitions and classification. Biosens. Bioelectron. **16**(1–2), 121–131 (2001). https://doi.org/10.1016/S0956-5663(01)00115-4
24. Scholz, F.: Voltammetric techniques of analysis: the essentials. ChemTexts **1**(4), 1–24 (2015). https://doi.org/10.1007/s40828-015-0016-y
25. He, Q., Sudibya, H.G., Yin, Z., Wu, S., Li, H., Boey, F., Huang, W., Chen, P., Zhang, H.: Centimeter-long and large-scale micropatterns of reduced graphene oxide films: fabrication and sensing applications. ACS Nano **4**(6), 3201–3208 (2010). https://doi.org/10.1021/nn100780v
26. Kudr, J., Adam, V., Zitka, O.: Fabrication of graphene/molybdenum disulfide composites and their usage as actuators for electrochemical sensors and biosensors. Molecules 24(18) (2019). https://doi.org/10.3390/molecules24183374
27. Sulleiro, M.V., Dominguez-Alfaro, A., Alegret, N., Silvestri, A., Gómez, I.J.: 2D materials towards sensing technology: from fundamentals to applications. Sens. Bio-Sensing Res. 38(July) (2022). https://doi.org/10.1016/j.sbsr.2022.100540

28. Kajale, S.N., Yadav, S., Cai, Y., Joy, B., Sarkar, D.: 2D material based field effect transistors and nanoelectromechanical systems for sensing applications. iScience 24(12), 103513 (2021). https://doi.org/10.1016/j.isci.2021.103513
29. Vargas-Bernal, R.: Electrical properties of two-dimensional materials used in gas sensors. Sensors (Switzerland) 19(6) (2019). https://doi.org/10.3390/s19061295
30. Kumar, Y., Hashmi, S.A., Pandey, G.P.: Lithium ion transport and ion-polymer interaction in PEO based polymer electrolyte plasticized with ionic liquid. Solid State Ionics **201**(1), 73–80 (2011). https://doi.org/10.1016/j.ssi.2011.08.010
31. Zhang, P., Xiao, Y., Zhang, J., Liu, B., Ma, X., Wang, Y.: Highly sensitive gas sensing platforms based on field effect transistor-a review. Anal. Chim. Acta **1172**, 338575 (2021). https://doi.org/10.1016/j.aca.2021.338575
32. Dai, C., Liu, Y., Wei, D.: Two-dimensional field-effect transistor sensors: the road toward commercialization. Chem. Rev. **122**(11), 10319–10392 (2022). https://doi.org/10.1021/acs.chemrev.1c00924
33. Hashtroudi, H., MacKinnon, I.D.R., Shafiei, M.: Emerging 2D hybrid nanomaterials: towards enhanced sensitive and selective conductometric gas sensors at room temperature. J. Mater. Chem. C **8**(38), 13108–13126 (2020). https://doi.org/10.1039/d0tc01968b
34. Wang, Y., Zhou, Y., Wang, Y., Zhang, R., Li, J., Li, X., Zang, Z.: Conductometric room temperature ammonia sensors based on titanium dioxide nanoparticles decorated thin black phosphorus nanosheets. Sens. Actuators B Chem. **349**, 130770 (2021). https://doi.org/10.1016/J.SNB.2021.130770
35. Dominguez-Alfaro, A., Jénnifer Gómez, I., Alegret, N., Mecerreyes, D., Prato, M.: 2D and 3D immobilization of carbon nanomaterials into pedot via electropolymerization of a functional bis-edot monomer. Polymers (Basel) **13**(3), 1–15 (2021). https://doi.org/10.3390/polym13030436
36. Zheng, L., Wang, X., Jiang, H., Xu, M., Huang, W., Liu, Z.: Recent progress of flexible electronics by 2D transition metal dichalcogenides. Nano Res. **153**, **15**(3), 2413–2432 (2021). https://doi.org/10.1007/S12274-021-3779-Z
37. Lei, Z.L., Guo, B.: 2D material-based optical biosensor: status and prospect. Adv. Sci. **9**(4), 2102924 (2022). https://doi.org/10.1002/ADVS.202102924
38. Subbanna, B.B., Choudhary, K., Singh, S., Kumar, S.: 2D material-based optical sensors: a review. ISSS J. Micro Smart Syst. **111** 2022 (2021), **11**(1), 169–177 (2022). https://doi.org/10.1007/S41683-021-00083-4
39. Chia, H.L., Mayorga-Martinez, C.C., Pumera, M.: Doping and decorating 2D materials for biosensing: benefits and drawbacks. Adv. Funct. Mater. **31**(46), 2102555 (2021). https://doi.org/10.1002/ADFM.202102555
40. Duan, Q., Liu, Y., Chang, S., Chen, H., Chen, J. H.: Surface plasmonic sensors: sensing mechanism and recent applications. Sensors 21(16), (2021). https://doi.org/10.3390/s21165262
41. Zhou, J., Yang, T., Chen, J., Wang, C., Zhang, H., Shao, Y.: Two-dimensional nanomaterial-based plasmonic sensing applications: advances and challenges. Coord. Chem. Rev. **410**, 213218 (2020). https://doi.org/10.1016/J.CCR.2020.213218
42. Ge, Y., Chen, S., Xu, Y., He, Z., Liang, Z., Chen, Y., Song, Y., Fan, D., Zhang, K., Zhang, H.: Few-layer selenium-doped black phosphorus: synthesis, nonlinear optical properties and ultrafast photonics applications. J. Mater. Chem. C **5**(25), 6129–6135 (2017). https://doi.org/10.1039/C7TC01267E
43. Luo, X., Qiu, T., Lu, W., Ni, Z.: Plasmons in graphene: recent progress and applications. Mater. Sci. Eng. R Reports **74**(11), 351–376 (2013). https://doi.org/10.1016/J.MSER.2013.09.001
44. Chen, M., Liu, D., Du, X., Lo, K.H., Wang, S., Zhou, B., Pan, H.: 2D materials: excellent substrates for surface-enhanced Raman scattering (SERS) in chemical sensing and biosensing. TrAC Trends Anal. Chem. **130**, 115983 (2020). https://doi.org/10.1016/J.TRAC.2020.115983
45. Qian, X.M., Nie, S.M.: Single-molecule and single-nanoparticle SERS: from fundamental mechanisms to biomedical applications. Chem. Soc. Rev. **37**(5), 912–920 (2008). https://doi.org/10.1039/B708839F

46. Wang, W., Wang, W., Liu, L., Xu, L., Kuang, H., Zhu, J., Xu, C.: Nanoshell-enhanced raman spectroscopy on a microplate for *Staphylococcal Enterotoxin* B sensing. ACS Appl. Mater. Interfaces **8**(24), 15591–15597 (2016). https://doi.org/10.1021/ACSAMI.6B02905/ASSET/IMAGES/LARGE/AM-2016-029052_0005.JPEG
47. Bradshaw, G.A.J., D. S., Resonance energy transfer: from fundamental theory to recent applications. Front. Phys. **7**(JULY), 1–19 (2019). https://doi.org/10.3389/fphy.2019.00100
48. Tian, F., Lyu, J., Shi, J., Yang, M.: Graphene and graphene-like two-denominational materials based fluorescence resonance energy transfer (FRET) assays for biological applications. Biosens. Bioelectron. **89**, 123–135 (2017). https://doi.org/10.1016/J.BIOS.2016.06.046
49. Yan, L., Shi, H., Sui, X., Deng, Z., Gao, L.: MoS2-DNA and MoS2 based sensors. RSC Adv. **7**(38), 23573–23582 (2017). https://doi.org/10.1039/C7RA02649H
50. Zhou, J., Li, Z., Ying, M., Liu, M., Wang, X., Wang, X., Cao, L., Zhang, H., Xu, G.: Black phosphorus nanosheets for rapid MicroRNA detection. Nanoscale **10**(11), 5060–5064 (2018). https://doi.org/10.1039/C7NR08900G
51. Zhou, J., Chen, J., Ge, Y., Shao, Y.: Two-dimensional nanomaterials for Förster resonance energy transfer-based sensing applications. Nanophotonics **9**(7), 1855–1875 (2020). https://doi.org/10.1515/NANOPH-2020-0065/ASSET/GRAPHIC/J_NANOPH-2020-0065_FIG_003.JPG
52. Neema, P.M., Tomy, A.M., Cyriac, J.: Chemical sensor platforms based on fluorescence resonance energy transfer (FRET) and 2D Materials. TrAC Trends Anal. Chem. **124**, 115797 (2020). https://doi.org/10.1016/J.TRAC.2019.115797
53. Devi, M.: Application of 2D nanomaterials as fluorescent biosensors. ACS Symp. Ser. **1353**, 117–141 (2020). https://doi.org/10.1021/BK-2020-1353.CH006
54. Zhang, L., Khan, K., Zou, J., Zhang, H., Li, Y.: Recent advances in emerging 2D material-based gas sensors: potential in disease diagnosis. Adv. Mater. Interfaces **6**(22), 1901329 (2019). https://doi.org/10.1002/ADMI.201901329
55. Di Francia, G., Alfano, B., La Ferrara, V.: Conductometric Gas Nanosensors. J. Sensors **2009** (2009). https://doi.org/10.1155/2009/659275
56. Korotcenkov, G., Cho, B. K.: Engineering approaches for the improvement of conductometric gas sensor parameters: part 1. improvement of sensor sensitivity and selectivity (Short Survey). Sens. Actuators B Chem. 188, 709–728 (2013). https://doi.org/10.1016/J.SNB.2013.07.101
57. Wang, C., Yin, L., Zhang, L., Xiang, D., Gao, R.: Metal oxide gas sensors: sensitivity and influencing factors. Sensors **10**, 2088–2106 (2010), **10**(3), 2088–2106 (2010). https://doi.org/10.3390/S100302088
58. Korotcenkov, G.: Metal oxides for solid-state gas sensors: what determines our choice? Mater. Sci. Eng. B **139**(1), 1–23 (2007). https://doi.org/10.1016/J.MSEB.2007.01.044
59. Perkins, F.K., Friedman, A.L., Cobas, E., Campbell, P.M., Jernigan, G.G., Jonker, B.T.: Chemical vapor sensing with monolayer MoS2. Nano Lett. **13**(2), 668–673 (2013). https://doi.org/10.1021/NL3043079/SUPPL_FILE/NL3043079_SI_001.PDF
60. Jakubik, W.P., Urbańczyk, M.W., Kochowski, S., Bodzenta, J.: Bilayer structure for hydrogen detection in a surface acoustic wave sensor system. Sens. Actuators B Chem. **82**(2–3), 265–271 (2002). https://doi.org/10.1016/S0925-4005(01)01061-9
61. Ricco, A.J., Martin, S.J., Zipperian, T.E.: Surface acoustic wave gas sensor based on film conductivity changes. Sens. Actuators **8**(4), 319–333 (1985). https://doi.org/10.1016/0250-6874(85)80031-7
62. Wohltjen, H.: Mechanism of operation and design considerations for surface acoustic wave device vapour sensors. Sens. Actuators **5**(4), 307–325 (1984). https://doi.org/10.1016/0250-6874(84)85014-3
63. Kumar, B., Min, K., Bashirzadeh, M., Farimani, A.B., Bae, M.H., Estrada, D., Kim, Y.D., Yasaei, P., Park, Y.D., Pop, E., Aluru, N.R., Salehi-Khojin, A.: The role of external defects in chemical sensing of graphene field-effect transistors. Nano Lett. **13**(5), 1962–1968 (2013). https://doi.org/10.1021/NL304734G/SUPPL_FILE/NL304734G_SI_001.PDF
64. Yang, W., Gan, L., Li, H., Zhai, T.: Two-dimensional layered nanomaterials for gas-sensing applications. Inorg. Chem. Front. **3**(4), 433–451 (2016). https://doi.org/10.1039/C5QI00251F

65. Arsat, R., Breedon, M., Shafiei, M., Spizziri, P.G., Gilje, S., Kaner, R.B., Kalantar-zadeh, K., Wlodarski, W.: Graphene-like nano-sheets for surface acoustic wave gas sensor applications. Chem. Phys. Lett. **467**(4–6), 344–347 (2009). https://doi.org/10.1016/J.CPLETT.2008.11.039
66. Xuan, W., He, M., Meng, N., He, X., Wang, W., Chen, J., Shi, T., Hasan, T., Xu, Z., Xu, Y., Luo, J.K.: Fast response and high sensitivity ZnO/glass surface acoustic wave humidity sensors using graphene oxide sensing layer. Sci. Rep. **4**, 7206 (2014). https://doi.org/10.1038/srep07206
67. Davis, C.E., Ho, C.K., Hughes, R.C., Thomas, M.L.: Enhanced detection of m-xylene using a preconcentrator with a chemiresistor sensor. Sens. Actuators B Chem. **104**(2), 207–216 (2005). https://doi.org/10.1016/J.SNB.2004.04.120
68. Hubble, L.J., Cooper, J.S., Sosa-Pintos, A., Kiiveri, H., Chow, E., Webster, M.S., Wieczorek, L., Raguse, B.: High-throughput fabrication and screening improves gold nanoparticle chemiresistor sensor performance. ACS Comb. Sci. **17**(2), 120–129 (2015). https://doi.org/10.1021/CO500129V/SUPPL_FILE/CO500129V_SI_001.PDF
69. Liu, Y., Parisi, J., Sun, X., Lei, Y.: Solid-state gas sensors for high temperature applications – a review. J. Mater. Chem. A **2**(26), 9919–9943 (2014). https://doi.org/10.1039/C3TA15008A
70. Ammu, S., Dua, V., Agnihotra, S.R., Surwade, S.P., Phulgirkar, A., Patel, S., Manohar, S.K.: Flexible, all-organic chemiresistor for detecting chemically aggressive vapors. J. Am. Chem. Soc. **134**(10), 4553–4556 (2012). https://doi.org/10.1021/JA300420T/SUPPL_FILE/JA300420T_SI_002.PDF
71. Fowler, J.D., Allen, M.J., Tung, V.C., Yang, Y., Kaner, R.B., Weiller, B.H.: Practical chemical sensors from chemically derived graphene. ACS Nano **3**(2), 301–306 (2009). https://doi.org/10.1021/NN800593M/ASSET/IMAGES/LARGE/NN-2008-00593M_0007.JPEG
72. Cho, B., Hahm, M. G., Choi, M., Yoon, J., Kim, A. R., Lee, Y. J., Park, S. G., Kwon, J. D., Kim, C. S., Song, M., Jeong, Y., Nam, K. S., Lee, S., Yoo, T. J., Kang, C. G., Lee, B. H., Ko, H. C., Ajayan, P. M., Kim, D. H.: Charge-transfer-based gas sensing using atomic-layer MoS2. Sci. Reports **51** 2015 (2015), 5 (1), 1–6. https://doi.org/10.1038/srep08052
73. Yang, H., Heo, J., Park, S., Song, H. J., Seo, D. H., Byun, K. E., Kim, P., Yoo, I. K., Chung, H. J., Kim, K.: Graphene barristor, a triode device with a gate-controlled Schottky barrier. Science (80) 2012, 336 (6085), 1140–1143. https://doi.org/10.1126/SCIENCE.1220527/SUPPL_FILE/YANG.SM.PDF
74. Luther, B.P., Wolter, S.D., Mohney, S.E.: High temperature Pt Schottky diode gas sensors on N-Type GaN. Sens. Actuators B Chem. **56**(1–2), 164–168 (1999). https://doi.org/10.1016/S0925-4005(99)00174-4
75. Singh, A., Uddin, M.A., Sudarshan, T., Koley, G.: Tunable Reverse-Biased Graphene/Silicon Heterojunction Schottky diode sensor. Small **10**(8), 1555–1565 (2014). https://doi.org/10.1002/SMLL.201302818
76. Raza, A., Zhang, X., Ali, S., Cao, C., Rafi, A. A., Li, G.: Photoelectrochemical energy conversion over 2D materials. Photochem. **2**, 272–298 (2022), **2**(2), 272–298 (2022). https://doi.org/10.3390/PHOTOCHEM2020020
77. Wang, J.: Nanomaterial-based electrochemical biosensors. Analyst **130**(4), 421–426 (2005). https://doi.org/10.1039/B414248A
78. Chen, A., Chatterjee, S.: Nanomaterials based electrochemical sensors for biomedical applications. Chem. Soc. Rev. **42**(12), 5425–5438 (2013). https://doi.org/10.1039/C3CS35518G
79. Guo, S., Dong, S.: Graphene nanosheet: synthesis, molecular engineering, thin film, hybrids, and energy and analytical applications. Chem. Soc. Rev. **40**(5), 2644–2672 (2011). https://doi.org/10.1039/C0CS00079E
80. Hong, X., Tan, C., Chen, J., Xu, Z., Zhang, H.: Synthesis, properties and applications of one- and two-dimensional gold nanostructures. Nano Res. **8**(1), 40–55 (2015). https://doi.org/10.1007/S12274-014-0636-3/METRICS
81. Wu, S., He, Q., Tan, C., Wang, Y., Zhang, H.: Graphene-based electrochemical sensors. Small **9**(8), 1160–1172 (2013). https://doi.org/10.1002/SMLL.201202896

82. Kannan, P.K., Late, D.J., Morgan, H., Rout, C.S.: Recent developments in 2D layered inorganic nanomaterials for sensing. Nanoscale **7**(32), 13293–13312 (2015). https://doi.org/10.1039/C5NR03633J
83. Wu, S., Zeng, Z., He, Q., Wang, Z., Wang, S.J., Du, Y., Yin, Z., Sun, X., Chen, W., Zhang, H.: Electrochemically reduced single-layer MoS2 nanosheets: characterization, properties, and sensing applications. Small **8**(14), 2264–2270 (2012). https://doi.org/10.1002/SMLL.201200044
84. Raza, A., Rafiq, A., Qumar, U., Hassan, J.Z.: 2D hybrid photocatalysts for solar energy harvesting. Sustain. Mater. Technol. **33**, e00469 (2022). https://doi.org/10.1016/J.SUSMAT.2022.E00469
85. Liu, H., Duan, C., Yang, C., Shen, W., Wang, F., Zhu, Z.: A novel nitrite biosensor based on the direct electrochemistry of hemoglobin immobilized on MXene-Ti3C2. Sens. Actuators B Chem. **218**, 60–66 (2015). https://doi.org/10.1016/J.SNB.2015.04.090
86. Wang, Y., Zhao, M., Ping, J., Chen, B., Cao, X., Huang, Y., Tan, C., Ma, Q., Wu, S., Yu, Y., Lu, Q., Chen, J., Zhao, W., Ying, Y., Zhang, H., Wang, Y., Zhao, M., Ping, J., Chen, B., Cao, X., Huang, Y., Tan, C., Ma, Q., Yu, Y., Lu Chen, Q.J., Zhao, W., Zhang, H., Ying, Y.: Bioinspired design of ultrathin 2D bimetallic metal–organic-framework nanosheets used as biomimetic enzymes. Adv. Mater. **28**(21), 4149–4155 (2016). https://doi.org/10.1002/ADMA.201600108
87. Liu, J., Su, D., Wang, J., Wu, K.: High-moment magnetic nanoparticles (2020)
88. Alhabeb, M., Maleski, K., Anasori, B., Lelyukh, P., Clark, L., Sin, S., Gogotsi, Y.: Guidelines for synthesis and processing of two-dimensional titanium carbide (Ti3C2Tx MXene). Chem. Mater. **29**(18), 7633–7644 (2017). https://doi.org/10.1021/ACS.CHEMMATER.7B02847/ASSET/IMAGES/LARGE/CM-2017-02847C_0005.JPEG
89. Chen, C., Xie, X., Anasori, B., Sarycheva, A., Makaryan, T., Zhao, M., Urbankowski, P., Miao, L., Jiang, J., Gogotsi, Y.: MoS2-on-MXene heterostructures as highly reversible anode materials for lithium-ion batteries. Angew. Chem. - Int. Ed. **57**(7), 1846–1850 (2018). https://doi.org/10.1002/anie.201710616
90. Lu, W., Yu, P., Jian, M., Wang, H., Wang, H., Liang, X., Zhang, Y.: Molybdenum disulfide nanosheets aligned vertically on carbonized silk fabric as smart textile for wearable pressure-sensing and energy devices. ACS Appl. Mater. Interfaces **12**(10), 11825–11832 (2020). https://doi.org/10.1021/ACSAMI.9B21068/ASSET/IMAGES/LARGE/AM9B21068_0005.JPEG
91. Wang, Z., Dong, S., Gui, M., Asif, M., Wang, W., Wang, F., Liu, H.: Graphene paper supported mos2 nanocrystals monolayer with cu submicron-buds: high-performance flexible platform for sensing in sweat. Anal. Biochem. **2018**(543), 82–89 (2017). https://doi.org/10.1016/j.ab.2017.12.010
92. Wang, Z., Gui, M., Asif, M., Yu, Y., Dong, S., Wang, H., Wang, W., Wang, F., Xiao, F., Liu, H.: A facile modular approach to the 2D oriented assembly MOF electrode for non-enzymatic sweat biosensors. Nanoscale **10**(14), 6629–6638 (2018). https://doi.org/10.1039/c8nr00798e
93. Yoon, J., Lee, S.N., Shin, M.K., Kim, H.W., Choi, H.K., Lee, T., Choi, J.W.: Flexible electrochemical glucose biosensor based on GOx/Gold/MoS2/Gold nanofilm on the polymer electrode. Biosens. Bioelectron. **140**(March), 111343 (2019). https://doi.org/10.1016/j.bios.2019.111343
94. Shin, M., Yoon, J., Yi, C., Lee, T., Choi, J. W.: Flexible HIV-1 biosensor based on the Au/MoS2 nanoparticles/Au nanolayer on the PET substrate. Nanomaterials **9**(8) (2019). https://doi.org/10.3390/nano9081076
95. Silvestri, A., Wetzl, C., Alegret, N., Cardo, L., Hou, H.L., Criado, A., Prato, M.: The era of nano-bionic: 2D materials for wearable and implantable body sensors. Adv. Drug Deliv. Rev. **186**, 114315 (2022). https://doi.org/10.1016/j.addr.2022.114315
96. Yin, F., Li, X., Peng, H., Li, F., Yang, K., Yuan, W.: A highly sensitive, multifunctional, and wearable mechanical sensor based on RGO/synergetic fiber bundles for monitoring human actions and physiological signals. Sens. Actuators, B Chem. **285**(January), 179–185 (2019). https://doi.org/10.1016/j.snb.2019.01.063
97. Pang, Y., Zhang, K., Yang, Z., Jiang, S., Ju, Z., Li, Y., Wang, X., Wang, D., Jian, M., Zhang, Y., Liang, R., Tian, H., Yang, Y., Ren, T.-L.: Epidermis microstructure inspired graphene pressure

sensor with random distributed spinosum for high sensitivity and large linearity. ACS Nano **12**(3), 2346–2354 (2018). https://doi.org/10.1021/acsnano.7b07613
98. Tao, L.-Q., Zhang, K.-N., Tian, H., Liu, Y., Wang, D.-Y., Chen, Y.-Q., Yang, Y., Ren, T.-L.: Graphene-paper pressure sensor for detecting human motions. ACS Nano **11**(9), 8790–8795 (2017). https://doi.org/10.1021/acsnano.7b02826
99. Ramírez, J., Rodriquez, D., D. Urbina, A., M. Cardenas, A., J. Lipomi, D.: Combining high sensitivity and dynamic range: wearable thin-film composite strain sensors of graphene, ultrathin palladium, and PEDOT:PSS. ACS Appl. Nano Mater. **2**(4), 2222–2229 (2019). https://doi.org/10.1021/acsanm.9b00174.
100. He, J., Xiao, P., Shi, J., Liang, Y., Lu, W., Chen, Y., Wang, W., Théato, P., Kuo, S.-W., Chen, T.: High performance humidity fluctuation sensor for wearable devices via a bioinspired atomic-precise tunable graphene-polymer heterogeneous sensing junction. Chem. Mater. **30**(13), 4343–4354 (2018). https://doi.org/10.1021/acs.chemmater.8b01587
101. Guo, H., Lan, C., Zhou, Z., Sun, P., Wei, D., Li, C.: Transparent, flexible, and stretchable ws2 based humidity sensors for electronic skin. Nanoscale **9**(19), 6246–6253 (2017). https://doi.org/10.1039/c7nr01016h
102. Borini, S., White, R., Wei, D., Astley, M., Haque, S., Spigone, E., Harris, N., Kivioja, J., Ryhänen, T.: Ultrafast graphene oxide humidity sensors. ACS Nano **7**(12), 11166–11173 (2013). https://doi.org/10.1021/nn404889b
103. Yoon, J.W., Lee, J.H.: Toward breath analysis on a chip for disease diagnosis using semiconductor-based chemiresistors: recent progress and future perspectives. Lab Chip **17**(21), 3537–3557 (2017). https://doi.org/10.1039/c7lc00810d
104. Pang, Y., Jian, J., Tu, T., Yang, Z., Ling, J., Li, Y., Wang, X., Qiao, Y., Tian, H., Yang, Y., Ren, T.L.: Wearable humidity sensor based on porous graphene network for respiration monitoring. Biosens. Bioelectron. **116**(March), 123–129 (2018). https://doi.org/10.1016/j.bios.2018.05.038
105. Mendes, F.A.R., França-Pinto, A., Martins, M.A., Cukier, A., Stelmach, R., Giavina-Bianchi, P., Carvalho, C.R.F.: Seasonal changes influence the improvement in asthma symptoms by exercise training in subjects with asthma. J. Asthma **56**(6), 674–679 (2019). https://doi.org/10.1080/02770903.2018.1484131
106. Jung, Y., Moon, H. G., Lim, C., Choi, K., Song, H. S., Bae, S., Kim, S. M., Seo, M., Lee, T., Lee, S., Park, H.-H., Jun, S. C., Kang, C.-Y., Kim, C.: Humidity-tolerant single-stranded DNA-functionalized graphene probe for medical applications of exhaled breath analysis. Adv. Funct. Mater. **27**(26), 1700068 (2017). https://doi.org/10.1002/adfm.201700068.
107. Wang, L., Jackman, J.A., Park, J.H., Tan, E.L., Cho, N.J.: A flexible, ultra-sensitive chemical sensor with 3D biomimetic templating for diabetes-related acetone detection. J. Mater. Chem. B **5**(22), 4019–4024 (2017). https://doi.org/10.1039/c7tb00787f
108. Du, B., Yang, D., She, X., Yuan, Y., Mao, D., Jiang, Y., Lu, F.: MoS2-based all-fiber humidity sensor for monitoring human breath with fast response and recovery. Sens. Actuators B Chem. **251**, 180–184 (2017). https://doi.org/10.1016/j.snb.2017.04.193
109. Cho, B., Yoon, J., Kwan Lim, S., Ra Kim, A., Kim, D.-H., Park, S.-G., Kwon, J.-D., Lee, Y.-J., Lee, K.-H., Hun Lee, B., Cho Ko, H., Gwan Hahm, M.: Chemical sensing of 2D graphene/MoS2 heterostructure device. ACS Appl. Mater. Interfaces **7**(30), 16775–16780 (2015). https://doi.org/10.1021/acsami.5b04541
110. Sahatiya, P., Kadu, A., Gupta, H., Thanga Gomathi, P., Badhulika, S.: Flexible, disposable cellulose-paper-based MoS2/Cu2S hybrid for wireless environmental monitoring and multifunctional sensing of chemical stimuli. ACS Appl. Mater. Interfaces **10**(10), 9048–9059 (2018). https://doi.org/10.1021/acsami.8b00245
111. Zhang, D., Sun, Y., Li, P., Zhang, Y.: Facile Fabrication of MoS2-modified SnO_2 hybrid nanocomposite for ultrasensitive humidity sensing. ACS Appl. Mater. Interfaces **8**(22), 14142–14149 (2016). https://doi.org/10.1021/acsami.6b02206
112. Li, N., Chen, X.D., Chen, X.P., Ding, X., Zhao, X.: Ultra-high sensitivity humidity sensor based on mos2/ag composite films. IEEE Electron Device Lett. **38**(6), 806–809 (2017). https://doi.org/10.1109/LED.2017.2699332

113. Xu, H.; Xiang, X.J.; Fei Lu, Y.; Kun Zhang, M.; Jia Li, J.; Bing Gao, B.; Jin Zhao, Y.; Ze Gu, Z.: Multifunctional wearable sensing devices based on functionalized graphene films for simultaneous monitoring of physiological signals and volatile organic compound biomarkers. ACS Appl. Mater. Interfaces **10**(14), 11785–11793 (2018). https://doi.org/10.1021/acsami.8b00073
114. Choi, H., Choi, J.S., Kim, J.S., Choe, J.H., Chung, K.H., Shin, J.W., Kim, J.T., Youn, D.H., Kim, K.C., Lee, J.I., Choi, S.Y., Kim, P., Choi, C.G., Yu, Y.J.: Flexible and transparent gas molecule sensor integrated with sensing and heating graphene layers. Small **10**(18), 3685–3691 (2014). https://doi.org/10.1002/smll.201400434
115. Cho, S.-Y., Koh, H.-J., Yoo, H.-W., Kim, J.-S., Jung, H.-T.: Tunable volatile-organic-compound sensor by using au nanoparticle incorporation on MoS2. ACS Sensors **2**(1), 183–189 (2017). https://doi.org/10.1021/acssensors.6b00801
116. Cho, B., Ra Kim, A., Jae Kim, D., Chung, H.-S., Young Choi, S., Kwon, J.-D., Won Park, S., Kim, Y., Hun Lee, B., Hwan Lee, K., Kim, D.-H., Nam, J., Gwan Hahm, M.: Two-dimensional atomic-layered alloy junctions for high-performance wearable chemical sensor. ACS Appl. Mater. Interfaces **8**(30), 19635–19642 (2016). https://doi.org/10.1021/acsami.6b05943
117. Mayorga-Martinez, C. C., Ambrosi, A., Eng, A. Y. S., Sofer, Z., Pumera, M.: Metallic 1T-WS2 for selective impedimetric vapor sensing. Adv. Funct. Mater. **25**(35), 5611–5616 (2015). https://doi.org/10.1002/adfm.201502223
118. Kim, J.-S., Yoo, H.-W., Ouk Choi, H., Jung, H.-T.: Tunable volatile organic compounds sensor by using thiolated ligand conjugation on MoS2. Nano Lett. **14**(10), 5941–5947 (2014). https://doi.org/10.1021/nl502906a
119. Tao, L.Q., Tian, H., Liu, Y., Ju, Z.Y., Pang, Y., Chen, Y.Q., Wang, D.Y., Tian, X.G., Yan, J.C., Deng, N.Q., Yang, Y., Ren, T.L.: An intelligent artificial throat with sound-sensing ability based on laser induced graphene. Nat. Commun. **8**, 1–8 (2017). https://doi.org/10.1038/ncomms14579
120. Lei, Y., Butler, D., Lucking, M.C., Zhang, F., Xia, T., Fujisawa, K., Granzier-Nakajima, T., Cruz-Silva, R., Endo, M., Terrones, H., Terrones, M., Ebrahimi, A.: Single-atom doping of MoS2 with manganese enables ultrasensitive detection of dopamine: experimental and computational approach. Sci. Adv. **6**(32), 1–10 (2020). https://doi.org/10.1126/sciadv.abc4250
121. Taylor, I.M., Robbins, E.M., Catt, K.A., Cody, P.A., Happe, C.L., Cui, X.T.: Enhanced dopamine detection sensitivity by PEDOT/graphene oxide coating on in vivo carbon fiber electrodes. Biosens. Bioelectron. **89**, 400–410 (2017). https://doi.org/10.1016/J.BIOS.2016.05.084
122. Liu, T.-C., Chuang, M.-C., Chu, C.-Y., Huang, W.-C., Lai, H.-Y., Wang, C.-T., Chu, W.-L., Chen, S.-Y., Chen, Y.-Y.: Implantable graphene-based neural electrode interfaces for electrophysiology and neurochemistry in in vivo hyperacute stroke model. ACS Appl. Mater. Interfaces **8**(1), 187–196 (2016). https://doi.org/10.1021/acsami.5b08327
123. Dai, X., Le, K., Wang, F., Wei, R., Liu, J., Jiang, Y., Li, H.: Single-molecule detection of acetylcholine by translating the neuronal signal to a single distinct electronic peak. ACS Appl. Bio Mater. **3**(10), 6888–6896 (2020). https://doi.org/10.1021/acsabm.0c00797
124. Jang, J., Kim, J., Shin, H., Park, Y.G., Joo, B.J., Seo, H., Won, J.E., Kim, D.W., Lee, C.Y., Kim, H.K., Park, J.U.: Smart contact lens and transparent heat patch for remote monitoring and therapy of chronic ocular surface inflammation using mobiles. Sci. Adv. **7**(14), 1–11 (2021). https://doi.org/10.1126/sciadv.abf7194
125. Kim, J., Kim, M., Lee, M.-S., Kim, K., Ji, S., Kim, Y.-T., Park, J., Na, K., Bae, K.-H., Kyun Kim, H., Bien, F., Young Lee, C., Park, J.-U.: Wearable smart sensor systems integrated on soft contact lenses for wireless ocular diagnostics. Nat. Commun. **8**(1), 14997 (2017). https://doi.org/10.1038/ncomms14997
126. Torrente-Rodríguez, R.M., Tu, J., Yang, Y., Min, J., Wang, M., Song, Y., Yu, Y., Xu, C., Ye, C., IsHak, W.W., Gao, W.: Investigation of cortisol dynamics in human sweat using a graphene-based wireless MHealth system. Matter **2**(4), 921–937 (2020). https://doi.org/10.1016/j.matt.2020.01.021

127. Wang, Z., Shin, J., Park, J.-H., Lee, H., Kim, D.-H., Liu, H.: Engineering materials for electrochemical sweat sensing. Adv. Funct. Mater. **31**(12), 2008130 (2021). https://doi.org/10.1002/adfm.202008130
128. Yun, Y. J., Ju, J., Lee, J. H., Moon, S.-H., Park, S.-J., Kim, Y. H., Hong, W. G., Ha, D. H., Jang, H., Lee, G. H., Chung, H.-M., Choi, J., Nam, S. W., Lee, S.-H., Jun, Y.: Highly elastic graphene-based electronics toward electronic skin. Adv. Funct. Mater. 27(33), 1701513 (2017). https://doi.org/10.1002/adfm.201701513
129. Lei, Y., Zhao, W., Zhang, Y., Jiang, Q., He, J.-H., Baeumner, A.J., Wolfbeis, O.S., Wang, Z.L., Salama, K. N., Alshareef, H.N.: A MXene-based wearable biosensor system for high-performance in vitro perspiration analysis. Small **15**(19), 1901190 (2019). https://doi.org/10.1002/smll.201901190
130. Li, G., Wen, D.: Sensing nanomaterials of wearable glucose sensors. Chin. Chem. Lett. **32**(1), 221–228 (2021). https://doi.org/10.1016/j.cclet.2020.10.028
131. Lee, H., Choi, T.K., Lee, Y.B., Cho, H.R., Ghaffari, R., Wang, L., Choi, H.J., Chung, T.D., Lu, N., Hyeon, T., Choi, S.H., Kim, D.-H.: A Graphene-based electrochemical device with thermoresponsive microneedles for diabetes monitoring and therapy. Nat. Nanotechnol. **11**(6), 566–572 (2016). https://doi.org/10.1038/nnano.2016.38
132. Lorencova, L., Bertok, T., Filip, J., Jerigova, M., Velic, D., Kasak, P., Mahmoud, K. A., Tkac, J.: Highly stable Ti3C2Tx (MXene)/Pt nanoparticles-modified glassy carbon electrode for H_2O_2 and small molecules sensing applications. Sens. Actuators B Chem. **263**, 360–368 (2018). https://doi.org/10.1016/j.snb.2018.02.124
133. Hao, Z., Luo, Y., Huang, C., Wang, Z., Song, G., Pan, Y., Zhao, X., Liu, S.: An intelligent graphene-based biosensing device for cytokine storm syndrome biomarkers detection in human biofluids. Small **17**(29), 2101508 (2021). https://doi.org/10.1002/smll.202101508
134. Goldoni, R., Farronato, M., Connelly, S. T., Tartaglia, G. M., Yeo, W.-H.: Recent advances in graphene-based nanobiosensors for salivary biomarker detection. Biosens. Bioelectron. **171**, 112723 (2021). https://doi.org/10.1016/j.bios.2020.112723
135. Mannoor, M.S., Tao, H., Clayton, J.D., Sengupta, A., Kaplan, D.L., Naik, R.R., Verma, N., Omenetto, F.G., McAlpine, M.C.: Graphene-based wireless bacteria detection on tooth enamel. Nat. Commun. **3**(1), 763 (2012). https://doi.org/10.1038/ncomms1767
136. Yan, W., Fuh, H.-R., Lv, Y., Chen, K.-Q., Tsai, T.-Y., Wu, Y.-R., Shieh, T.-H., Hung, K.-M., Li, J., Zhang, D., Ó Coileáin, C., Arora, S. K., Wang, Z., Jiang, Z., Chang, C.-R., Wu, H.-C.: Giant gauge factor of van der Waals material based strain sensors. Nat. Commun. **12**(1), 2018 (2021). https://doi.org/10.1038/s41467-021-22316-8
137. Park, M., Park, Y.J., Chen, X., Park, Y.-K., Kim, M.-S., Ahn, J.-H.: MoS2-based tactile sensor for electronic skin applications. Adv. Mater. **28**(13), 2556–2562 (2016). https://doi.org/10.1002/adma.201505124
138. Zhu, M., Du, X., Liu, S., Li, J., Wang, Z., Ono, T.: A review of strain sensors based on two-dimensional molybdenum disulfide. J. Mater. Chem. C **9**(29), 9083–9101 (2021). https://doi.org/10.1039/D1TC02102H
139. Kedambaimoole, V., Kumar, N., Shirhatti, V., Nuthalapati, S., Sen, P., Manjunatha Nayak, M., Rajanna, K., Kumar, S.: Laser-induced direct patterning of free-standing Ti3C2–MXene films for skin conformal tattoo sensors. ACS Sensors **5**(7), 2086–2095 (2020). https://doi.org/10.1021/acssensors.0c00647
140. Chun, S., Son, W., Wan Kim, D., Lee, J., Min, H., Jung, H., Kwon, D., Kim, A.-H., Kim, Y.-J., Kyoo Lim, S., Pang, C., Choi, C.: Water-resistant and skin-adhesive wearable electronics using graphene fabric sensor with octopus-inspired microsuckers. ACS Appl. Mater. Interfaces **11**(18), 16951–16957 (2019). https://doi.org/10.1021/acsami.9b04206
141. Mak, K.F., Shan, J.: Photonics and optoelectronics of 2D semiconductor transition metal dichalcogenides. Nat. Photonics **10**(4), 216–226 (2016). https://doi.org/10.1038/nphoton.2015.282
142. Wu, W., Wang, L., Li, Y., Zhang, F., Lin, L., Niu, S., Chenet, D., Zhang, X., Hao, Y., Heinz, T.F., Hone, J., Wang, Z.L.: Piezoelectricity of single-atomic-layer MoS2 for energy conversion and piezotronics. Nature **514**(7523), 470–474 (2014). https://doi.org/10.1038/nature13792

143. Yang, Y., Huo, N., Li, J.: Sensitized monolayer MoS2 phototransistors with ultrahigh responsivity. J. Mater. Chem. C **5**(44), 11614–11619 (2017). https://doi.org/10.1039/C7TC03476H
144. Yin, Z., Li, H., Li, H., Jiang, L., Shi, Y., Sun, Y., Lu, G., Zhang, Q., Chen, X., Zhang, H.: Single-layer MoS2 phototransistors. ACS Nano **6**(1), 74–80 (2011). https://doi.org/10.1021/nn2024557
145. Choi, C., Choi, M.K., Liu, S., Kim, M., Park, O.K., Im, C., Kim, J., Qin, X., Lee, G.J., Cho, K.W., Kim, M., Joh, E., Lee, J., Son, D., Kwon, S.-H., Jeon, N.L., Song, Y.M., Lu, N., Kim, D.-H.: Human eye-inspired soft optoelectronic device using high-density MoS2-graphene curved image sensor array. Nat. Commun. **8**(1), 1664 (2017). https://doi.org/10.1038/s41467-017-01824-6

SPRINGER NATURE

GPSR Compliance

The European Union's (EU) General Product Safety Regulation (GPSR) is a set of rules that requires consumer products to be safe and our obligations to ensure this.

If you have any concerns about our products, you can contact us on ProductSafety@springernature.com

In case Publisher is established outside the EU, the EU authorized representative is:

Springer Nature Customer Service Center GmbH
Europaplatz 3
69115 Heidelberg, Germany

The manufacturer's authorised representative in the EU is Springer
Nature Customer Service Centre GmbH, Europaplatz 3, 69115 Heidelberg,
Germany. If you have any concerns regarding our products, please
contact ProductSafety@springernature.com

Printed and bound by CPI Group (UK) Ltd, Croydon, CR0 4YY
01/10/2026
02225906-0002